ENTERPRISE INTER- AND INTRA-ORGANIZATIONAL INTEGRATION

Building International Consensus

IFIP - The International Federation for Information Processing

IFIP was founded in 1960 under the auspices of UNESCO, following the First World Computer Congress held in Paris the previous year. An umbrella organization for societies working in information processing, IFIP's aim is two-fold: to support information processing within its member countries and to encourage technology transfer to developing nations. As its mission statement clearly states,

IFIP's mission is to be the leading, truly international, apolitical organization which encourages and assists in the development, exploitation and application of information technology for the benefit of all people.

IFIP is a non-profitmaking organization, run almost solely by 2500 volunteers. It operates through a number of technical committees, which organize events and publications. IFIP's events range from an international congress to local seminars, but the most important are:

- The IFIP World Computer Congress, held every second year;
- open conferences;
- working conferences.

The flagship event is the IFIP World Computer Congress, at which both invited and contributed papers are presented. Contributed papers are rigorously refereed and the rejection rate is high.

As with the Congress, participation in the open conferences is open to all and papers may be invited or submitted. Again, submitted papers are stringently refereed.

The working conferences are structured differently. They are usually run by a working group and attendance is small and by invitation only. Their purpose is to create an atmosphere conducive to innovation and development. Refereeing is less rigorous and papers are subjected to extensive group discussion.

Publications arising from IFIP events vary. The papers presented at the IFIP World Computer Congress and at open conferences are published as conference proceedings, while the results of the working conferences are often published as collections of selected and edited papers.

Any national society whose primary activity is in information may apply to become a full member of IFIP, although full membership is restricted to one society per country. Full members are entitled to vote at the annual General Assembly, National societies preferring a less committed involvement may apply for associate or corresponding membership. Associate members enjoy the same benefits as full members, but without voting rights. Corresponding members are not represented in IFIP bodies. Affiliated membership is open to non-national societies, and individual and honorary membership schemes are also offered.

ENTERPRISE INTER- AND INTRA-ORGANIZATIONAL INTEGRATION

Building International Consensus

IFIP TC5 / WG5.12 International Conference on Enterprise Integration and Modeling Technology (ICEIMT'02) April 24–26, 2002, Valencia, Spain

Edited by

Kurt Kosanke
CIMOSA Association e.V.
Germany

Roland Jochem
Fraunhofer Institute for Production Systems and Design Technology (IPK)
Germany

James G. Nell
National Institute of Standards and Technology (NIST)
USA

Angel Ortiz Bas
Polytechnic University of Valencia
Spain

SPRINGER SCIENCE+BUSINESS MEDIA, LLC

Library of Congress Cataloging-in-Publication Data

A C.I.P. Catalogue record for this book is available from the Library of Congress.

Enterprise Inter– and Intra–Organizational Integration:
Building International Consensus
Edited by Kurt Kosanke, Roland Jochem, James G. Nell and Angel Ortiz Bas

DOI 10.1007/978-0-387-35621-1

Printed on acid-free paper.
www.springer.com/mycopy

Contents

Committees of the EI3-IC ix

Acknowledgements xi

Foreword – The European Commission xiii

Foreword – The US National Institute of Standards and Technology xv

Preface xvii

PART 1. Overview and Results 1

EI3-IC Overview
Kurt Kosanke 3

ICEIMT: History and Challenges
H. Ted Goranson 7

Accomplishments of the ICEIMT'02
James G. Nell, and H. Ted Goranson 15

Enterprise Modelling and Integration
Francois B Vernadat 25

PART 2. Knowledge Management in Inter- and Intra-Organisational Environments 35

A Merged Future for Knowledge Management and Enterprise Modeling
H. Ted Goranson, (Ed.), Michael N. Huhns, James G. Nell, Hervé Panetto, Guillermina Tormo Carbó, and Michael Wunram 37

Anchoring Knowledge in Business-Process Models to support Interoperability of Virtual Organizations
Peter Heisig, (Ed.), Martine Callot, Jan Goossenaerts, Kurt Kosanke, John Krogstie, and Nenad Stojanovic 51

Managing Processes and Knowledge in
Inter-Organisational Environments
*David Chen, (Ed.), Frank Lillehagen, Niek du Preez,
Raul Poler Escoto, and Martin Zelm* 61

Ontologies and their Role in Knowledge Management and
E-Business Modelling
Hans Akkermans 71

Semantic Bridging of Independent Enterprise Ontologies
Michael N. Huhns, and Larry M. Stephens 83

Active Knowledge Models and Enterprise Knowledge Management
Frank Lillehagen, and John Krogstie 91

Synthesising an Industrial Strength Enterprise Ontology
Chris Partridge, and Milena Stefanova 101

**PART 3. Enterprise Inter- and Intra-Organisational
Engineering and Integration** 111

Agents and Advanced Virtual Enterprises: Needs and an Approach
*H. Ted Goranson, (Ed.), Guillermina Tormo Carbó, Yoshiro Fukuda,
Lee Eng Wah, James G. Nell, and Martin Zelm* 113

Virtual Enterprise Planning Methods and Concepts
*Richard H. Weston, (Ed.), Cheng Leong Ang, Peter Bernus,
Roland Jochem, Kurt Kosanke, and Henry Ming* 127

Quality of Virtual Enterprise Reference Models
Peter Bernus 135

The Business Process (Quiet) Revolution;
Meir H. Levi 147

Enterprise Architecture and Systems Engineering
Peter Webb 159

Proposal of a Reference Framework for
Manufacturing Systems Engineering
*Gregor von Cieminski, Marco Macchi, Marco Garetti, and
Hans-Peter Wiendahl* 167

The Users View of Enterprise Integration and the
Enterprise Process Architecture
Juan Carlos Mendez Barreiro 177

Matching Teams to Business Processes
Nikita Byer, and Richard H. Weston 183

Analysis of Perceptions of Personnel at Organisational Levels on the Integration of Product, Functional and Process Orientations
Ruth Sara Aguilar-Savén 195

Challenges to Multi-Enterprise Integration
William J. Tolone, Bei-tseng Chu, Gail-Joon Ahn, Robert G Wilhelm, and John E. Sims 205

Practices in Knowledge Management in Small and Medium Firms
Raul Poler Escoto, Angel Ortiz Bas, Guillermina Tormo Carbó, and David Gutierrez Vañó 217

Component-Based Automotive Production Systems
Richard H. Weston, Andrew A. West, and Robert Harrison 225

The MISSION Project
Markus Rabe, and Frank-Walter Jaekel 235

PART 4. Interoperability of Business Process and Enterprise Models 243

System Requirements: Products, Processes and Models
James G. Nell, (Ed.), Em delaHostria, Richard, L. Engwall, Myong Kang, Kurt Kosanke, Juan Carlos Mendez Barreiro, Weiming Shen 245

Ontologies as a New Cost Factor in Enterprise Integration
H. Ted Goranson, (Ed.), Bei-tseng Chu, Michael Gruninger, Nenad Ivezic, Sern Kulvatunyou, Yannis Labrou, Ryusuke Masuoka, Yun Peng, Amit Sheth, and David Shorter 253

From Integration To Collaborative Business
Mike Payne 265

Enterprise Interoperability: A Standardisation View
David Chen, and François B. Vernadat 273

Interoperability of Standards to Support Application Integration
Em delaHostria 283

MultiView Program Status: Data Standards for the Integrated Digital Environment
Richard L. Engwall, and John W. Reber 295

Workflow Quality of Service
Jorge Cardoso, Amit Sheth, and John Miller 303

Improving PDM Systems Integration Using Software Agents
Yinsheng Li, Weiming Shen, and Hamada H. Ghenniwa 313

Ontologies for Semantically Interoperable Electronic Commerce
Leo Obrst, Howard Liu, Robert Wray, and Lori Wilson 325

PART 5. Common Representation of Enterprise Models 335

Steps in Enterprise Modelling
Ioannis L. Kotsiopoulos, (Ed.), Torsten Engel, Frank-Walter Jaekel, Kurt Kosanke, Juan Carlos Mendez Barreiro, Angel Ortiz Bas, Michaël Petit, and Patric Raynaud 337

New Support Technologies for Enterprise Integration
H. Ted Goranson, (Ed.), Roland Jochem, James G. Nell, Hervé Panetto, Chris Partridge, Francesca Sempere Ripoll, David Shorter Peter Webb, and Martin Zelm 347

Some Methodological Clues for Defining a Unified Enterprise Modelling Language
Michaël Petit 359

Common Representation through UEML – Requirements and Approach
Roland Jochem 371

UML Semantics Representation of Enterprise Modelling Constructs
Hervé Panetto 381

Language Semantics
Ioannis L. Kotsiopoulos 389

Modeling of Distributed Business Processes
H. Grabowski, and Torsten Engel. 399

Needs and Characteristics of Methodologies for Enterprise Integration
Marc Hawa, Angel Ortiz Bas, and Francisco-Cruz Lario Esteban 407

Argumentation for Explicit Representation of Control within Enterprise Modelling and Integration
Bruno Vallespir, David Chen, and Guy Doumeingts 417

Authors Index........................ 425

Committees of the EI3-IC Ininitiative

Scientific Committee

Ang, Cheng Leong, Gintic, Singapore
Berio, Giuseppe, Univ. Torino, Italy
Bernus, Peter, Griffith University, Australia
Brandl, Dennis, self-employed, USA
Bremer, Carlos, EESC-Univ. of Sao Paulo, Brazil
Browne, Jim, CIMRU, Ireland
Camarinha-Matos, Luis, New Univ. of Lisbon, Portugal
Doumeingts, Guy, GRAISOFT/Univ. of Bordeaux 1, France
Engwall, Richard L. R.L. Engwall and Associates, USA
Ferreira, Joao J.P. INESC Porto, Portugal
Fox, Marc, University of Toronto, Canada
Fukuda, Yoshiro, Hosei University, Japan
Goossenaerts, Jan, Eindhoven University, The Netherlands
Goranson, H. Ted, Old Dominion University, USA
Guilbert, Gerard, EADS, France
Hawa, Marc, DMR Consulting Group, Spain
Huhns, Michael, University of South Carolina, USA
Katzy, Bernhard, CeTim/ BW University Munich, Germany
Lario Esteban, Francisco-Cruz, Polytechnic Univ. of Valencia, Spain
Lillehagen, Frank, COMPUTAS, Norway
Matsuda, Michiko, Kanagawa Inst. of Technology, Japan
Molina, Arturo, ITESM Campus, Monterrey, Mexico
Neal, Richard, NGM Progr. Office, USA
Preez, Niek. D. du, Univ. of Stellenbosch, South Africa
Reyneri, Carla, Data Consult, Italy
Rhodes, Tom, NIST, USA
Scheer, August W., Univ. of Saarbrücken, Germany
Schuh, Günther, University of St Gallen, Switzerland
Segarra, Gerard, Renault DIO-EGI, France
Solte, Dirk, FAW Ulm, Germany
Vernadat, Francois B, ENIM/University of Metz/EC-Eurostat affiliate, France/Luxembourg
Weston, Richard H., Loughborough University, UK
Wortmann, Hans, Eindhoven University, Netherlands

Technical Committee

Jochem, Roland, FhG-IPK, Germany
Kosanke, Kurt, CIMOSA Association, Germany
Nell, James G. NIST, USA
Ortiz Bas, Angel, Polytechnic Univ. of Valencia, Spain
Poler Escoto, Raul, Polytechnic Univ. of Valencia, Spain
Zelm, Martin, CIMOSA Association, Germany

Acknowledgements

We sincerely thank all of the workshop and conference participants and all the authors for their valuable contributions. We appreciate the efforts of the scientific and the technical committee that helped to plan and organise the workshops and conference and provided their time to review and improve the quality of the papers published herein. Special commendation is due to Ted Goranson and Martin Zelm, both of whom have been committed to the success of the ICEIMT activities this time and in the previous initiatives.

We greatly appreciate the efforts of EADS, Gintic, NIST, and IPK for providing venues for the enormously productive workshops preceding the conference. We are grateful to the Polytechnic University of Valencia, especially to Professor Francisco-Cruz Lario Esteban, Director of the Centre for Investigation and Production Management and Engineering (CI-GIP). He and his team provided the resources and facilities to create an environment that made the conference a great forum for learning and exchanging new ideas to further inter- and intra-organisational interoperability.

Finally we thank the European Commission and NIST for their financial support, which enabled us to involve key people and host the workshops in Asia, Europe and the USA enabling a more global participation in the initiative. We are happy to acknowledge the support of the International Federation of Information Processing through the IFIP TC5 WG 5.12, leading to the publishing of these proceedings as an IFIP Publication.

The organisers

Kurt Kosanke
CIMOSA Association
Böblingen, Germany

Roland Jochem
Fraunhofer Institute for Production Systems and Design Technology (IPK),
Berlin, Germany

James G. Nell
National Institute of Standards and Technology (NIST),
Gaithersburg,
Maryland, USA

Angel Ortiz Bas
Polytechnic University of Valencia,
Spain

2002-June 30

Foreword – The European Commission

"The ICT Industry and the Information Society Technologies (IST, http//www.cordis.lu/ist) are radically transforming the economy and our daily lives, whether at work, at home or while on the move. Within the European IST Programme of the European Commission, the initiative in New Methods of Work and Electronic Commerce (IST- Key Action II, http//www.cordis.lu/ist/ka2) is playing a key role in developing leading edge research and development, which makes possible the creation of dynamic innovative and competitive organisations.

In this context Interoperability and Standardisation are critical issues for organisations that have started or are looking to do business electronically. For organisations to do business over the Internet, they must communicate and effect electronic exchanges with a wide range of business partners. E-business is much more than just buying and selling transactions over the Internet. It involves new forms of collaboration in which business processes, resources, skills and eventually knowledge have to be shared. Today, the insufficient level of interoperability of business applications impedes the adoption of this new form of collaboration. It is within this context that the European Commission is supporting relevant international initiatives dealing with Interoperability and Standardisation, e.g. the "Enterprise Inter- and Intra-organisational Integration" initiative and the working group on "Interoperability of Enterprise Software", thereby enabling European industry to take a leading role in this field and building industrial consensus to launch a large scale action capable of creating impact."

Rosalie Zobel,
Director,
Information Society Technologies: New Methods of Work and Electronic Commerce,
Directorate-General Information Society,
European Commission

Foreword – The US National Institute of Standards and Technology

The National Institute of Standards and Technology, NIST, welcomes the Enterprise Inter- and Intra-organizational Integration efforts of 2001 and 2002. The four workshops produced some very advanced thinking, and in-depth discussions achieved some new understanding and consensus in areas of enterprise engineering and integration. The primary focus this year was interoperability of business processes within an enterprise and in support of globally oriented electronic commerce. Some key projects to further technology were proposed to enable better communication where there is connectivity and to better understand and create useful knowledge from the data that is transferred.

The NIST mission is to develop and promote measurements, standards, and technology to enhance productivity, facilitate trade, and improve the quality of life. Therefore, NIST feels that consensus in the approach to enterprise engineering and integration coupled with technology development will do much to improve the climate for productive commerce in the world-wide, electronic-based marketplace. The results of the aforementioned workshops and International Conference on Enterprise Integration and Modeling Technology are presented in this book. As it has in 1992 and 1997, we are confident that this process has added key knowledge to the field of process interoperability, and has enabled enterprise integration to progress as a result.

Dale Hall,
Director,
Manufacturing Engineering Laboratory
National Institute of Standards and Technology
United States Department of Commerce

Preface

The international initiative on Enterprise Inter- and Intra-Organisational Integration (EI3-IC) had the objective to increase both international consensus (IC) and public awareness on enterprise integration. In these proceedings we intend to present the current status in inter- and intra-organisational integration for electronic commerce and thereby to further increase awareness and consensus within academia and industry about enterprise inter-and intra-organisational integration.

The conference proceedings contain the papers presented at the ICEIMT conference in Valencia, Spain, selected papers presented at the different workshops and three papers on the initiative itself: overview, history and results. The proceedings follow the conference structure with each section (Parts 2 to 5) starting with the workgroup reports, followed by a particular view on the section theme and additional papers either presented at the conference or during the related workshop. Section editorials discuss the different contributions.

As stated in the paper by Nell and Goranson in section 1 the results from all workshops indicate the important role of business processes in the area of e-commerce and virtual enterprises. Sharing relevant knowledge between co-operating partners and making it available for decision support at all levels of management and across organisational boundaries will significantly enhance the trust between the partners on the different levels of partner operations (strategy, policy, operation and transaction). Clearly business process modelling can significantly enhance establishment, operation and decommission of the required collaboration.

Merging knowledge management and business process modelling will provide synergy and improve efficiency of enterprise collaborations (Part 2 and Workshop 1). However, the benefits of knowledge sharing between collaborators can only be exploited if interoperability of business processes and business-process models can be assured (Part 4 and Workshop 3). This is especially important during the enterprise establishment phase where the required and provided capabilities have to be matched under the time constraints of the usually rather short market window.

But interoperability has not only an information technology aspect, but a human aspect as well. Only if the business-process model representation is commonly understood, will the people involved in the collaboration be able to build and maintain the needed trust in each other's capabilities (Part 5 and Workshop 4). Emphasis has been placed on the need for user oriented business process modelling, which is seen as a prerequisite for model based decision support. Specific aspects of virtual enterprise planning have been addressed (Part 3 and Workshop 2). Agent technology has been a subject in all four workshops and several proposals for further work have been made. The same is true for the concept of ontologies, which will play an important role in solving the interoperability issues through the harmonisation of business knowledge semantics.

The Editors

Kurt Kosanke
CIMOSA Association
Böblingen, Germany

Roland Jochem
Fraunhofer Institute for Production Systems and Design Technology (IPK), Berlin, Germany

James G. Nell
National Institute of Standards and Technology (NIST), Gaithersburg, Maryland, USA

Angel Ortiz Bas
Polytechnic University of Valencia, Spain

2002-June 30

PART 1. OVERVIEW AND RESULTS

EI3-IC Overview
The Initiative on Enterprise Inter- and Intra-Organisational Integration – International Consensus

Kurt Kosanke
CIMOSA Association, Germany, kosanke@cimosa.de

Abstract The initiative is the third one aimed on building international consensus on enterprise integration. With the focus on virtual enterprises the aspects of enterprise engineering, knowledge management and business process modelling, interoperability of business processes and models and model representation were addresses in the workshops and results presented at the conference.

1 INTRODUCTION

This third international initiative had again the objective to increase both international consensus and public awareness on enterprise integration. Following the two previous initiatives in 1992 and 1997 (Goranson, 2002 Kosanke, 1997, Petrie, 1992), the focus of the third initiative was on Enterprise Inter- and Intra-Organisational Integration (EI3). This included the recognition of competitive benefits, as well as organisation and infrastructure implications. Drivers, barriers and enablers for electronic commerce in general and the virtual enterprise in particular, as well as potential benefits from the application of integration supporting information and communication technology have been addressed. Application areas include business-to-business, e-business, e-commerce, extended and virtual enterprises, and supply chains.

The initiative is supported by the European Commission, DG Information Society, IST 2001-92039 and the International Federation of Information Processing - IFIP TC5/WG 5.12. The organisation has been done jointly by the CIMOSA Association, Germany, the Fraunhofer Institute IPK (Institute for Production Systems and Design Technology), Germany, the National

Institute of Standards and Technology (NIST), USA and the Polytechnic University of Valencia, Spain

The EI3-IC initiative has provided a basis for an international discourse on the subject of enterprise inter- and intra-organisational co-operation with emphasis on virtual enterprises and Business-to-Business type e-commerce. Inviting experts in the field has enabled pulling in insights and results from other projects hence enabling a consolidation of this fragmented know-how, and thereby contributing to an international consensus in the field. The community built during the EI3-IC initiative will continue as an international forum to further identify and eliminate barriers to utilisation of inter- and intra-organisational integration technology (Nell, Goranson, 2002).

2 RATIONALE

Globalisation combined with the emergence of powerful information and communication technologies drive enterprises towards new forms of co-operation. Electronic commerce and virtual enterprises are a new way for small and medium enterprises (SMEs) to unite forces, increase their competitiveness and meet today's market needs and jointly behave as one producer towards the customer. Up to now application of relevant ICT support has been hampered by a lack of business justification, by a plethora of seemingly conflicting solutions and confusing terminology (Kosanke, 2001), and by an insufficient understanding of the technology by the end-user community. These barriers inhibit, or at least delay, the use of relevant methods and tools in the industry, especially in small-to-medium-sized enterprises (SMEs).

One of the main concerns in the required collaborations is the need for information on partner capabilities. Such information will also help to establish the much needed trust between the partners. Partner capabilities can best be described through relevant business processes and associated resources. Linking compatible processes into an overall business process would allow evaluating the collaboration prior to its real implementation.

3 METHODOLOGY AND ACTIVITIES

The initiative on enterprise inter- and intra-organisational integration (EI3-IC) consists of two parts:

1. Four workshops with international experts reviewing and consolidating a set of issues in enterprise inter and intra-organisational integration.

2. The ICEIMT'02 (International Conference on Enterprise Integration and Modelling Technologies) that is aimed on state of the art overview and presentation of the workshop results.

A scientific committee (see below) guided and supported the initiative. It acted as advisory committee for reviewing workshop and conference agendas and papers and helped to identify the experts to be invited to the workshops.

The workshops have been organised with plenary sessions for all participants and a number of parallel working group sessions. The first plenary session held in all workshops provided time for the participants to present their own work as it relates to the predefined set of issues. This methodology has led again to very good results. It enables the members of the working group to have a common understanding of each other's position leading to much better focusing on the issues to be discussed.

During the first plenary session the experts will usually amend the set of predefined issues. Working groups have then worked on subsets of the issues of the particular workshop. Presentation of working group results and discussions of the topic with all working groups have been done during subsequent plenary sessions.

Papers on workshop results were prepared co-operatively by the working groups and presented at the– ICEIMT'02 by a group member.

3.1 Workshops and Conference

Four thematic workshops with international experts in the field have been organised. The workshop themes have been selected according to their importance for the management of business collaborations. The following workshops have been held:

- Workshop 1, Knowledge management in inter- and intra-organisation environments (EADS, Paris, France, 01-12-05/06)
- Workshop 2, Enterprise inter- and intra-organisation engineering and integration (Gintic, Singapore, 02-01-23/25)
- Workshop 3, Interoperability of business processes and enterprise models (NIST, Gaithersburg, MD, USA)
- Workshop 4, Common representation of enterprise models (IPK, Berlin, Germany, 02-02-20/22)

Workshop editorials are presented as part of the related sections in these proceedings.

The ICEIMT'02 was held at the Polytechnic University of Valencia, Spain, on 2002-04 24/26. It was structured following the themes of the workshops. In addition to an opening session with keynote papers, a special session on international projects provided information on actual work done on an international level.

4 CONCLUSIONS

International consensus on the contents of enterprise intra- and inter-organisation integration is a prerequisite for real industry acceptance and application of the technology. A more common terminology is expected to be one result of the initiative. With its particular focus on e-commerce the third initiative identified major players in this field both in industry and academia and thereby has continued to build the community on enterprise integration. A community that will continue the drive for consensus beyond this initiative and towards a follow-on ICEIMT (Nell, Goranson, 2002).

However, significant efforts are still needed to gain awareness and acceptance in the industry. Large-scale demonstrations of pilot applications as well as professional training of potential users would be means to convince the user community of the benefits of the technology.

5 REFERENCES

Goranson, H. T, (2002), *ICEIMT: History and Challenges,* these proceedings.

Kosanke, K., Nell, J.G. (Eds.), (1997), *Enterprise Engineering and Integration: Building International Consensus;* Proceedings of ICEIMT'97 Intern. Conference on Enterprise Integration and Modelling Technology; Springer-Verlag.

Kosanke, K. de Meer, J. (2001), *Consistent Terminology - A Problem in Standardisation – State of Art Report of Enterprise Engineering,* Proceedings SITT'01, Boulder, Col, USA, October, 3/5.

Nell, J.G. Goranson, H. T. (2002), *Accomplishments of the ICEIMT'02 Activities,* these proceedings.

Petrie, Jr. C.J. (Ed.), (1992), *Enterprise Integration Modelling,* Proceedings of the First International Conference, MIT Press.

ICEIMT: History and Challenges

H. Ted Goranson
Old Dominion University, USA, tedg@sirius-beta.com

Abstract ICEIMT was initiated as a multiactivity event to meaningfully address and partially solve a precise set of problems. The history behind the original activity is reviewed in the context of the problem set. Now, as ICEIMT continues ten years later, the problem set has evolved to be immensely more difficult. That new agenda is defined.

1 HISTORY OF ICEIMT: THE PROBLEM

Around 1986, the highest levels in the United States government recognized a problem with profound economic consequences, and resolved to address it. American competitiveness was considered low and decreasing, especially in the manufacturing sector. Most of the preliminary work on identifying underlying causes was done at the research consortium SEMATECH. That body was funded at two billion dollars to try to rescue the strategically important semiconductor industry from the Japanese threat. SEMATECH sponsored a supplier's working group, representing the majority of the world's information technology infrastructure suppliers. The group had a special legal exclusion from antitrust restrictions and high-level visibility in both the administration and board rooms.

The problem was seen as revolving around enterprise integration, the ability to quickly and cheaply get all the models, metrics and control software interfaced, collaborating and optimized. The threat from the Japanese was not they did this well, but that their tight, stable vertical enterprises made the components far less dynamic, so each new enterprise just reused the infrastructure from the older one. There was far less innovation in the Japanese approach, but that was more than compensated by an ability to focus on continuous improvement of processes. Some industries, like semiconductors, automobiles and consumer electronics especially benefited from this structurally competitive asymmetric advantage.

The goal was set to build infrastructure capability that would accommodate the "American" business model of many relatively independent suppliers, each innovating in processes and underlying software. The notion was to task the U. S. research establishment to develop integration frameworks and methods that could match the Japanese approach in speed but also accommodate diversity and innovative dynamism. That innovativeness was seen as the societally rooted competitive advantage that would save the day.

A second order problem was that the US research establishment was demonstrably unable to deliver this sort of result for structural reasons. The primary agency for such research was the (then named) Advanced Research Projects Agency (ARPA). ARPA had been established in 1957 as the owner of the nation's high risk, high payoff research problems. Some notable successes had resulted, and ARPA was the focus for information technology research. But by 1986, ARPA had had some spectacular failures, costing billions directly with astounding opportunity costs. Several studies in industry and the intelligence community suggested a structural barrier in the highly inbred advisory "invisible college," named ISAT, which influenced research directions and rough distribution of funds. ISAT (Information Science and Technology panel) was — and still is — populated by power brokers from major research universities and dominated by a mutual back-scratching protocol. The collapse of artificial intelligence, robotics and high performance computing initiatives was directly traced to this structural problem.

The National Science Foundation (NSF) was unable to address the problem. NSF's mission at the time was to subsidize university research under heavy peer review at the graduate student level. But the peer review process is highly departmentalized. The enterprise integration problem was seen as large and interdisciplinary, not amenable to decomposition into graduate student sized portions.

The suppliers and intelligence community raised the visibility of these problems to an attentive White House.

Meanwhile, the European Union was struggling with a similar problem and structural inadequacy.

European enterprises of the time tended to be in between those of the American and Japanese types in terms of centralization and stability, but with no inherent advantage over either other than guaranteed home markets (and at the time, subsidies). No European firm was a significant player in information infrastructure; all such infrastructure was controlled by American-owned multinationals.

The European Union's research initiatives had been spectacular failures, more publicly so than in the US. The structural problem here was that research initiatives were intended as comprehensive, interdisciplinary efforts. But in practice, the teams (composed by law of universities and business from different countries) divided the work up and proceeded independently. Each partner "owned" the results of the effort. Sharing even among partners was poor. Reporting and commercialization was usually below a useful critical mass.

Senior policy makers of the USA and EU decided to explore breaking these research barriers by creating a Joint Research Organization of some kind. ARPA, the NSF and the National Institute of Standards and Technology (NIST) represented the US in developing collaboration protocols, which were established in 1988. In 1990, meetings were held in San Francisco and Washington to explore the tactics of enterprise integration collaboration. Key US and EU projects were selected and "action officers" designated: Kurt Kosanke for the EU and Ted Goranson for the USA. Large working meetings to shape the collaboration were held in Daytona Beach in January 1990 and in Berlin the next July. A final planning meeting was held in Brussels in January of 1992.

Traction for collaboration was built around the general superiority of modeling theory in the EU. Several hundreds of millions were planned. Deep relationships with suppliers were committed. A tacit agreement was reached to develop a European infrastructure industry

A multi-tiered initiative was constructed under the rubric of ICEIMT. Significant consensus work was done within the suppliers' working group to define problems, candidate solutions and reasonable commercialization strategies. To address futures, four facilitated workshops were held among the research community; two in Austin, Texas, USA in February, two in April in Nice, France.

Products of the effort were a book edited at some cost by a US based consortium, a conference in Hilton Head, South Carolina in June, and detailed, closed briefings to the supplier and defense communities. These latter had profound impact on the future of information infrastructure. A major step toward object-oriented infrastructures directly resulted. The origin of the enterprise resource planning market can be traced to associated decisions and technology transfer. And major technical alliances that persist today were initiated.

The intended pan-Atlantic joint research organization did not emerge.

2 NEW PROBLEMS IN THE MARKET

A second ICEIMT was initiated by the EU at the five-year mark. This ICEIMT is the third.

Ten years after ICEIMT defined the enterprise integration problem it will be worthwhile to revisit how things have changed in the decade since the problem was defined as a matter of national survival.

The Japanese are no longer the threat they once were. Their system has collapsed because of a structural weakness in the banking system that should have been anticipated. They are now joining the rest of the world in their enterprise structures and inheriting the enterprise integration problem. The drivers for enterprise integration today are primarily ones of company rather than national survival, though manifest destiny is that much less complex manufacturing will flow to emerging economies. So domestic manufacturing needs to be more agile than lean, more niche than mass oriented and more proximity and service centered.

The bottom line of this dimension is that the situation is worse today than ten years ago. A decade ago, the problem was at the center of a national emergency. Now it is not. ARPA has become DARPA, the "Defense" ARPA, with a narrowly focused operationally military focus. No one in the USA government owns or openly cares about the problem. (Some well-funded intelligence and experimentation agencies are working identical problems but to date have steered clear of the civil industrial marketplace for reasons noted below.)

The supplier situation is different and far, far more complex. Europe is now a major player in Enterprise Resource Planning and CAD-led Product Data Management. Microsoft was not a player a decade ago, but is now, bringing to the sector a level of monopolistic rancor not part of the prior scene. In particular, the internet and web are central parts of the environment now. Vendors are significantly less driven by customer satisfaction than by winning strategic positions. Most major architectural decisions are now guided by strategic advantage for the supplier than the user.

The users have less clout than before for another reason as well, the balkanization of enterprise integration communities. Back in the days before computers and models, expertise was stored in implicit ways, largely in tacit knowledge, rules of thumb and in trusted managers. Once models became a way of making these explicit, the various communities in the enterprise tended to coalesce around what knowledge they "owned" and could use as leverage to do their job better. Suppliers identified niches within the enterprise based on these functions. As a result, we now have enterprises that consist of warring infrastructures, methods and metrics. We now have Enterprise Resource Planning, Customer Relations Management, Supply Chain

Management, Activity Based Costing Management, Knowledge Management, Product Data Management and on and on.

The voice of the user to integrate these systems has been muted to essentially nothing. And now the problem of enterprise integration is not merely to integrate functions by their processes, we have to integrate enterprise integration infrastructures as well. On the supplier side, this introduces new competitive dynamics. After all, an enterprise does not buy and champion enterprise integration tools, senior managers do. Market forces drive the supplier to speak to that manager's concerns. We are in the unhappy state that the very existence of effective models has increased the dis-integration of the enterprise. The existence of functional frameworks for model integration has had the unexpected result of fragmented and cannibalized integration markets.

At the same time, enterprises have become enormously more complex in the past decade, the products much more sophisticated and interdisciplinary and the speed of change is at unparalleled levels.

One other success of enterprise integration has engendered new problems. The original impetus came from the operation side of the enterprise to balance the management of production with the management of capital and capital-driven assets. Since then, emerging business models (fluid supply chains and agile virtual enterprises) have allowed for the independent management of capital and production. Unfortunately, the legacy of enterprise modeling in industrial engineering has unduly influenced the targeting of the frameworks. Instead of growing to handle both functions (capital and production), it has inexplicably stayed with the latter.

In short, the difficulty of the problem has grown in complexity and difficulty faster than the solution has evolved.

3 NEW TECHNICAL PROBLEMS

In addition to the problem set increasing, there are new technical barriers as well.

The first ICEIMT looked at integration strategies in general. An enterprise can be integrated at the level of basic services, at the level of applications, or at the level of models. The first was the default at the beginning of ICEIMT and was deemed inadequate. The goal is to integrate at the level of models — in fact this can be used as a definition of enterprise integration. A baseline for model-centric integration was that component of the CIMOSA architecture that related different models, model views and generic types.

But the market at that point was obsessed with application integration. The reason was straightforward: most of the research and vendor attention had responded to the so-called "software crisis," wherein most applications or application synthesis projects failed. The result was a collection of techniques for software engineering through modularity, encapsulation and reuse under the aegis of "object orientation." The market is sustained by selling applications, not models, and the supplier working group guessed that something like model integration could be accomplished by dual use of application integration technologies. The standards community was quite ready to respond because application integration standards were well understood with an established constituency.

The industry knew that the compromise would result in immediate progress but serious barriers in the longer term. In fact, the situation today is far worse than ten years ago because of this compromise. Markets and many business practices blur the concepts of process and object. Because of encapsulation, object-oriented models by definition lack the sort of visibility and "zoomable auditability" which formed the original desiderata. So today, enterprise modelers have to work around an unfriendly legacy that they helped create.

And the goals of enterprise integration have escalated in several dimensions:

- The original ICEIMT scope concerned coordination and optimization of operations and related resources. Businesses now are used to thinking in terms of strategic planning integrated to operations. Some rudimentary integration of this type exists in terms of qualitative metrics (accounting dollars) in the form of activity based costing. Models are very much more complex than flat numbers, so integration from strategic to operational domains is a tough problem, but one expected by astute managers. The gains would be substantial if such a thing could be accomplished. This called the "vertical integration" issue. (The current ICEIMTs first workshop touched on this issue in a targeted way by addressing the merger of knowledge management and enterprise integration. See the report from that workshop for some concrete recommendations.)
- The original ICEIMTs range of the business life cycle made assumptions that you knew what you were going to make, and how and to whom you were going to sell it. The engineered system only addressed how it was made in most cases. Today's ICEIMT agenda must address the whole life cycle of operations, from discovering markets, designing products and services, and creating and supporting them. This is the "horizontal" expansion of the integration scope. (Both horizontal and vertical expansions sweep in a greater variety of model types, view and uses. But they also necessitate for the first time the explicit modeling of soft items: uncertainties, unknowns, unknow-

ables, social and cultural collaborative dynamics, and certain types of trust. These are difficult problems)

- Originally, the ICEIMT user community was content with "batch" engineering. In the assumption that the world would not change very much, one would model, integrate and optimize an enterprise. Then it would be operated in that mode for some long period without change. After some period, a re-engineering would occur for some other static period. Almost no one will accept that today. The world is dynamic. Conditions change, you discover mistakes you made in the original models and assumptions, you improve your processes, you change and evolve products, and you swap your partners at will. The need for continually evolving systems has redefined enterprise integration problem in a more ambitious, demanding way.

4 NEW APPROACHES

The original ICEIMT defined a spectrum of approaches that ranged from model-centric to language-centric. The model-centric approach was deemed less capable but more realistic at the time. Since then, significant work has been done on ontologies and ontology languages, and the language-centric approach seems to now dominate the agenda. Examples are the process specification language and the unified enterprise modeling language.

As noted above, the modularity-by-object philosophy was adopted as a compromise with existing market trends. It is a manifestly inadequate approach for the expanded agenda (and perhaps even its original, smaller scope). Since the first ICEIMT, workable notions of "features" are used in product data management versions of enterprise modeling. And even within the programming community, features are being grafted onto object oriented programming through the new strategy of "aspect"-oriented programming. Quite probably, some abstraction of models into enterprise value features (or something similar) will be developed as the language-based mechanism for enterprise model integration.

The old ICEIMT agenda was satisfied to stick to process features that can be explicitly, unambiguously represented. The new agenda requires modeling of partial, uncertain or unknown facts. There are few techniques for accomplishing this, but they are well known and all the subjects of experiments.

In response to the need for dynamism and distributed federation, the "activity" of models is likely to change. In the original ICEIMT vision, it was sufficient to have "passive" models, representations of processes that simply captured some superficial behavior. The new generation will certainly use some notion of agents (active models) that reflect some of the cause and effect mechanics of the underlying processes.

Ten years ago, three types of repository strategies were defined, with a simple unified approach at one end and a more difficult federated one at the other. First generation integration relied on everyone using the same methods, the models being all collected in the same location under single control. Next generation integration is expected to relax that somewhat, with the models being distributed, the methods being varied somewhat, and controlled more locally — ideally by the same person that owns the process of interest — the federated model.

5 SUMMARY

The original ICEIMT defined a response to an extremely important problem that was not being adequately addressed by market forces or government agencies. It did a good job, engaging with suppliers and users finding a practical balance among emerging trends, valuable benefits and tolerable tradeoffs. The world changed as a result.

Now, ten years later, market forces and government agencies depend even more heavily on the reborn ICEIMT. The situation is very much more difficult, and solutions likely more valuable. Some technical barriers and challenges exist that did not before.

Almost certainly, the same practical radicalism is required. Smooth evolution from first generation solutions will be inadequate.

6 REFERENCES

Petrie, C.J. Jr. (Ed.), *Enterprise Integration Modeling,* Proceedings of the First International Conference, MIT Press 1992.

Kosanke, K., Nell, J.G. (Ed's), *Enterprise Engineering and Integration: Building International Consensus;* Proceedings ICEIMT'97, Springer-Verlag 1997.

Accomplishments of the ICEIMT'02

Summary Paper

James G. Nell[1], and H. Ted Goranson[2]
[1]National Institute of Standards and Technology, *USA* [2]*Old Dominion University, USA,*
nell@cme.nist.gov

Abstract: The purpose of this paper is to analyze the activities of this initiative on Enterprise Intra- and Inter-organizational Integration--International Consensus (EI3-IC) and especially of its conference the ICEIMT'02. We have extracted the major accomplishments, identified how the discussions have furthered our knowledge about enterprise integration, and attempted to show how the information was parlayed into better knowledge about the topic. In addition to the analysis of the initiative as a whole and of its results, we report results from a plenary discussion, held as the closing session of the conference.

1 ICEIMT'02

ICEIMT'02 strove to improve international consensus on issues in enterprise engineering, modeling, and integration technologies with emphasis on inter-organizational relations. The conference identified barriers, proposed solutions, and communicated results, thereby helping to justify the technology to industry so that key technology can be moved profitably from the international R&D domain to broadly based implementation.

The conference agenda comprised reports from workshops and invited papers. The papers were intended to communicate status and a sampling of the many different views on enterprise integration. The program especially emphasized results from the four workshops that preceded the conference. The workshops produced recommendations on research directions and a number of proposals for R&D projects.

Selected experts in the fields of engineering, business administration, and computer science attended the workshops. About 75 persons from 18 countries on 5 continents attended the ICEIMT'02, coming from academic institutions, government, industry, and consortia. The majority was aligned with academia. These conference proceedings provide about 40 papers that offer a very comprehensive overview of the state-of-the-art in of enterprise integration as well as providing directions for further research.

2 PRESENT STATE OF INTEGRATION

In *ICEIMT: History and Challenges*, in the introductory section of this book, Ted Goranson has detailed the things we thought we knew about integration in 1992 and 1997. In fact, we knew enough to implement solutions in some cases. These solutions have produced mixed benefits. Some of them have served us well. Others have performed as planned but have created significant barriers to further improvement.

For example, as the chunks of enterprise that are able to share information in an electronic format get larger, the more diverse the context, the more varied the granularity of the concepts, and the more difficult the subject is to model. In addition, the level of executive necessary to approve improvement projects rises as the scope of improvement increases. Often this level is beyond the technical ability of the executive so there is increasing trepidation about such large improvements.

Some of the large integration products, such as enterprise-resource planning, are expensive and to achieve the interoperability needed, they force the enterprise to conform to the model of the software. This sets up a larger barrier within the enterprise. But also, with the advent of the Internet, global commerce, and virtual enterprises these systems also become formidable barriers to electronic commerce and, so-called supply-chain integration becomes more frustrating to accomplish.

Past attempts to integrate migrated from installing better information-transmission conduits, to mandating similar hardware, common software packages, and common languages. We have not succeeded because we have avoided the real problem, the hard problem, the semantics involved. Regardless of investments in mandating the above assets to be common, for there to be information exchange in which the receiver acts as the sender intended, connoting an understanding, the sign and symbols in the syntax must convey the same meaning, no more or no less. We now have concluded that a tool called ontology is the key to congruence of meaning between the sender and receiver. Five years ago we were just beginning to appreciate the importance of a good ontology.

The purpose of ICEIMT'02 has been to address the difficult parts of transferring information among applications and organizations.

3 VISION

As in the past, cultural inertia continues to limit the effectiveness and use of standards to constrain enterprises. Enterprise management will insist upon retaining the freedom to tinker with the enterprise and process design. This will be a survival maneuver to improve operational efficiency and to differentiate products. Therefore, attempts to standardize enterprises, processes, or reference architectures will most probably be ignored.

A primary focus of the conference was on similar visions for a new enterprise environment that would use a combination of standards and tools to allow inter-operating business processes to determine the best way to communicate information among those processes in a way that elicits desired behavior. These are called self-integrating processes and they may be thought of as software that adapts to its environment with no human assistance. We assume an environment in which not all interfaces are known, but the nature of the interfaces can be discovered, such as by, querying, learning and guessing. The new environment would operate for intra-organizational transactions and inter-organizational transactions on a global scale.

3.1 Model-Driven Enterprises

Another part of the vision is to have the enterprise model, to the most extent practicable, operate enterprises. This EI3-IC spent considerable time in its working groups exploring how these model-driven processes could work, and discovering that very little, if any, new technology is required to make that happen. We determined that the engines that accomplish the model-driven enterprises, software agents, need knowledge of enterprise goals, process goals, a system to trigger and accomplish the work, and a system that oversees the work that also knows the enterprise and process goals. These agents can adjust the enterprise to benefit the enterprise, and, perhaps, at the expense of the good of individual processes. To act this way the agents will need some sense of self, and information that humans use to form patterns of knowledge, with which it can assess value. This is real-world information with which humans create what we call tacit, implicit, or unspoken knowledge. Setting this up will require skill in knowing what to model, what to include in knowledge bases, and knowing what is unnecessary to include.

3.2 Interoperability

One of the important changes from prior ICEIMT endeavors is the focus on process interoperability. There were attempts to define exactly what we mean by interoperability. Francois Vernadat, in his paper that is presented in the first-session section of this book: *Enterprise Modeling and Integration: from Fact Modeling to Enterprise Interoperability* has provided a flexible and extendible definition of system interoperability. We have extended that definition toward the part of the system that will provide better information transfer.

- System interoperability = Ability of a system to use parts of another system
- Enterprise interoperability = Ability of an enterprise to use parts of another enterprise
- Process interoperability = Ability of a process to use parts of another process
- Information-aspect interoperability = Ability of the information aspect of one process to use part of the information aspect of another process

4 INTEROPERABILITY SITUATION

Processes that communicate information were the focus of the last two ICEIMT efforts even though the problem was identified as an enterprise-integration problem. The earlier ICEIMT vision foresaw that true integration would be next to impossible and advocated a concept called federation. Federation led to the concept of self-integration, in which processes learn to deal with the situation they encounter on an ad hoc basis.

In his history paper, Goranson analyzed the changing integration situation, which can be summarized as:

- Worse now than in 1992 and 1997. Because the scope of what is needed is larger, and because of the inflexible nature of some solutions, the problem is now more difficult to solve
- We are now dealing with a supply-chain problem that is larger than the enterprise, rather than dealing merely with intra-enterprise processes.
- The Internet on a global basis has allowed the problem focus to migrate from inside the company border, past the country border, and back to the process border, however now, the processes may be on different continents
- The process-border issue has been reduced to the meaning of information, or semantics--the other problems have been solved largely, leaving the difficult ones yet to be solved.

In other words, the core problem always was meaning of information, and we chose not to recognize it or solve it, because the solution is difficult and expensive.

5 RESULTS OF THE ICEIMT

The Conference ended with an open, interactive discussion on what to do next. Noting that some key leaders and advocates of the process are exiting the scene, or at least reducing their activity, the starting question was if and how to continue the initiative on enterprise integration.

Of the three ICEIMTs, this conference had the fewest attendees, but attendance at the workshops in 2001/2002 was higher than in the two earlier initiatives, 1992 and 1997. People in attendance were mostly from the research aspect of academia. Others attended from consortia--a very few were from individual industries. Some at the conference believed the core research community was as large as always with the loss apparently being among corporate users. One conclusion could be that our workshop-based ICEIMT model provided experts in small working groups with opportunities for detailed discussions and ensuing insight. This process probably provides much more satisfaction to the participants than the paper presentations at the conference. Apparently, because of this perceived value, approval to travel to the workshops is easier to obtain than approval for a conference; plus, attendance is possible on a regional basis when the workshops are conducted in various locations.

The planning for next steps identified discrete areas of need. These are listed below in no particular order. Discussants had diversity of opinion on priorities. Everyone agreed that the core focus of the community on enterprise integration must be maintained, and that the bias toward manufacturing applications be shifted more to the enterprise as a whole. Also, some participants expressed unanimous agreement that the interdisciplinary nature of the work in terms of approaches be emphasized. Most participants seemed to agree that the "demand" side of the equation, attracting users, should not be a focus. Rather, emphasize the "supply" side by consistently improving solutions and the supply of practitioners.

In supporting the supply side, all agreed that the future of the community should be taken into its own hands as an international concern; this is opposed to counting on forthcoming support from the EU as the driving factor. Independently, the immense influence of the Web indicates that the EI community needs to be more active in relevant Web developments.

The group also noted that because EI concentrates on a core set of problems, and because EI is also interdisciplinary in approach, that we should be careful about the "buzzwords" used to characterize the problem and community. Specifically mentioned was the preponderance at this workshop of the term "interoperability". The discussant questioned whether the dilution of focus is worth the discrimination of the concept. Apparently not everyone shared the concern, and no specific recommendation ensued.

6 ACTION LIST

6.1 Create an educational program

This is the core action proposed to create a science of enterprise integration. The notion is that there needs to be a "seed" set of education materials and consonant support for creating (probably graduate) curricula. All agreed on the general need for this as a top priority. Other than including the need for a textbook(s), no details were discussed other than this important action: Mark Fox, University of Toronto, took responsibility for hosting a working session to develop a plan for bringing this about. At least one of these details is the understanding of the academic positioning, whether in industrial engineering, business management, computer science, or other departments. The theme of an article, *Who Needs a Whole MBA?* Business Week, March 25, 2002, was that a specialized master's degree may be more relevant than an MBA. Mentioned is the Massachusetts Institute of Technology's System Design and Management program that combines engineering and business courses. This model seems to be attractive to industry. There could be enterprise engineering included in such a program.

6.2 Create a publication venue

Mark Fox also offered to host the first EI conference in Toronto in 2004, which is seen as the earliest practical time. The planning for this will need a substantial committee, many of whom seemed pre-committed to help. The rationale for the conference primarily is to provide a publication forum, feeding existing journals, or perhaps creating a new one. A substantial majority of the group believes that the key value is in preserving the EI focus of the community. But some voices were raised in support of attaching to one of the large IT conferences instead.

6.3 Continue the workshop process

Many people spoke of the value of the ICEIMT workshops as the primary vehicle for sharing ideas, getting reality checks, and establishing consensus research directions. The preference was for more frequency, perhaps yearly. No action leader volunteered, and the default position now is that this might be folded into the conference, at least initially. However, some murmuring of proposing for EU support for a series of ICEIMT-type workshops was heard, and this might emerge under the banner of proposed projects. The

European Union Sixth Framework has identified an "instrument" that can support these. The EU IST representative and James G. Nell of NIST were impressed by the quality of the workshop attendees, discussions, and reports.

6.4 Support EI standards

The reasoning behind this suggestion was that the EI community depends on standards as a key strategy and focus for interaction. Yet few standards are rationalized from a focused EI perspective and those that do, conflict and compete with others. The idea was that coherent, strong support would help define and center the community.

6.5 Form, propose and execute research projects

Obviously, there was much interest in proposing projects to the EU IST Sixth Framework Program, which seems well disposed to the EI agenda. Probably, this will proceed in the old-fashioned manner, with small affinity groups forming alliances for specific projects. Very likely, the CIMOSA Association may propose for support for some near-term ICEIMT-like workshops.

6.6 Support the vendor base

A vendor spoke up on the reliance of his sector on being exposed periodically to the state-of-the-art. He suggested a *Handbook of Enterprise Integration* that is frequently updated. Peter Bernus, Griffith University, spoke of work underway to fill this need partly. Many others mentioned the importance of feeding the vendor base, but seemed to believe that other actions listed here would be sufficient. One suggestion was that when vendors are invited to another event to give them all a single case study so that comparisons can be made,

6.7 Outreach to parallel communities

As EI is highly interdisciplinary, the importance of outreach to other disciplines is essential. The Web connection has been noted, and also the cogent academic disciplines. Of special interest is the human-factors community. Presumably, this outreach is not a separate activity (no one volunteered) and will be supported as a philosophy underlying the other actions.

6.8 Develop a clear business case

Only one voice articulated this concern, but the voice was forceful. How this can be supported as an individual activity is not known, so it probably will be satisfied through practical user consciousness in the above listed actions.

7 MAJOR FINDINGS FROM ICEIMT'02 AND EI3-IC

7.1 On agent-oriented solutions

- No additional enterprise modeling languages or methodologies are required to represent the information used to simulate knowledge as used by agents and enterprise models.
- There is a need to plan carefully the amount of tacit knowledge to be made available; that is, provide no more or no less than is necessary to support autonomous agents that perform reasoning-type tasks.

7.2 On process-model concepts

- To improve model reuse and to reduce the complexity and cost of enterprise models, move the semantics-intensive content from the enterprise models to the ontology of the application being modeled.
- Find a way to match up the global, more soft, less deterministic, enterprise-level models with more deterministic, process-level models by modeling the process-level material in the enterprise-level format.
- The ontology is the place to resolve verb-oriented process models and noun-oriented enterprise- and product-object models. These resolutions are necessary to permit such activities as computer-based simulations.
- If we are talking about improving the information sharing and interactions in an enterprise, and the enterprise is a system, then enterprise engineering must use a systems-engineering approach when re-engineering an enterprise.

7.3 On Enterprise Integration community building

- There is a need to meet and discuss these integration issues in a workshop format more often than every five years.
- Pursuant to that end, the group welcomed Mark Fox's proposal to organize the next ICEIMT in Toronto in 2004.
- Formalize an academic curriculum around a to-be-created science of enterprise integration.

8 SUMMARY

The group concluded that we need to create a single focus in the enterprise-integration and enterprise-engineering fields, and publish papers, articles, and success stories in some venue. Implementing this action should be on the agenda for the educational meeting proposed by Mark Fox of University of Toronto. A business case is very complicated, but important to justify significant investment; therefore, a progressive approach would be beneficial if it was created, planned, and undertaken by a group of vendors, users, and R&D types. Planning should begin immediately to preserve the momentum apparent at the ICEIMT'02 in Valencia.

Enterprise Modelling and Integration
From Fact Modelling to Enterprise Interoperability

Francois B. Vernadat
EC/EUROSTAT, Luxemburg & LGIPM, ENIM/University of Metz, France,
Francois.vernadat@cec.eu.int

Abstract: Enterprise Modelling and Integration has evolved over the last decades from entity-relationship and activity modelling to object and flow modelling as well as from pier-to-pier system integration to inter-organisational exchanges enabling various forms of electronic commerce. The next challenge is Enterprise Interoperability, i.e. seamless integration in terms of service and knowledge sharing. The paper discusses modelling and integration issues to progress towards Enterprise Interoperability and shows how the CIMOSA architecture can be revised to host these emerging techniques and standards.

1 INTRODUCTION

Enterprise Modelling (EM) is the art of externalising enterprise knowledge, which adds value to the enterprise, be it a single enterprise, a private or government organisation, or a networked enterprise (e.g. extended enterprise, virtual enterprise or smart organisation). *Enterprise Integration* (EI) deals with facilitating information flows, systems interoperability and knowledge sharing among any kind of organisation. *Enterprise Interoperability*, as one of the many facets of EI, provides two or more business entities (of the same organisation or from different organisations and irrespective of their location) with the facility to exchange or share information (wherever it is and at any time) and to use functionalities of one another in a distributed and heterogeneous environment (Kosanke, Nell, 1997, OAG. OAGIS, 2001, Petrie 1992, Vernadat, 1996).

With the emergence of A2A (application-to-application) and X2X technologies in business (B2B: business-to-business, B2C: business-to-customer,

C2C: customer-to-customer...) as well as in governments (G2B: government-to-business, G2C: government-to-citizen, G2G: government-to-government, G2N: government-to-non government organisations), there is a need for sound and efficient methods and tools to design and operate efficient integrated systems made of autonomous units.

In this context, EM provides a *semantic unification space* at the corporate level where shared concepts can be properly defined, mapped to one another and widely communicated in the form of enterprise models (Goranson, 1992).

This position paper first briefly reviews the current state of EM and EI and then probes the future in terms of their evolution before indicating how the CIMOSA framework can be revised to cope with these evolutions.

2 ENTERPRISE MODELLING & ENGINEERING

What it is: *Enterprise Modelling* is concerned with representing the structure, organisation and behaviour of a business entity, be it a single or networked organisation, to analyse, (re-)engineer and optimise its operations to make it more efficient. Enterprise Modelling is a crucial step both in Enterprise Engineering and Enterprise Integration programmes (Vernadat. 1996).

Enterprise Engineering (EE) is concerned with designing or redesigning business entities. It concerns all activities, except enterprise operation, involved in the enterprise life cycle, i.e. mission identification, strategy definition, requirements definition, conceptual design, implementation description, installation, maintenance and continuous improvement as defined in PERA and GERAM (IFAC-IFIP Task Force, 1999, Williams 1992). It mostly concentrates on engineering and optimising business processes in terms of their related flows (materials, information/decision and control), resources (human agents, technical agents, roles and skills) as well as time and cost aspects. EM techniques for EE should therefore support at least representation and analysis of *function*, *information*, *resource* and *organisation* aspects of an enterprise (AMICE, 1993, IFAC-IFIP Task Force, 1999, Vernadat. 1996).

As advocated in the Zachman Framework (Sowa, Zachman, 1992), the objective of EM is to define the six perspectives of *what, how, where, who, when* and *why* of the *Enterprise Model, System Model, Technology Model* and *Component* level of an enterprise. The what defines entities and relationships of the business entity, the how defines the functions and processes performed, the where defines the network of locations and links of entities and agents, the who defines agents and their roles, the when defines time aspects and the schedule of events, and the why defines the strategy of the enterprise.

What needs to be modelled: The following aspects are concerned (AMICE, 1993, IFAC-IFIP Task Force. 1999).

- Function aspects: functional domains, triggering events, business processes (or control flows), enterprise activities (or process steps)
- Information aspects: enterprise objects, object relationships (semantic and user-defined links), object flows, object states
- Resource aspects: doers (human and technical agents), resource components, resource capabilities and/or competencies, roles
- Organisation aspects: organisation units, organisation cells (or decision centres), responsibilities, authorities
- Temporal and causal constraints

These are the usual modelling constructs found in prominent EM languages (ARIS, CIMOSA, GRAI, IDEF, IEM...) as reviewed in (Vernadat. 1996).

What for: The enterprise models must provide abstract representations of the things of the organisation being analysed with enough precision and in a way which lends itself to computer processing to support:

- Enterprise Reengineering / Process Improvement (establishing the business-process map, simplifying and reorganising some processes, optimising use of resources, simulating enterprise behaviour)
- Workflow design and management (to automate critical processes)
- Tuning enterprise performances (mostly in terms of costs and delays but also quality, reactivity and responsiveness)
- Management decision support ("what if" scenarios, simulating planned situations, forecasting, etc.)
- Enterprise integration (i.e. seamless exchange across the system to provide the right information at the right place at the right time)

Enterprise Knowledge Management: Enterprise modelling is a form of enterprise knowledge representation method in the sense that it captures, represents and capitalises basic facts and knowledge about the way the enterprise is structured, organised and operated (mostly surface knowledge).

According to G. Mentzas, Enterprise Knowledge Management (Tiwana, 2000) is a new discipline for enabling individuals, teams and the entire organisation to *collectively* and *systematically* create, share and apply *corporate knowledge* to better achieve organisational *efficiency*, *responsiveness*, *competency* and *innovation*. Thus, there is a need to also address deep knowledge.

Within an enterprise, knowledge is exhibited at various levels. It is in the mind of people (individual level), within team structures (team level), encapsulated in business processes and rules (organisational level) and linked to inter-organisational interactions (environment level).

Knowledge is usually classified as *tacit* (formalised as a theory or expressed in a structured language/notation) or *implicit* (individual feeling or known by humans but not formalised in a theory or in a structured model).

Nonaka has proposed a cyclic model of knowledge emergence and consolidation within an organisation (Tiwana, 2000). The model is a cycle made of four steps: *socialisation* (tacit know-how becomes shared know-how), *externalisation* (shared tacit know-how becomes codified knowledge), *combination* (codified knowledge becomes enterprise knowledge), and *internalisation* (enterprise knowledge becomes individual tacit know-how).

Evolution of Enterprise Modelling and future issues: The origins of Enterprise Modelling can be set back to the mid-70's when several diagrammatic methods were proposed for information system analysis and software development. The early methods can be qualified as *fact modelling* methods in the sense that little or poor semantics of the enterprise was captured. Pivot concepts taken into account were the concepts of enterprise *entities*, *relationships* among entities and *activities* made of sub-activities. The models produced only represent static facts. Pioneering methods are the entity-relationship model of P.P.S. Chen and the SADT method of D.T. Ross, also known as IDEF0, (Vernadat, 1996).

They were soon followed in the 80's by *flow-charting* methods combining ideas of the two previous ones but in addition depicting the flow of processing activities (SSAD by Gane and Sarson, Yourdon's notation, DeMarco's notation, MERISE in French spheres) (Martin, McClure, 1985). For CIM, IDEF and GRAI methods appeared (Vernadat, 1996). Time aspects were missing in such models.

At the same period, a lot of more fundamental work was carried out on (1) *semantic models* (e.g. extended entity-relationship model, semantic networks, frames, binary model) to capture more of the semantics of data or for knowledge representation, and (2) on *formal models* to analyse system behaviours (e.g. Petri nets, timed Petri nets, coloured Petri nets, state-charts).

The 90's have been dominated by two complementary trends, which have seriously impacted and boosted EM: business process (BP) modelling and object-oriented (OO) modelling. *BP modelling* focuses on business processes and related concepts: *events*, *activities*, *roles*, *resources* and *object flows*. Many of the common EM tools and approaches have emerged from this trend (CIMOSA, IDEF3, ARIS, IEM and the workflow technology). *OO modelling* focuses on the abstract concept of objects and brings structuring modelling principles, e.g. object uniqueness, property inheritance, aggregation mechanisms, and reusability. The prominent method in the field is UML (Unified Modelling Language), which has become an OMG and ISO standard and has supplemented OMT (Object Modelling Technique) (ISO/IEC DIS 19501-1, 2000).

Current modelling tools are quite good at modelling structured business processes, i.e. deterministic sequences of activities with related object flows and associated resources (e.g. ARIS Tool Set, FirstSTEP, etc.). However, they need to be extended in several ways. Among these, we can cite:

- *Socio-organisational aspects*: More research work and extensions to commercial tools are required in terms of modelling human roles, individual and collective competencies, decision centres. To this end, a competency model has recently been validated in industry and proposed to extend CIMOSA constructs (Berio, Vernadat, 1999, Harzallah, Vernadat, 2002).
- *Weakly structured workflow*: Structured business process and workflow system implementations tend to rigidify the enterprise, i.e. to automate processes in an inflexible way. Modern tools should be able to cope with weakly or ill-structured processes, i.e. processes for which the exact control flow sequence is not fully known. Three essential constructs have been proposed to this end but not yet implemented in commercial tools: AND construct (the process step is made of *n* activities that must all be done but the execution order of which will be decided at run-time), XOR construct (there are *n* activities in the process step but only one will be executed, the choice of which will be decided at run-time), and the OR construct (*k* among *n* activities will be done in the process step at run-time but the selection will be decided at run-time) (Berio, Vernadat, 1999). Another interesting problem concerns the modelling of the decision knowledge associated to each case, which is also a research issue (El Mhamedi, et al, 2000).
- *Inter-organisational Interaction and Co-ordination* aspects: The modelling of networked organisations and supply chains requires that new constructs be proposed to cope with such structures.
- *EM ontologies*: Because there are different ways of representing the same concepts, there is the need to have an ontology of enterprise modelling concepts (specialised by industrial sectors, application domains, tools, and so on) (ACM, 2002). Examples of such ontologies for enterprise modelling are the TOVE ontology (Fox, Gruninger, 1998) or the ontology for PSL (Process Specification Language) (Schelenoff, et al, 2000). The UEML (Unified Enterprise Modelling Language) initiative of the IFAC-IFIP Task Force on Enterprise Integration is another one (Vernadat, 2001). EM ontologies have a crucial role to play to make Enterprise Interoperability a reality in the next decades.

3 ENTERPRISE INTEGRATION

Enterprise Integration: Since the early 90's, EI has drastically evolved from specialised communication protocols (e.g. MAP, TOP, field-buses), diverse dedicated standard data exchange formats (e.g. IGES, STEP, EDI/EDIFACT, HTML...) and complex monolithic integration infrastructures for distributed computing environments (e.g., OSF/DCE in the Unix world, OLE and DCOM in the MS Windows world and OMG/CORBA in the OO world) proposed at that time (Vernadat, 2001). Regarding Enterprise Application Integration (EAI), the state of the art is now to use *Message-Oriented Middleware* (MOM) (either in stateless or state-full mode as well as in synchronous or asynchronous mode) on top of computer networks compatible with TCP/IP (Linthicum, 2000). The middleware must provide sufficient scalability, security, integrity and reliability capabilities. Messages are more and more in the form of HTML and XML documents. The most recent trend is to switch to Java programming (JSP, EJB) and apply the J2EE (Java to Enterprise Edition and Execution) principles to build integrated collaborative systems.

On top of these, large applications are implemented according to the 3-tier client-server architecture using the web architecture and a standard protocol (HTTP). A client user can access the application on his/her PC via HTTP using a standard HTML browser. The request is sent to a web server, which concentrates all requests and passes the request to the application server (AS). The AS processes the request using its local database server.

A new trend for the development of application servers is to build them as a set of remote services accessible via the web, called *web services*. The client does not need to know where they are located on the web but can request their use at any time. Services need to be declared via WSDL (Web Service Description Language) and registered in a common web repository, called UDDI

Concerning message exchange, the trend is to make wide use of XML (eXtensible Mark-up Language) (XML, http) to neutralise data because of the ability of XML to separate the logic of documents as well as data formatting from data itself. This means that well-known data exchange formats used in industry (e.g. EDI, STEP, etc.) will soon have to be reworked in the light of XML (e.g. cXML, ebXML...).

Finally, concerning transport of messages, new protocols are being proposed including SOAP (Simple Object Access Protocol) (http://), RosettaNet (http://), Bolero.net (http://), Biztalk (http://) among others.

Towards Enterprise Interoperability: Broadly speaking, interoperability is a measure of the ability of performing interoperation between two or more different entities (be they pieces of software, processes, systems, organisations...). Thus, Enterprise Interoperability is concerned with interoperability between organisational units or business processes either within a large distributed enterprise or within a network of enterprises (e.g. supply chain, extended enterprise or virtual enterprise). The challenge relies in communication, co-operation and co-ordination of these processes.

4 CIMOSA REVISION

CIMOSA (AMICE, 1993), a pioneering Enterprise Integration architecture designed in the late 80's – early 90's, is made of three main components, namely the *Modelling Framework* (MFW), the *Integrating Infrastructure* (IIS, made of distributed computer services) and the *System Life Cycle* (SLC or deployment methodology). This architecture can be revisited as follows.

Concerning the EM Modelling Framework, it is proposed to add a modelling view to CIMOSA, called *Interaction View,* to deal with inter-organisational aspects, mostly interaction and co-ordination mechanisms between business entities making a networked organisation or supply chain. Constructs of this modelling view would include (Fig. 1):

- *Business Entity*, used to define the components (or nodes) of a networked organisation or supply chain. They can represent External Suppliers, Manufacturing Units, Warehouses, Final Assembly Units, Distribution Centres and Customers)
- *Interface*, used to define the corporate competencies and services offered by each Business Entity and the protocol to access them
- *Channel*, used to define exchange mechanisms between two Business Entities in terms of frequency, exchange mode, exchange rate, carrier, exchange cost, availability, reliability and alternatives). Two types of Channels need to be distinguished: *Communication Channels* for data/information exchanges (information flows) and *Transportation Channels* for goods exchanges (material flows).

Concerning the Integrating Infrastructure (IIS) the recommendation is to develop IIS services as Web services on top of a Message-Oriented Middleware where messages would be encapsulated in XML format and exchanged in a secured SOAP-like envelope.

Concerning the System Life Cycle, currently CIMOSA uses the life cycle defined in GERAM and approved by ISO TC 184/SC5 (IFAC-IFIP Task Force, 1999). However, this life cycle has a linear layout, which might confuse the business user because it does not show the principles of Continuous Process Improvement currently prevailing in industry and based on the Deming's Wheel philosophy (Deming, 1982). We suggest the adoption of a more cyclic view of the SLC, presented on Fig. 2 and based on modern iterative prototyping methods used in software engineering as well as in system design and implementation.

Interaction View	Function View	Information View	Resource View	Organisation View
Bus. Entity	Event	Ent. Object	Resource	Org. Unit
Interface	Bus. Proc.	Obj. View (or State)	Component	Org. Cell
Channel	Activity	Integrity rules	Capability/ Competency	Authority/ Responsibility

Figure 1: Revised CIMOSA MFW

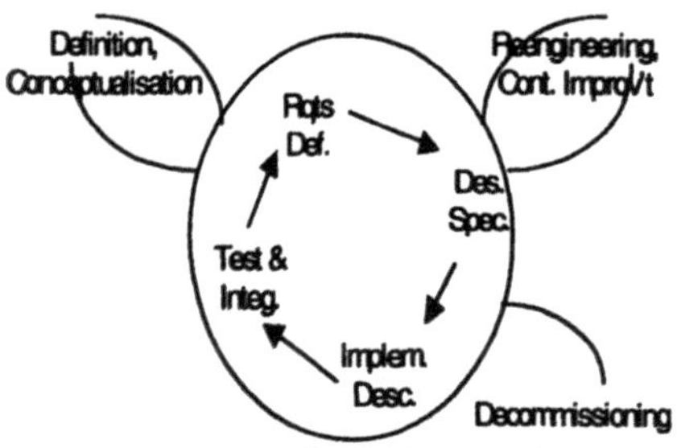

Figure 2: Revised CIMOSA SLC

5 CONCLUSION

Enterprise Modelling has evolved over the last three decades from fact modelling to Knowledge Management while at the same time Enterprise Integration has evolved from computer systems integration and CIM to Enterprise Interoperability and e-commerce.

This paper has provided a short overview of the field in terms of where we stand and what has to be done next. It also proposes an extension of the CIMOSA framework to host extended principles for Enterprise Modelling and Integration.

6 REFERENCES

ACM, (2002), *Special issue on Ontology Applications and Design*, Communications of the ACM, 45(2).

AMICE, (1993), *CIMOSA: Open System Architecture for CIM*, second revised and extended edition, Springer-Verlag.

Berio, G. Vernadat, F.B. (1999), *New developments in enterprise modelling using CIMOSA*, Computers in Industry, 40 (2-3): 99-114.

Biztalk (http://www.biztalk.org)

Bolero.net, http://www.bolero.net

Deming, E.W. (1982), *Quality, Productivity and Competitive Position*, The MIT Press.

El Mhamedi, A. Sonntag, M. and Vernadat, F.B. (2000), *Enterprise engineering using functional and socio-cognitive models*, Engineering Cost and Valuation Analysis.

Fox, M.S. Gruninger, M. (1998), *Enterprise Modelling*, AI Magazine, Fall, 109-121.

Goranson, H. T. (1992), *Dimensions of Enterprise Integration, in Enterprise Integration Modeling* (Petrie, C. Ed.), The MIT Press, pp. 101-113.

Harzallah, M. Vernadat, F.B. (2002), *IT-based competency modeling and management: From theory to practice in enterprise engineering and operations*, to appear in Computers in Industry.

IFAC-IFIP Task Force, (1999), *GERAM: Generalised Enterprise Reference Architecture and Methodology*, Version 1.6, in ISO IS 15704, TC 184 SC5 WG1.

ISO/IEC DIS 19501-1, (2000), *Information Technology – Unified Modelling Language (UML)* - Part 1: Specification.

Kosanke, K. Nell, J.G. (Eds.), (1997), *Enterprise Engineering and Integration: Building International Consensus*, Springer-Verlag.

Linthicum, D.S. (2000), *Enterprise Application Integration*, Addison-Wesley.

Martin, J. McClure, C. (1985), *Structured Techniques for Computing*, Prentice-Hall.

Nonaka, I. A. (1994), *dynamic theory of organizational knowledge creation*, Organization Science, 5, 4-37.

OAG. OAGIS, (2001), Open Applications Group Integration Specification, Open Application Group, Inc., Release 7.2.1, Doc. No. 20011031.

Petrie, C. (Ed.) (1992), *Enterprise Integration Modeling*, The MIT Press.

RosettaNet, http://www.rosettanet.org

Schelenoff, G. Gruninger, M, Tissot, F. Valois, J. Lubell, J., Lee, J. (2000), *The Process Specification Language (PSL): Overview and Version 1.0 Specification*, National Institute of Standards and Technology (NIST), Gaithersburg, MD. USA.

SOAP, http://www.w3.org/TR/SOAP

Sowa, J.F., Zachman, J.A. 1992, *A logic-based approach to Enterprise Integration, in Enterprise Integration Modeling* (Petrie, C. Ed.), The MIT Press, pp. 152-163.

Tiwana, A. (2000), *The Knowledge Management Toolkit. Practical Techniques for Building a Knowledge Management System*, Prentice-Hall.

Vernadat, F.B. (1996), *Enterprise Modeling and Integration: Principles and Applications*, Chapman & Hall.

Vernadat, F.B. (2001), *UEML: Towards a Unified Enterprise Modelling Language*. Proc. 3ème Conférence Francophone de Modélisation et Simulation (MOSIM'01), Troyes, France, 25-27 April, pp. 3-13.

Williams, T.J. (1992), *The Purdue Enterprise Reference Architecture*, Instrument Society of America.

XML. http://www.w3.org/TR/1998/REC-xml-19980210.

PART 2. KNOWLEDGE MANAGEMENT IN INTER- AND INTRA-ORGANIZATIONAL ENVIRONMENTS

Knowledge Management (KM) has been gaining significant momentum within enterprise organisations and is considered an important success factor in the enterprise operation. However, wide differences exist in the understanding of what a knowledge management system is and does. Perception ranges from using the enterprise-wide database or expert systems with various ontologies and roles to enterprise modelling and integrated communication systems supported with Internet technology. Generally accepted guidelines or standards are missing to support the design and implementation of a knowledge management system in an organisation or between organisations. Capturing knowledge and using it across organisational boundaries with a satisfactory acceptance of the human user is another major challenge.

Three workgroup reports address the relations between knowledge management and enterprise modelling concluding that joining in some form could be possible and synergy would bring additional benefits. One focus was on possible combined futures and the research roadmap these futures require (Goranson). Three different levels of potential work have been identified: near term, medium term and longer term oriented. At each level problems and limits have been identified and potential solutions are proposed.

Discussing the mapping of enterprise modelling onto knowledge management similarities and differences as well as solutions have been identified (Heisig). Thereby focus has been also on ontologies, which will play an important role in this mapping. A role that would become intensified with the move towards inter-organisational collaboration or virtual enterprises.

Concentrating on guidelines for enterprise modelling to cover scope and goals, architectures, infrastructures and approaches to implementation, the third workgroup looked at examples of industrial solutions and tool strategies (Chen). Potential synergies and solutions have been identified with emphasis on the human role in future environments.

Ontologies are conceptual reference models that formally describe the consensus about a domain and that are both human-understandable and machine processable. Akkermans in his overview paper gives an overview of recent developments, issues, and experiences in Semantic Web research, and especially discusses the role of ontologies in innovative intelligent e-applications, using the On-To-Knowledge project for ontology-based knowledge management as a particular example.

The paper by Huhns describes a methodology by which information from many independent sources and with different semantics can be associated, organised, and merged. A preliminary evaluation of the methodology has been conducted by relating 53 small, independently developed ontologies for a single domain.

Lillehagen in his paper presents a novel approach for integrating enterprise modelling and knowledge management in dynamic networked organisations. The approach is based on the notion of active knowledge models (AKM™). An AKM is a visual model of enterprise aspects that can be viewed, traversed, analysed, simulated, adapted and executed by industrial users.

The last paper in this section presents a report on work in progress of a synthesis of selected state of the art enterprise ontologies, which aims to produce a Base Enterprise Ontology (Partridge). The synthesis is intended to harvest the insights from the selected ontologies, building upon their strengths and eliminating – as far as possible – their weaknesses. Early results of this work are reported.

The Editors

Kurt Kosanke
CIMOSA Association,
Böblingen, Germany

Martin Zelm
CIMOSA Association,
Stuttgart, Germany

A Merged Future for Knowledge Management and Enterprise Modeling

Report Workshop 1/Workgroup 1

H. Ted Goranson[1], (Ed.), Michael N. Huhns[2], James G. Nell[3], Hervé Panetto[4], Guillermina Tormo Carbó[5], and Michael Wunram[6]

[1]Old Dominion University, USA; [2]University of South Carolina, USA; [3]National Institute of Standards and Technology, USA; [4]CRAN - Research Center For Automatic Control, France; [5]Universidad Politecnica De Valencia, Spain; [6]Universität Bremen, Germany,
tedg@sirius-beta.com

Abstract: see Quad Chart on page 2

1 INTRODUCTION

The workgroup examined the relationship between knowledge management (KM) and enterprise modeling (EM). The specific focus was on possible combined futures and the research roadmap these futures require. The workgroup concluded that a combination of techniques from KM and EM shows promise in addressing the limitations of each.

The following Quad-Chart (Table 1) summarizes the work of the group that addressed those requirements. It identifies the approach taken and proposes a concept for integrating the KM and BPM technologies.

7 BACKGROUND

Enterprise modeling and knowledge management should be key contributors to decision making in an enterprise. Managers, engineers, and technicians all need knowledge and expertise in order to be most effective. Whether the necessary knowledge is internal or external to an enterprise, it needs to be located, reconciled, and focused on problems at the very moment when it can have the greatest benefit. From an idealistic viewpoint, the entire corporate expertise should be brought to bear on each problem or decision. For this to happen, the knowledge must be organized to be locatable and understandable: this can be provided by EM, with the result that the knowledge is isomorphic to the enterprise itself.

Table 1: Working Group Quad-Chart

EI3-IC Workshop 1 ***KM in inter- and intra-organization environments***	**Workgroup 1** ***A merged future for knowledge management and enterprise modeling***	**2001-December-5/7** **EADS, Paris, France**
Abstract: Both knowledge management and enterprise modeling have strong interest communities; each has a sustainable market in the enterprise supporting practitioners and theorists. Both have structural barriers at fulfilling early promise.		**Major problems and issues:** – Institutionalize individual knowledge – Support education at individual, team, and enterprise levels – KM metrics for financial and cost-benefit analyses – Auditability of intellectual property – Promote self-awareness in automated enterprise agents – Manage diverse corporate culture in virtual and merged enterprises – Softness of many KM topics – Lack of enterprise-wide continuity in KM systems
Approach: – Define what is the nature of "knowledge" in KM systems – Determine where the knowledge is used and how it is accessed – Examine the problem of collecting the necessary knowledge that is tacit and "assumed" by humans – Analyze how human and software agents would apply contents of a knowledge base – Consider how small-to-medium sized enterprises can access this capability		**Results and Future work**: – Knowledge exists only in human minds--stuff stored electronically is information – No new techniques are needed to model information relating to knowledge – Need methods for representing information about "soft" enterprise activities such as strategic planning and decision making – Need metrics for measuring the adequacy of soft information – Need methodology to define what we know, need to know, do not know, cannot know. – Need methodology to define what we should forget either permanently or for the subject analysis

Both knowledge management and enterprise modeling are well established in enterprises today. In both one can find:

- A robust vendor and consulting community.
- Well-established university research groups, funded in different ways.
- An active press, targeting both managers and technicians.
- Enough promise — supported by case studies — to fuel continued investment and implementation.

However, there have been some spectacular failures and some vexing limits in successful implementations.

7.1 Enterprise Modeling and Integration: Background

Enterprise modeling is done for a purpose, and an important one is to support the optimization of operations, through what is termed Enterprise Integration (EI). This is a fundamental business need with direct and measurable benefit. For some time, there have been many techniques to model processes and other elements of the enterprise. Modeling in this context means creating an explicit representation, usually computable, for the purposes of understanding the basic mechanics involved. One often uses that understanding to measure, manage, and improve the process or element.

A basic problem is that there are many types of elements to be modeled in an enterprise, and many perspectives and contexts in which those models would be "viewed." Enterprise integration in this context combines models and their uses in such a way that the whole system can be seen in various coherent ways and for multiple purposes. EI provides a model framework in which components can be interrelated.

Some EM and EI systems are wholly computable. Enterprise Resource Planning (ERP) is one that focuses on specific tasks, delivering planning and control functions. It generally requires a constrained modeling approach and heavy use of generic models, thus restricting the processes for better or worse. The more general EI philosophy is framework based, such frameworks supporting:

- Levels of model genericity to enable model and best practice reuse.
- Relationships among different views (for instance views needed to see organizational linkages versus information flows).
- Relationships among different types of basic entities in the enterprise; for instance, activities need to be modeled differently than roles or resources.

CIMOSA is a strong example of such an integrating framework, major elements of which are standardized internationally as ISO 15704. EI frameworks are widely used, especially in the subset of ERP noted above and a similarly focused subset of Product Data Management (PDM), which supports activities centered on the evolution of product features as they are transformed by processes in the enterprise. Enterprises that use computer-aided design heavily implement EM in this fashion.

1.1 Enterprise Modeling: Problems and Limits

The major problems of EM are of two types. First, EM assumes that one knows what should be made or done, who will do it, and a precise notion (perhaps to change later) about how each element of work will be done. Because the primary leverage from the approach is the system view, some substantial part of the system must be included in the model. But those enterprises desiring a system view might wish to include strategic marketing and product design elements, if applicable. Such processes aren't as easily captured as process models however: they have "soft" elements like unknown futures, tacit knowledge, and poorly understood cultural and collaborative dynamics.

Second, EM usually deals with the normative, stable, deterministic case. In other words, managers expect their world to remain as it is because they are going to great lengths to engineer an operational enterprise. Dynamic environments, evolving processes, shifting partnerships and changing products are a way of life for many enterprises. So if EI is employed, it must be more federated than unified. That means the EI system must ideally be cheap to assemble, must change the source models and process in little or no way, be responsive to change, even indicate change, and be to some extent self-organizing and adapting.

Adding KM techniques to the mix can mitigate these two problems, possibly in a revolutionary manner.

1.2 Knowledge Management: Background

Knowledge Management solutions address several needs that all share the underlying notion that enterprises depend heavily on individual and institutional knowledge, and the knowledge must be better understood and managed. KM is a set of philosophies, tools, and techniques to support various functions within this need. While both KM and EM address pressing business needs, EM originates from the industrial engineering and operational perspective and is technique-centric; KM originates from the management perspective and is needs centric. The two communities have a poor history of deep collaboration, which may explain why such an apparent synergy has been hitherto unexploited.

The discrete problems addressed by the KM community are:

- A need to capture individual knowledge to make it "institutional" knowledge so that it can be reused in the enterprise, and persist when an expert leaves.
- A second order intent to use standardized knowledge elements and communication methods to develop and support corporate culture for competitive benefit.
- Support for the "learning organization:" education at the individual, team and enterprise level.
- The development of knowledge "metrics." Significant investment is wrapped up in knowledge, and there is currently no good way to quantify the value of the result. Metrics are needed by financial accountants to evaluate capital knowledge assets, and planners using simple cost-benefit analyses in decision-making.
- Auditability of intellectual property. Tracking the initiation of an idea and the various inputs can reveal who contributed what and when and prove it in court.
- Self-awareness. The better you "know" yourself and your relationship to the world, the better you can change and manage yourself. This notion is the very same driver as in EI, where it is focused on operations, but in the KM world it is more focused on strategic planning.
- KM is often invoked as the backbone around which diverse corporate cultures will be combined after a merger or acquisition.

Because the needs of KM are more diffuse, the tools and implementations are too. Many tools are simply ways of aiding collaboration by structuring the way information is stored, indexed, and shared. Also, many of the techniques are "soft" and merely philosophical, motivational or concerned with building awareness.

1.3 Knowledge Management: Problems and Limits

The general problems with KM systems are of two types:

- KM systems are "soft," almost by definition. They deal with intellectual property for which no good value metrics exist; they deal with collaborative contexts that are not well modeled; and they implicitly address the slippery reality of "tacit" knowledge. Many KM systems deal with strategic planning, which means they address uncertain futures, but without extrapolating from the current situation. The current situation is often described only by an EI or other operational system, whether or not formalized and automated.
- KM systems deal with both "know-what" and "know-how," but with little emphasis on the "how." In other words, the knowledge is not sufficiently bound to the work of the enterprise, or what that work

might become. One part of this problem is the age-old lack of linkages between strategic planning and operational management — it is not just an impedance mismatch between functions, but between methods and basic representations as well. This mismatch frequently produces strategic decisions that make little sense.

Just from this brief overview, the reader may already be anticipating suggestions from the working group on how the strengths of one approach could strengthen the weaknesses of the other. KM needs formalisms (which might help with metrics) and anchoring in the enterprise's actual work; EM needs ways of dealing with knowledge about context and other soft elements, specifically including tacit knowledge.

8 NEAR TERM FUTURE: DEDUCTIVE TRUST AND PROCESS SITUATING

The workgroup recognized a few near term synergies between EM and KM.

"Knowledge" in the KM context is "justified true belief." Each of those three words conveys different dimensions of trust in the information. Usually that trust is "inductive;" the trust is based on (in ascending order of "closeness" to your own judgment):

- Authority: Someone in the enterprise represents that the knowledge is to be trusted. This person might be trusted by you, in which case you trust that person as a certifier of sorts; but usually you are delegating trust.
- Votes: The second case above involves a certifying agent that has the authority of the enterprise, which can be seen as a case of enough votes of the right kind. This type involves votes directly on the information itself. You might not have cause yourself to trust the information, but some group dynamic provides additional confidence, by aggregated authority or broadened depth. (There are likely several group mechanisms involved here, but the workgroup did not exhaustively explore them.)
- Experience: You have seen this case before with enough similarity and enough times to have confidence that it will turn out the same way the next time.

But there is a different basis on which one might base trust, a "deductive" basis that involves understanding the cause and effect mechanics behind the situation in sufficient detail to determine the outcome. For example, one may have experienced many sunrises so have inductive confidence that the sun will rise again tomorrow. Or that person may have deductive trust based on knowledge of the planetary mechanics that produce sunrises. Deductive trust produces a better foundation for justified true belief.

In the business enterprise, deductive trust is much preferred, because it is auditable: decision makers can — if so inclined — "audit" the trust behind the knowledge by zooming in on the underlying physics. Most knowledge in an enterprise is of the inductive type and this is reflected in current KM systems, whereas managers want most to be of the deductive type. Enterprise models capture cause-and-effect dynamics within the enterprise, so a marriage would seem manifest destiny. In this case, each element of knowledge in the KM system is linked to modeled processes (representing activities) in the EM.

Such a linkage can be made during the (already costly) modeling and knowledge capture processes without unduly extending the difficulty of either. The benefit to KM would be rather profound: some significant portion of the knowledge will be (or be expected to be) deductively auditable by linkage to actual processes. Another way of putting this is that knowledge in a KM system is know-how; current KM approaches focus on the "know," but not the "how." Linkage of KM to EM provides the how. And that "how" linkage provides a significant benefit: maintaining knowledge costs money — maintaining vitality in that knowledge base costs more.

Knowledge managers need to know which knowledge to "forget." If there is not a robust linkage to processes (current and future), the knowledge has no apparent relevance to the business. That should prompt an examination with one of the following results:

- The EM is incomplete and needs to be extended. In this case, the existing knowledge indicates what processes need to be better modeled or added. Experience indicates that this can be a powerful technique for modeling processes that have "soft" mechanics, such as many marketing processes.
- The knowledge is determined to be not relevant, and can therefore be deliberately forgotten. The ability to know what is not relevant is an important step in a system's knowledge of itself, which in turn is a necessary condition for being a "learning organization." Knowledge should be deliberately forgotten because it is out of date; because of machine constraints on storage or search time; or because it can be more robustly handled by a collaborating agent.
- The knowledge is determined to be relevant, but poorly supported by processes in the existing enterprise. This would indicate modifying the enterprise. Often the solution in this instance is to develop business partnerships with entities that can support the knowledge process linkage either by supplementing the source enterprise, or maintaining that knowledge itself.

- The situation is the complement of the first case, where an EM is more complete than the knowledge base. This can be used as an indicator of knowing what you don't know within the universe of interest.

This rounds out the four likely conditions for full KM: knowing what you know in a trusted way; knowing what you can forget; knowing what you do not know; and knowing what you can delegate. Knowledge resides in the individual, but has value in the context of the enterprise. KM can be seen as the management of pieces of knowledge, while EM can be seen as the compositional framework for those pieces.

Another way of understanding the problem is to consider a breakdown of KM into four elements: revealing information; forming and managing facts; forming and managing relationships and contexts among facts; and applying that knowledge to effect. Today's KM systems do the first two well enough, but need help with the other two.

EM may help with understanding contexts. The basic idea behind EM is taking fragments of information within the enterprise and placing them in a larger context. EM provides a registration framework for the parts that relate one to another. But this framework relies on artifacts of the modeling process that capture local interdependencies. KM systems based on *ontologies* can allow global registration. Ontologies are formal descriptions of elements and behaviors, originally devised to help share knowledge between systems employing different representations.

A focus on ontologies should provide a bridge between EM and KM, but the leverage is likely to come more from the EM side, because enterprise models are based on the notion of activities and outcomes, which automatically captures a notion of local dependencies among information elements. This notion is what — at root — allows compositions into larger context and systems. The state of the art in process ontologies is the Process Specification Language, developed at the U. S. National Institute of Standards and Technology [PSL citation] and proposed as an international standard.

To provide a bridge between KM and EM, PSL is the likely starting point. In particular, the combination of a PSL-like ontology structure and CIMOSA-like composition strategies can be overlain on existing KM tools and theories to provide for system behavior and business context. Both PSL and CIMOSA (or substitutes) will have to be examined carefully for needed extensions. Neither was designed for this larger, more ambitious role.

The "effect" problem in KM is the problem of linking each piece of justified knowledge to a business role. The workgroup believes EM can help if there is a slight shift of emphasis from the normative notion of "task" in EM. EM is concerned with doing work, and processes that perform tasks are the logical currency. But knowledge is more naturally seen as being applied to solve problems. So a "problem-centric" notion of the basic unit is proposed as a bridging strategy. A problem is seen as a combination of a task (or set of tasks) together with an element (or elements) of knowledge.

At first glance, this seems an immediately implementable strategy to take short-term advantage of synergies between existing EM and KM tools and techniques. The workgroup proposes serious research focused on this likely "low hanging fruit."

There is a precedent for the sort of merger suggested here, and an example of how quickly the result can spread and become the normal way of doing things. Financial management is a matter of collecting many pieces of information and managing them in much the same way that KM intends to manage knowledge. In fact, financial knowledge is a simple case — the qualitative case — of general knowledge; so KM is a generalization of financial management.

About two decades ago, accounting reached a crisis very similar to the KM crisis today: (financial) knowledge was collected but not relevantly "situated." All of the problems noted above existed in some form. The response was Activity Based Costing (ABC), which simply uses a reduced form of enterprise model to ground individual costs and provide a way of intelligently assembling and relating them. ABC went from a proposal to standard practice in less than a decade; substantial benefits resulted. The near-term EM/KM proposal simply extends this logical evolution. As with the ABC revolution, a key strategy is to continue the same basic tools already in place; in this case, that means to continue using the operational and business process modeling methods that are already part of the management toolkit.

In the KM context, most KM is non-formalized and non-managed so of course it is non-computable. Informal KM is a human-to-human phenomenon based on personal networks. So this end of the merged KM/EM system must leverage and ride on top of the human infrastructure.

9 MEDIUM TERM FUTURES: FACT BASED DECISION MAKING

EM is generally focused on tactical optimization and similar types of self-examination. But many enterprises have their most pressing needs in strategic planning in the context of uncertain futures. The more uncertain the future, the more significant the threats and opportunities, but the less valid are simple extrapolations from the past.

The importance of thinking about the future is paramount for many enterprises, and for these real resources must be committed for designing processes to be able to respond in an agile way. Decisions are weighty and should be deductive where possible. Often this is termed "fact based decision making," and it is frequently supported by iterative simulations of what-if situations.

The connection of this task with both EM and KM is straightforward and obvious. "Traditional" EM structures processes so that systems can be optimized. EM for simulation (though not recognized as such) does precisely this with the twist that the processes are executable representatives of the processes. Models in most conventional EI systems don't have this character, they are representatives used to understand, not control processes. But the extension to control is not so great in many cases, and indeed modern EI systems perform substantial but limited control. The further extension to simulatable elements is also not so great, generally involving substituting synthetic stimuli for real ones. So it seems quite logical and cost effective to speak of EM in the context of strategic simulation, especially when the basic unit is the problem as suggested above.

(It should be noted that the advantage does not flow the other way. Most built-from-scratch simulation systems use "models" that cheaply emulate the behavior of processes. This cheapness is usually achieved by not modeling the underlying "physics" of the system; also the granularity is not determined by the unit of work as seen at the level of the work, but at some coarser subsystem granularity. As a result, simulation-derived models cannot easily be adapted for wider purpose.)

The workgroup has three recommendations to make at this medium term horizon and in the context of strategic, fact-based simulation.

- The merger of EM and KM should be extended (and justified by) the use of the combined, structured knowledge/process base for simulation. The advantages are potentially profound because of the reuse of information, the running start in well-founded infrastructure that works, and the hard-won existing, practical binding to the way things are really done. The technical challenges seem to be in "packetizing" knowledge elements from the KM side and adding a few new expressions to modeling methods on the EM side.
- Notions of reuse should be better exploited. The advantages of this are seen as similarly profound. The basic problem is that KM systems are generally case-based, meaning that the knowledge and its representation are bound in specific cases containing details that are irrelevant artifacts of how the information appeared. It is hard work to wade through cases to find relevant insights, extrapolate what is needed, and apply it in a specific new context. The preferred alternative is to build analogy-based KM systems, which index and manage information at a more generic and reusable abstract level.

Analogy-based systems are hard to build, and certainly not expected in the near term. But the first step toward such systems may not be so far away. It concerns clear guidelines about what is generic and what is specific to a task, problem or application. As it happens, EI frameworks are nearly universal in dealing with this problem in some way. Unfortunately, the solution is a matter of art specific to the expert who is the source for the knowledge being modeled. It probably is the case that every practical determination of what is generic must be captured in this manner. In other words, it is a type of meta-knowledge that is captured at the same time and using the same methods as the "base" knowledge. The format comes from the integrating framework.

The bottom line is that KM systems can take a large step toward identifying generic analogies by adopting EM methods when collecting knowledge from experts.

The final medium term recommendation concerns knowledge feedback, or self-reinforcing truths. An example is when a prominent stock analyst predicts a stock will rise. It does in part because of her recommendation, which further reinforces confidence in her "analytical" ability. It turns out that many dynamics in an enterprise may be of this type. For example a quality metric may indicate quality because second order dynamics may have adjusted or grown up around it to promote quality results. For instance, a quality metric may be related to number of inspections, and the precision of those inspections adjusted to the fact that the system drives to many inspections. In fact, the same quality could be achieved with fewer inspections, but only by breaking the cycle of driving toward many, promoted by the "truth" feedback.

Both EM and KM systems have this problem. Usually it is concealed in so-called "tacit" knowledge, which is the concern of many KM systems. But tacit knowledge is a famously black hole, not exhaustible. Good KM practices will help identify which tacit knowledge needs to be captured and why, and (sometimes) at what cost. But these truth feedback loops are best identified when they are deliberately broken as experiments, for instance actually trying to reduce the number of inspections while taking concurrent action elsewhere. One can practically do this only in simulated enterprises, which brings us back to the merger of EM, KM and strategic simulation.

The workgroup did not have time to make specific recommendations of steps and research issues toward solving this problem. But there is a general feel that opportunities are available when the problem is well stated and the more near term steps noted above are taken.

10 LONGER TERM FUTURES: SELF ORGANIZING ENTERPRISES

The workgroup considered the next generation of EI systems. These are likely to exhibit federating behavior and to do so using an agent system. They are also likely to cover much more of the enterprise. The new scope will include at minimum some strategic planning and product definition as one dimension of expansion, and some human, knowledge, and collaboration dynamics in the other dimension.

Agents in this context would likely be the result of evolution from first generation models that represent the superficial behavior of a process, and the second generation noted above where the models capture underlying physics and can be exercised in a simulation environment. Third generation models will be agents, small pieces of software code that include the model and have the ability to negotiate among themselves to optimize the system.

The result will be federated enterprise integration on the model level - not the enterprise proper - where the system self-integrates. But since these models have the ability to control, the effect is much the same.

This vision of EI was already identified in the second ICEIMT when creating a capability model for integrated systems. A high level of integration was when a process had the ability to see itself, see its context in the system, and change itself to optimize the system — perhaps in collaboration with others — even when it would apparently "harm" the agent. Presumably, the risk-reward environment would be structured to reward this behavior, and even reward an earnest but unsuccessful search for such optimization.

A higher level of integration is achieved when an agent has the ability to see into the system — following a relationship chain of some sort — discern a change in the system that would optimize the system, and effect that change. In this scenario, all of the agents involved would be rewarded in some way. For example, you may have a set of processes that do nothing but search and optimize for agility against a likely general change. If the enterprise were a virtual enterprise, this agent would be looking at processes involved in the work and others not currently engaged. All processes are in different formats, use only partially integrated applications, and cross business and cultural boundaries. Agents in these companies would be expected to enthusiastically support simulations that could eliminate them from the partnership. In fact, each company is expected to devise novel notions to support this process. This was considered an achievable goal.

In this case, distinctions among knowledge bases, operational process models, business processes, financial metrics and simulation agents will have all but disappeared. But there clearly are barriers. Perhaps the key barrier concerns realities of agent mechanics. As noted above, these agents need to know themselves and what they know, know what they do not, know where to get trusted information remotely, know what to forget, and know the system's goals and associated metrics. Perhaps it will collaboratively determine those goals.

Knowledge managed by these agents will include soft elements such as unknown futures, tacit knowledge and collaborative (cultural) dynamics. The system will integrate (in addition to factors currently handled by EI frameworks) product features, process features, and system features. (This latter incorporates the system optimization metrics.) The managing context will be through bounding constructs (for instance discretely supervised profit centers), practical constraints and financial and implementation motivations.

The good news is that lots of work by bright people is going into the general case. The business case provides a much simpler universe than "real life" because businesses (not necessarily their employees) are presumably motivated by financial rewards that are quantifiable. There are only complications about deferred rewards (market share, stock price, increased capability, new markets and the like). Moreover, the business application can justify significant investments in research and products — a repeatable improvement of only a few percent means hundreds of billions a year. Moreover, an agent-based system seems inevitable because it is the only scalable strategy for either knowledge or model management.

Agents are introduced to mitigate complexity, so agents themselves will be engineered for simplicity. One strategy will be to devise agents that all behave the same. The reason is that each agent has to know how the others will behave; if they are all the same and the agent "knows itself" (or has recourse to examine itself), it can predict how others will behave.

There are likely to be many thorny research issues, but the workgroup focused on two related ones that are key. The first involves harmonizing the notion of uniform agents with the wild variety of models likely to be involved. Recall that at this level of federated integration, diversity of methods is expected, even encouraged. Obviously, some sort of agent wrapper must be devised. The work indicated in the near and midterm agenda sketched above indicates that this wrapper structure will almost certainly be designed at the ontology level, built on extensions to PSL. This work will begin on a firm basis because the first extensions to the existing PSL base will be known agent needs. The most prevalent approach would be to use "speech acts" which have several formal advantages and the elegant property of being intuitively related to processes as they are currently modeled.

The second challenge indicated for attention by the workgroup is the so-called multilevel agent problem. This problem has an analog in the real world: not all processes need or want the same level of freedom. Some collection of processes or organizational elements will be bound more tightly within the enterprise. For instance, several processes will typically be collected in a partner company. The processes act as agents, but the company does too, and one is not a simple sum of its constituents. Similar aggregations may occur by functions and many aggregations may overlap.

The research challenge is to design the wrapper so that it can both support the aggregation process and accommodate the agency of these higher-level agents. Clearly, this strategy will be framework-based, by methods extended from today's EI frameworks.

11 EXTRA CONSIDERATIONS

In addition to the ambitious agenda noted above, the workgroup raised three issues to be considered by the EI and KM communities.

The first is a common suggestion that needs to be underscored. EI and KM are generally thought of as something that large firms do to preserve their way of doing things, which is maintaining centralized control. The agenda above adds the clear alternative of smaller companies or profit centers opportunistically aggregating to act as large enterprises. That means that a future merged strategy must be devised with sensitivities to small and medium enterprises. Flexibility and tailorability must increase and complexity and cost must decrease from current practice.

The second is the complement. Implementing a new infrastructure with the level of cleverness outlined will change some fundamentals of how business is done. Some optimization must be considered at a higher level than the larger enterprise, beyond to national and societal interest. This is especially cogent, as the initiating research will likely be funded by governments.

The final concern extends that notion in a structural way. Some technologies seem inherently abusable, while others seem self-correcting by design. For example, the Internet will likely be an inherently democratizing force despite the best efforts of large companies to "own" it or repressive governments to co-opt it. The workgroup recommends a project to study how to ensure that this new direction for merged EI/KM is inherently "good" and designed in a way that prevents capturing by inevitable corporate attempts to bend it one way or another for selfish purposes that compromise other elements of society.

Anchoring Knowledge in Business-Process Models to support Interoperability of Virtual Organizations

Report Workshop 1/Workgroup 2

Peter Heisig[1], (Ed.), Martine Callot[2], Jan Goossenaerts[3], Kurt Kosanke[4], John Krogstie[5], and Nenad Stojanovic[6]
[1]FhG-IPK, Germany, [2]EADS, France, [3]Eindhoven Univ. of Technology, Netherlands, [4]CIMOSA Association, Germany, [5]SINTEF, Norway, [6]University Karlsruhe, Germany
Peter.Heisig@ipk.fhg.de

Abstract: see Quad Chart on page 2

The only function of knowledge is to enable right decisions
(Chinese wisdom – Neo-Mohism about 200 BC)

1 INTRODUCTION

With the emphasis shifting to global markets and inter-organizational co-operation, complexity of enterprise systems is further increasing and with it the importance of real time information and knowledge for decision support. In these complex relationships management acting and reacting must be based on a blend of relevant knowledge and up-to-date information. It is this need for information that becomes of paramount importance in the decision-making processes at all management levels of inter-organisational enterprises.

The following Quad-Chart (Table 1) summarises the work of the group that addressed those requirements. It identifies the approach taken to resolve the issues and proposes a concept for integrating the KM and BPM technologies and ideas for future work for testing and enhancing the proposed solutions.

1.1 Background on Knowledge Management

The concept of knowledge management has been used in different disciplines, previously mostly in knowledge management and engineering (Skyrme, Amidon, 1997, De Hoog, 1997, Schreiber, et al. 2000) and artificial intelligence (Göbler, 1992, Forkel, 1994).

Table 1: Working Group Quad-Chart

<table>
<tr><td colspan="2">EI3-IC Workshop 1
Knowledge management in Inter- and Intra-organis. Envir's</td><td colspan="2">Workgroup 2:
Integrating KM and BPM to support interoperability in VEs</td><td colspan="2">2001-December-05/07
EADS, France</td></tr>
<tr><td colspan="3">Abstract:
The working group investigated the relations between KM and BPM to increase the efficiency of enterprise collaborations in the virtual environment.
The report presents a concept for connecting both knowledge management (KM) and business process modelling (BPM), and thus enhancing model based decision support.</td><td colspan="3">Major problems and issues:
– How to create and exploit synergy between KM and BPM to increase efficiency of enterprise engineering in the virtual environment?
– How to integrate general knowledge into business-process models and thereby enhance model based decision support?
– How to identify critical knowledge in business processes?
– What is the role of ontologies in KM and BPM?
– How to establish a common domain or even enterprise ontology?</td></tr>
<tr><td colspan="3">Approach:
– Review KM and BPM technologies and selected applications to identify commonalities and differences
– Focus on the process view of both technologies
– Discuss ontologies and their role in KM and BPM and potential contribution to decision support in establishing, exploiting and closing virtual enterprises
– Map KM onto BPM using representations of current technologies
– Categorise the knowledge needed in business process based decision support</td><td colspan="3">Results:
– KM and BPM are very similar and have some common objectives (capture knowledge, structure knowledge, provide knowledge for decision making
– Proposal for mapping the two technologies onto each other for enhancing decision making in the virtual environment
Future work:
– Establish a formal base for enterprise ontologies
– Define domain and enterprise ontologies
– Analyse the potential contributions of semantic web technologies
– Explore methodologies for knowledge structuring in addition to business process based structuring</td></tr>
</table>

IT-based approaches towards knowledge management are dominant. However, knowledge management is mainly understood by practitioners from manufacturing and the service industry as part of corporate culture and a business-oriented method as "The sum of procedures to generate, store, distribute and apply knowledge to achieve organizational goals".

All approaches to knowledge management emphasise the process character with inter-linked tasks or activities. The wording and the number of knowledge management tasks mentioned by each approach differ markedly. They extend from the four activities mentioned above to an approach in Germany with eight building blocks: Identify, Acquire, Develop, Share, Utilise, Render, Assess and Manage knowledge and knowledge goals. The close relationship between processes and knowledge management is underscored by the feedback from companies identifying the design of structures and processes as a critical factor for the success of knowledge management, indicating their focus on the core competence business processes to implement knowledge management.

1.2 Background on Business Process Modelling

Business process modelling is usually done for very specific goals, which partly explains the great diversity of approaches found in literature (Vernadat, 1996) and practice. The main reasons for doing BPM are:

a) To improve human understanding and communication: to make sense of aspects of an enterprise and communicate with other people
b) To guide system development
c) To provide computer-assisted analysis through simulation or deduction
d) To enable model deployment and activation for decision making and operation monitoring and control

A number of modelling frameworks have been developed (e.g. ARIS, CIMOSA, GRAI, IEM, PERA) that provide business process modelling languages allowing description of business processes with various degrees of details and for different points of view on the process itself. The GERAM framework work developed by the IFAC/IFIP Task Force (Bernus, et al, 1996) has become the base for international and European standards (pre EN ISO 19439, 2002). The work is still in progress.

The major application area of BPM is still Business-Process Reengineering (BPR) and Business-Process Optimisation. The real potential of BPM - real time decision support – is barely exploited.

1.3 Background on Ontologies

The task of the ontologist is described as: "to recognise, analyse and interrelate those concepts enabling him to produce a unified picture of reality" (Bunge, 1977). With reality understood as being the concrete world, but not including the concepts that words may designate. Ontology joins the natural and social sciences as a discipline concerned with concrete objects. It has the task to construct the most general theories concerning these concrete objects, their being and becoming. In contrast, common "scientific" knowledge domains such as ergonomics, logistics and many others, each define concepts and relationships, and connect them to some area of investigation. Whereas the practitioner of a discipline has a strong awareness of the concrete-world things as the anchors and purposes of the analysis, the heavy conceptual bias of the knowledge engineer or information analyst has given rise to several so-called ontologies, which are void of the being and becoming of the object of study.

Focussed ontologies have been defined and used in several domains including medicine, chemistry, and legal knowledge representation. In the area of enterprise modelling, early work that would nowadays be classified under the name enterprise ontology is the REA Accounting Model (McCarthy, 1982). Quite a few "enterprise" ontologies do not emphasise the distinction between things and their changes on the one hand and concepts on the other hand. These ontologies therefore have more fundamental concepts than strictly necessary. Examples are the Enterprise Ontology project (Ushold, et al, 1998) and TOVE (Toronto Ontology for Virtual Enterprise) (Fox, et al, 1998).

2 APPROACHES TO INTEGRATE KM AND BPM

Both KM and BPM aim at improving the results of the organisation, delivering a product or/and service to a client. The related business processes use knowledge as a resource. Nevertheless, only very few approaches to knowledge management have explicitly acknowledged this relation. And even fewer approaches have tried to develop a systematic method to integrate knowledge management activities into the business processes. Three forms of KM-BPM integration can be found (Mueller, et al, 2001):

a) BPM as the basis for the knowledge management is based on treating knowledge management as a specific business process in which an organisation creates and uses individual and collective knowledge (Macintosh et al, 1998, Mentzas, Apostolou, 1998).
b) KM as a basis for the Business-Process Improvement/Reengineering can provide knowledge for modelling, optimisation and automation of business processes.
c) KM integrated in the process- or workflow-management systems to provide access to the knowledge that is relevant for the current task.

In this paper we focus on the last form since it is the most reliable approach for integrating KM and BPM in the virtual organisation.

Following is a list of selected approaches:

- **CommonKADS methodology** (Schreiber, et al, 2000) integrates an organizational model, critical success factors and the KM cycle with seven activities: *Identify, plan, acquire and/or develop, distribute, foster the application, control and maintain, dispose.*
- **Business KM** (Bach, et al, 1999) tries to relate KM activities to business objects and business processes. The approach distinguishes between business processes, the knowledge structure, and the knowledge base.
- **Knowledge value chain approach** (Weggeman, 1999) is a continuously repeated process, which is composed of six KM tasks on the operational level: *identify, document, develop, share, apply and evaluate knowledge.*
- **Model-based KM approach** (Allweyer, 1998) adds a new perspective especially for knowledge-intensive processes (less structured, not exactly foreseeable and, in most cases, not repeatable).
- **Reference-model for KM** (Warnecke, et al, 1998) is an approach of a model-based design of knowledge-oriented processes for KM. The reference model consists of an object model with system elements and activities (*identify, make explicit, distribute, apply and store*), a process model and an implementation model.
- **Process KM** (Jorgenson, Carlsen, 1999, Jorgensen, 2000) is defined as the collection of processes necessary for *innovation, dissemination, and exploitation* of knowledge in a co-operating ensemble where knowledge seekers are linked to knowledge sources and a shared knowledge base is cultivated.

3 PROPOSAL FOR INTEGRATING KM AND BPM

3.1 Assumptions and approach

Our approach to business process oriented knowledge management is based on the following assumptions:

- KM operative methods and procedures used to generate, store, distribute and apply knowledge have to be integrated and oriented towards particular business processes.

- KM has to consider the specific cultural conditions - the network of different professional cultures, functional cultures and underlying corporate traditions and values (Davenport, et al, 1996).
- KM has to accommodate the daily use of knowledge and know-how of our colleagues, suppliers, clients, competitors and other resources (Hansen, et al, 1999).
- The drivers for both the traditional business processes and the knowledge management processes are combined to fulfil the business needs (Bullinger, et al, 1997).

Our approach rests on identifying relations between KM and BPM, using the IPK approach on Knowledge Management shown in Fig. 1 (Heisig, 2001) and the Enterprise Modelling Framework identified by (pre EN/ISO 19439, 2002) and partly shown in Fig. 2. Business process related knowledge is being captured/generated, stored and applied during all phases of the model life cycle. Such knowledge is used in model-based enterprise engineering during most of the life cycle phases and is applied for operational use during the enterprise operation phase. Knowledge distribution beyond the area of the business processes is not covered in the modelling framework.

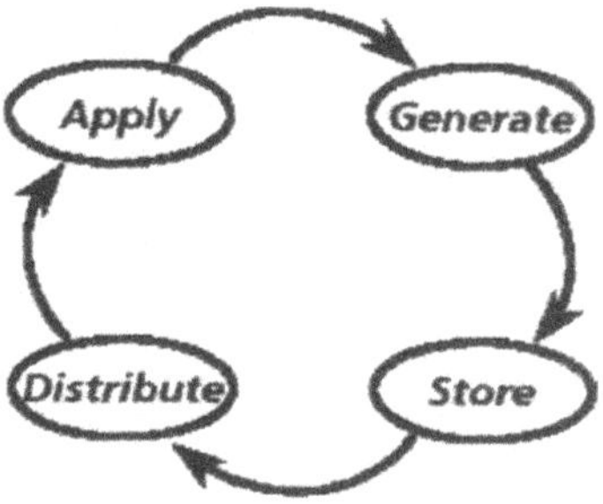

Figure 1: KM activities

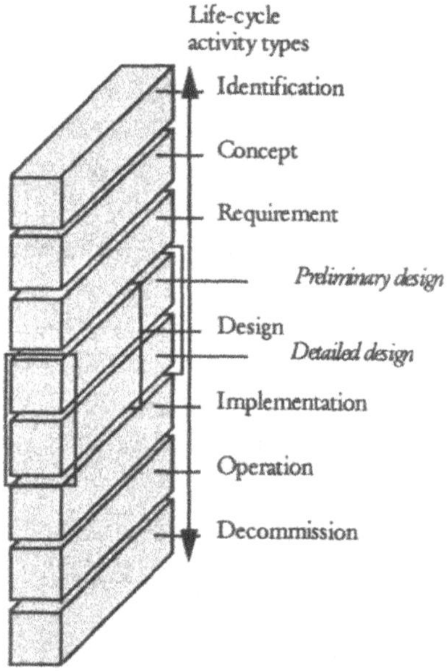

Figure 2: Modelling Framework

Therefore the KM activity *Distribute* has to be defined as being applicable during all life cycle phases identified in the modelling framework, providing for authorisations, promotion and exploitation of all the enterprise knowledge. This additional distribution needs might give rise for additional properties of the process model, i.e. meta-data specifically useful for reuse across the enterprise.

Establishing term (index) mappings between information and knowledge according to the structure of the business process has the advantage that knowledge distribution and application in the business process community is significantly improved since this structure is well known and accepted in the enterprise.

3.2 The role of ontology

Ontologies are a conceptualisation of a domain (Gruber, 1993). Thus, they provide mechanisms to structure knowledge sources according to the characteristics of the domain. It means that ontologies (or the vocabulary that one ontology provides) can be used for the creation of an indexing system, which is appropriate for the content description of the knowledge sources in order to make the sharing of this knowledge more efficiently (Stabs, et al, 2001). This is achieved by constraining the meaning of some indexes (terms) according to the axioms in the ontology. For example, it is possible to distinguish term chair as an organisational role from the term chair in the context of a business activity where chairs, as furniture, are assembled. Therefore, ontologies provide means for the semantic-based providing and access to knowledge, which is the crucial requirement for an efficient knowledge management system.

In order to anchor knowledge sources to the business processes, one needs two kinds of the indexes and term mappings between them – one index for each knowledge source pertaining to a problem domain (e.g. automobile industry, logistics, or ergonomics) and one index for the knowledge on the business process (e.g. assembling a product). In that way knowledge sources can be applied to each business process for which a mapping has been established. An efficient integration of KM and BPM need two kinds of ontologies: the Domain ontology that describes the knowledge sources of a problem domain (content) and the Enterprise ontology that corresponds to the business processes (creation and application context) (Abecker, et al, 1998).

From the virtual organisation point of view, the role of the ontologies in the knowledge sharing is even more important:

- Different vocabularies, used in geographically distributed organisational units, can be merged on the conceptual level (i.e. not on the syntax level, but on the level of the meaning of the terms) using a Domain ontology;
- Inputs and outputs of the business process can be described on the conceptual level (e.g. an input of a business activity is the Name of the customer, but not any string) using an Enterprise ontology; term (index) mappings existing between the Enterprise ontology and certain Domain ontologies then enable -semantic composition of the processes in a supplier-customer chain

- The comparison between similar business processes in different organisational units can be performed more accurately when the processes are described on the conceptual level, using an Enterprise ontology or one or another Domain ontology.

The presented arguments confirm the importance of the usage of ontologies in the KM-BPM integration and motivate our further research in this direction.

The PSIM environment (Goossenaerts, Pelletier, 2002) makes the distinction between the physical reality of the enterprise – it's being and becoming (context) – on the one hand, and the concepts and relationships (content) that knowledge domains use to analyse this reality. Within the Organization, which is the subject of various analyses in different knowledge domains, the business-process model serves as the pivotal core for term mapping and translation services in the organisation's knowledge engine. These services allow knowledge from various disciplines to be applied in the analysis of the organisation. The importance of reuse of past experience and solutions in organisational learning also justify anchoring the problem domain ontologies in the physical reality of the assembly operations.

3.3 Gaps and further work

Various methods and tools for Business-Process Reengineering (BPR) or Business-Process Optimisation, have been developed by academia and consulting companies. Despite these developments, a comparative study of methods for business process redesign completed by the University St. Gallen, Switzerland (Hess, Brecht, 1995) concludes: „ hidden behind a more or less standard concept, there is a multitude of the most diverse methods. A standardised design theory for processes has still not emerged.“

Adopting an ontology-based approach, further work must focus on how to define domain and enterprise ontologies and how to express term-mappings between the two ontologies. Also the combined application of KM and BPM in enterprise engineering (EE) especially in the area of virtual enterprises needs further investigations. The aim is to explore the relations between knowledge structuring and process structuring. Interoperability of virtual organisations is another area where BPR and EE will benefit from using such an ontology-based approach.

Semantic web technologies seem to have the potential to contribute to application of KM and BPM as well. However, basic research is needed in this area.

4 SUMMARY AND CONCLUSIONS

Knowledge management is currently one of the buzzwords on the agenda of Top-Management and of software providers and consulting companies. Knowledge is regarded as one or even the main factor for private and public organisations to gain competitive advantage.

With business process engineering, companies have focused their attention on eliminating non-value-adding process steps. In the future, companies will regard knowledge management activities as an integral part of their business processes. They will enhance their ability to deploy a significant source of competitive advantage - the know-how and learning of the people.

Behind the buzzword of knowledge management hide essential techniques for the systematic management of knowledge and experiences about operational processes. These techniques will not become superfluous as long as the economy remains dynamic. On the contrary, they will become part of services that add "ease of knowledge application" to the "ease of planning and operation" that has already revolutionised work in organisations.

5 REFERENCES

Abecker, A. Bernardi, A. Hinkelmann, K. Kuehn, O. Sintek, M. (1998), *Towards a Technology for Organizational Memories.* IEEE Intelligent Systems & Their Applications, 13(3).

Allweyer, Th. (1998), *Modellbasiertes Wissensmanagement.* In: Information Management, 1.

Bach, V. Vogler, P. Österle, H. (Eds.) (1999): Business Knowledge Management. Praxiserfahrungen mit Intranet-basierten Lösungen, Springer-Verlag.

Bernus, P, Nemes, L, Williams, T.J. (Eds.), (1996), *Architectures for Enterprise Integration,* The findings of the IFAC/IFIP Task Force, Chapman & Hall.

Bullinger, H.-J. Wörner, K. Prieto, J. (1997), *Wissensmanagement heute. Daten, Fakten, Trends*, Fraunhofer IAO, Stuttgart.

Bunge, M. (1977), *Ontology I: The Furniture of the World.* Treatise on Basic Philosophy Vol. 3, Reidel, Boston.

Davenport, Th. H. Jarvenpaa, S.L. Beers, M.C. (1996), *Improving Knowledge Work Processes.* Sloan Management Review.

De Hoog, R. (1997), *CommonKADS: Knowledge Acquisition and Design Support Methodology for Structuring the KBS Integration Process.* In: Leibowitz J. Wilcox, L.C. (Eds.), Knowledge Management and Its Integrative Elements. CRC Press, Boca Raton, New York.

F.B. Vernadat, (1996), *Enterprise Modelling and Integration, Principles and Applications;* Chapman and Hall.

Forkel, M. (1994), *Kognitive Werkzeuge – ein Ansatz zur Unterstützung des Problemlösens.* Hanser Verlag, München.

Fox, M.S. Barbuceanu, M. Gruninger, M. & Lin, J. (1998), *An Organisation Ontology for Enterprise Modeling.* In M. Prietula, K. Carley & L. Gasser (Eds.), Simulating Organizations: Computational Models of Institutions and Groups, Menlo Park CA: AAAI/MIT Press.

Göbler, Th. (1992), *Modellbasierte Wissensakquisition zur rechnerunterstützten Wissensbereitstellung für den Anwendungsbereich Entwicklung und Konstruktion*. Hanser Verlag, München.

Goossenaerts, J.B.M. Pelletier, C. (2002), *The PSIM Ontology and Enterprise Modeling*. In: Eijnatten van, F.M. (Ed.) Participative Simulation Environment for Integral Manufacturing Enterprise Renewal. TNO, Arbeid, Amsterdam, The Netherlands (forthcoming).

Gruber, T. R. (1993), *A Translation Approach to Portable Ontology Specifications*. Knowledge Acquisition, 5(2).

Hansen, M.T. Nohria, N. Tierney, T. (1999), *What's your Strategy for Knowledge Management*. In: Harvard Business Review, March-April.

Heisig, P. (2001), *Business Process Oriented Knowledge Management*. In: Kai Mertins, Peter Heisig, Jens Vorbeck (Eds.): Knowledge Management. Best Practices in Europe, Springer-Verlag.

Hess, Th. Brecht, L. (1995), *State of the Art des Business Process Redesign. Darstellung und Vergleich bestehender Methoden*. Gabler, Wiesbaden, Germany.

Jørgensen, H. D. (2000), *Software Process Model Reuse and Learning*, in Proceedings of Process Support for Distributed Team-based Software Development (PDTSD'00), Orlando, Florida. IIIS - International Institute of Informatics and Systemics.

Jørgensen, H. D. Carlsen, S. (1999), *Emergent Workflow: Integrated Planning and Performance of Process Instances*, Workflow Management '99, Münster, Germany.

Macintosh, A. Filby, I. & Tate, A. (1998), *Knowledge Asset Road Maps*. In Proceedings of The Second International Conference on Practical Aspects of Knowledge Management (PKM98), 29-30 October, Basel, Switzerland.

Mentzas, G. and Apostolou, D. (1998), *Towards a Holistic Knowledge Leveraging Infrastructure: The KNOWNET Approach*, Proc. Second International Conference on Practical Aspects of Knowledge Management, 29-30 October, Basel, Switzerland.

McCarthy, W.E. (1982), *The REA Accounting Model: A Generalized Framework for Accounting Systems in a Shared Data Environment*. The Accounting Review, Vol. LVII(3).

Mueller, H. J. Abecker, A. Maus, H. Hinkelmann, K. (2001), *Software-Unterstützung für das Geschäftsprozessorientierte Wissensmanagement*. In Proceedings des Workshops Geschäftsprozessorientiertes Wissensmanagement anlässlich der WM'2001 in Baden-Baden.

Pre EN/ISO 19439, (2002), *Enterprise integration - Framework for enterprise modelling*, CEN TC 310, WG1.

Schreiber A. Th. Hoog, R. Akkermans, H. Anjewierden, A. Shadbolt, N. Velde W. (2000), *Knowledge Engineering and Management. The CommonKADS Methodology*. The MIT Press.

Skyrme, D.J. Amidon, D.M. (1997), *Creating the Knowledge-Based Business*. Business Intelligence, London, New York

Staab, S. Schnurr, H.-P. Studer, R. Sure, Y. (2001), *Knowledge Processes and Ontologies*. IEEE Intelligent Systems. 16(1), Special Issue on Knowledge Management.

Uschold, W. King, M. Moralee, S. & Zorgios, Y. (1998), *The Enterprise Ontology*. The Knowledge Engineering Review, 13, Special Issue on Putting Ontologies to Use.

Warnecke, G. Gissler, A. Stammwitz, G. (1998), *Referenzmodell Wissensmanagement – Ein Ansatz zur modellbasierten Gestaltung wissensorientierter Prozesse*. Information Management 1.

Weggeman, M. (1999), *Kenntnismanagement. Inrichtigen besturing van kennisintensieve organisaties*. Scriptum, Schiedom German: *Wissensmanagement – Der richtige Umgang mit der wichtigsten Ressource des Unternehmens*. MITP-Verlag, Bonn, Germany.

Managing Processes and Knowledge in Inter-Organisational Environments

Report Workshop 1/Workgroup 3

David Chen[1], (Ed.), Frank Lillehagen[2], Niek du Preez[3], Raul Poler Escoto[4], and Martin Zelm[5]

[1]LAP/GRAI, University Bordeaux I, France, [2]Computas AS, Norway, [3]Stellenbosch University, South Africa, [4]Universidad Politecnica de Valencia, Spain, [5]CIMOSA Association, Germany

chen@lap.u-bordeaux.fr

Abstract: see Quad Chart on page 2

1 PROBLEMS

Knowledge Management (KM) has been gaining significant momentum within enterprise organisations. However, the differences in understanding of what a KM system is range from enterprise-wide database and information systems to generalised knowledge-based systems, via enterprise modelling and integration systems. This could be a barrier for promoting KM in industry and, consequently, the scope and goal of KM need to be better defined. The workgroup represented the business end user, vendor, consultant and researcher on KM with experiences in KM applications as for example METIS tool of Computas or the EDEN software of Indutech or the IMAGIM tool of GRAISOFT supporting the use of the GRAI Methodology.

Further, the problem of lack of guidelines to support the implementation of KM system in companies was raised. The view hold was that enterprise modelling techniques (e.g. constructs, templates, models…) could provide help to capture and represent knowledge in an appropriate form. Nevertheless the relationship between enterprise modelling and KM needs to be better clarified (for example through a mapping between business process and

KM). As more R&D work remains to be done to make KM a reality, the group also felt it important to identify future needs in this domain.

The following Quad-Chart (Table 1) summarises the work of the group that addressed those requirements. It identifies the approach taken to resolve the issues and proposes a concept for integrating the KM and BPM technologies and ideas for future work.

Table 1: Working Group Quad-Chart

<table>
<tr><td colspan="2">EI3-IC Workshop 1
KM in Inter- and Intra Organisational Environments</td><td colspan="2">Workgroup 3:
Managing Processes and Knowledge in Inter-Organisational Environments</td><td colspan="2">2001-December-05/07
Paris, France</td></tr>
<tr><td colspan="3">Abstract:
KM is considered an important success factor in enterprise operation, however, capturing knowledge and using it across organisational boundaries is still a major challenge. Starting from a comparison of KM and BPM, the paper elaborates on methodologies for integrating enterprise modelling and KM in dynamic networked organisations. Examples of KM/BPM applications in SMEs are provided and discussed.</td><td colspan="3">Major problems and issues:
– What are the definition of a KM system that cause issues like:
  – Lack of a common understanding?
  – A barrier for KM in industry?
– How to define the scope and goal of KM enabling to:
  – Grow with the (system) life cycle?
  – Adapt to evolving infrastructures?
– Why are existing standards not used?
– How to define guidelines for implementation and use of KM systems especially in SMEs?</td></tr>
<tr><td colspan="3">Approach:
– Define the scope of KM applying the GERAM life-cycle concept.
– Compare knowledge and business-process management.
– Discuss the requirements for a KM system infrastructure and the concept of the active knowledge model.
– Present and evaluate examples of actual KM application
– Refer to standards wherever possible, mainly ISO IS 15704, 14258 CEN ENV 40003, 12204, 13559 and others focussing on interoperability.
– Derive future needs from the above.</td><td colspan="3">Results:
– Realising KM with BPM by mapping the basic KM tasks onto BPM.
– Requirements for KM system infrastructures
– Synthesis from examples of Process and KM applications
Further work needed:
– Define methodologies for scalable KM systems for decentralised decision-making.
– Investigate dependencies and interoperation of (process) model management and KM
– Define an infrastructure consisting of IT and non-IT services to support KM across organisational borders
– Design modules for user guidance and training to implement KM systems especially between SMEs for the network of knowledge value chains.</td></tr>
</table>

2 ISSUES

This section presents the main issues discussed and reflects results of the work carried out by the group members independently of the workshop.

2.1 Scope and Goals

The scope of a KM should cover the full system life cycle. According to ISO 15704, (1998) and pre EN ISO 19439, (2002) the life cycle starts from domain and concepts definitions, to requirement identification, design and implementation down to operation and decommission. The scope should also be capable of growing dynamically as the understanding and infrastructure evolves. At each phase of the system life cycle, the main tasks of KM are: (1) identifies, structures, and activates information to become knowledge, (2) structures the mass of information, to make it efficiently useable and (3) supports the co-ordination of collaborative work (Davenport, Probst, 2001).

The KM differs from the enterprise modelling in the fact that the latter deals with the development of the modelling languages and methodology (Vernadat, 1996) while the former is concerned with the capturing, structuring, localising, distributing and utilising the knowledge. In other words, enterprise modelling will provide techniques (constructs and formalisms) and tools to represent knowledge from various viewpoints.

The goal of KM is the improvement of the organisational capabilities to achieve a better utilisation and sharing of its knowledge. An effective KM system may have the following characteristics: (1) Enterprise-wide decision-making support and performance evaluation, (2) Clear knowledge mapping and indexing structures that are well communicated throughout the organisation to facilitate efficient collection of (and effective reuse of) critical information needed for decision-making, (3) Portal-based context-preserving User Environments, (4) Work-process driven development, change and evolution, (5) Infrastructure supported reuse, cultivation and re-engineering of knowledge and solutions, (6) Model managed solutions design, problem-solving and learning, and (7) Knowledge-integrated processes, activities and actions.

2.2 Knowledge and Business-Process Management

Knowledge and business-process management are closely linked. Business processes themselves are a particular type of enterprise knowledge. Associate KM with BPM has many advantages. First it allows to identify operational information as business processes modelling focuses on daily enterprise operation procedures with its needed information and knowledge.

Secondly because people are usually familiar with process structures, the use of the concept of process may facilitate the capture of knowledge. Third it leads to direct use for decision support e.g. simulating alternative scenarios, business cases etc. Moreover process modelling enables to continuous capturing of new knowledge. According to the changing environment of business, processes need to be modified and changed continuously. Furthermore it allows authorisation according to the specific business processes (Need to know basis). Finally knowledge captured in process models allows business partners to pursue common interests. Table 2 shows a tentative mapping between KM tasks and business processes modelling.

Table 2: Mapping Knowledge Management to Business Process Modelling

KM Tasks	Business Process Modelling
Capture knowledge	Inputs and outputs of an activity are relevant information to capture knowledge
Structure knowledge	Inputs and outputs of an activity determine the content of the model views
Identify knowledge	Via navigation across the model views
Localise knowledge	Identification of the higher level Information e.g. processes, enterprise objects
Utilise knowledge	Use the Inputs/outputs of activities for decision support (Simulation of the model)
Manage knowledge	Distribute knowledge and control access rights in the Model or parts of it

It has been considered important that the implementation of an active KM system will influence the behaviour of the system's users. Starting from current working processes, the effects can be evaluated and measured. Information and knowledge in use are presented in various forms such as routines, databases, manual procedures and reports. The accessibility and forms will have an impact on their way of working, and their behaviour.

In summary: (1) BPM is supported by many languages and methods, but often only used for documentation, process analysis and BP re-engineering, (2) With generic, standardised modelling constructs most process types (management, planning, operation, ...) can be modelled and used for dynamic decision support, in virtual enterprises, with real time data etc. and (3) In addition, 'external knowledge' as market analysis and work laws and procedures can be included.

2.3 Infrastructure of Knowledge Management System

The WG3 believes that information and communication technology plays a crucial role to evolve KM pragmatically. Hence, it was considered very important to define and implement an infrastructure, which consists of both IT and non-IT services. In particular, when across-organisational co-operation moves beyond the buying and selling of goods and well defined services, there is need for a flexible infrastructure that supports not only information exchange, but also knowledge sharing, creation, utilisation and management within and across the traditional organisational borders.

The workgroup has studied the architecture for application methodology of operational Enterprise Engineering (EE) with METIS. It shows that different engineering teams can work on, use and modify collaboratively their solution model, each of which is derived from an underlying (solution) meta model. In addition, the teams can modify and adapt their modelling templates in the meta model. The meta models are based on and constrained by a common Meta Data repository representing the inner core of the infrastructure. The Meta Data repository enables exchange and interoperability of data, knowledge and solution models across the entire enterprise.

Then the group has considered the possible structure of an infrastructure being an architecture that provides a certain number of services with a layered structure: Layer 1 is the ICT platform, software architectures, tools, components and applications. Layer 2 is concerned with the knowledge content, representation, sharing and knowledge repositories access. Layer 3 is the model engineering and management layer, providing work processes and services. Layer 4 represents the solution modelling, meta-modelling and the work performance team environment.

2.4 The Active Model Concept

When implementing KM, there is a need to have some models and methodology to help to capture knowledge to perform enterprise modelling. There is also needs to use templates, constructs or meta-constructs to model/describe the knowledge. Enterprise modelling can provide such support, particularly, in capturing and managing the knowledge and using it in a dynamically changing environment and across organisational boundaries.

Constructs and relationships representing the so-called Active (knowledge) Model are implemented in the METIS tool and its constituents (Lillehagen, 2002). The purpose is to show the concept of multi-views and the instantiation from meta data of the active knowledge model. An Active Model is on the one hand composed of sub-models, which are presented with model views such as domain, scope or view style. On the other hand an Active Model consists of template objects derived from Meta Models. These Meta Models can be either viewed under system aspects with features like domain, scope etc. Or they can be represented under user aspects, which means that they are instantiated from processes, activities, resources and organisation units.

2.5 The Standardisation Issues

Standardisation will play an important role to implement interoperable KM systems, especially in inter-organisational environments. Standards will: (1) provide common understanding of (knowledge) content; (2) enable interoperability between models; and (3) insure investments of users and vendors. One problem is that many standards are not used in industry, particularly in SMEs. The possible reasons are: (1) these standards are not known at the industry level, they are ignored; (2) some of them are not developed to a sufficient detail or be operational. As a consequence there is a need for better communication and promotion of standardisation activities.

3 EXAMPLES OF INDUSTRIAL APPLICATION

A summary of knowledge management applications in industry involving WG3 members is given to derive future needs from this experience.

3.1 Process and Knowledge Management Application in Spanish SMEs

CEMENTOS LEMONA, a cement producer with 250 employees has applied KM to improve their processes. The project started from a realistic basis centred on the search for effectiveness in the application of KM as a discipline for process improvement and was implemented to a subset of selected processes. Hence, pilot experience was gained from these processes permitting, at a later date, to apply this tool to other areas of the company. The marked commitment of the management of the company and its strong bet on new management methods constitute a guarantee of the project's success.

VICINAY CADENAS has carried out an important effort to introduce the culture of individual and collective knowledge during the last 10 years. In this way, the relationship between the company and the worker is based more on what he does well rather than on how much he does and with what effort. Two projects were centred in areas in which generated knowledge is crucial for the sustained improvement of its operations, namely the transmission of knowledge between shifts and the management of improvement suggestions. After the implementation of the methodology and a few organisational changes, the company has achieved operational improvements with KM being one of its key competitive elements.

JAZ ZUBIAURRE is an example to apply KM in SMEs supported by a methodology. The company is making production systems for metals surface treatment and a national leader in the production of metallic brushes with 70 employees. JZ applies the administration of the knowledge through a methodology called RUMBO (developed by the foundation Tekniker). The principles of this methodology are: (1) Operability like integrated group, (2) Mobilisation of the human resources, materials and assets toward attaining the adopted strategies, (3) Foment of the conditions that facilitate the acquisition and diffusion of the knowledge, (4) New forms of allotment of the power, (4) Team work, (5) Development of capacities for the administration of the change with success.

Information technology can play an important role to make enterprise-wide KM a reality, in particular for large companies. UNION FENOSA is a managerial group with business in the generation and distribution of eclectic energy with a size of 25.000 employees. It has developed a model for intellectual capital management. The UF knowledge portal has as main components: (1) A standard model that integrates the key elements for business management (strategy, organisation, processes, systems and infrastructure), (2) Contents structured for the elements of the pattern, (3) Databases, experiences and suggestions associated to the elements of the pattern and (4) A computer tool of support integrated in the intranet of the company.

ARTECHE with 1000 people consists of several companies producing electronic measure and protection goods. KM has become a key tool for this managerial group to treat the processes related to innovation, the information system and to learning process, all operated via their intranet.

Besides the support from methodology and information technology, experiences show that we must not forget the people. Human should be at the centre of any KM system as shown in the case of IRIZAR - a 2.300 employees company and leader in the production of luxury buses - famous for comfort, safety and reliability. The company has a Knowledge Project based on people: (1) The knowledge workers should manage themselves (They must have autonomy), (2) The continuous innovation should be part of the work, task and responsibility of the knowledge workers, (3) The knowledge work requires continuous learning, but it demands continuous teaching on the other hand, and (4) The knowledge workers productivity is not only question of quantity but also of the produced quality.

3.2 Enterprise Modelling Based Knowledge Management Applications in South Africa

Several interesting and diverse case studies of Enterprise modelling using knowledge maps and route guides are currently deployed in the southern African region. In all cases the modelling methodology and associated route guides uses a comprehensive life cycle approach and in most cases a multiple lifecycle context. The EDEN framework and software is used, as a modelling environment and invariably the knowledge mapping of each company is a variant of the generic knowledge maps available in the EDEN software.

The single most important common denominator is that of maturity of KM culture and the unqualified non-negotiable need to compete globally. A second very strong success factor is the presence of a project champion to persist in deploying the ICT enabled innovation support efforts. Some examples are listed below: (A) Innovation modelling in the product development process of wine for the global markets, (B) Innovation and deployment modelling of the Rapid ERP implementation process, (C) Capturing of the IP and Process modelling of the Product Development Process in a specialised vehicle manufacturing enterprise, (D) Strategy Deployment modelling of a Health Care and industrially limiting pandemic, (E) Deployment of ISO 9001 within a company, (F) Modelling of a component based supply chain 3D model simulator, (G) Deployment of a Rapid Product development process.

All of these diverse applications share a number of common characteristics. These are: (1) Speed and efficiency of innovation is crucial for success, (2) Multidisciplinary teams are essential for the integration and deployment, (3) Knowledge based innovation processes requires a common knowledge map, excellent storage categorisation and retrieval functionality, (4) Structure and flexibility in the innovation process, (5) An ICT facilitated use of a large variety of modelling, simulation and communication tools and (6) Most important of all a culture to innovate.

In conclusion, the universe of available and accessible Knowledge components varies widely from one company to the next. Mature companies like SCANIE invest a substantial portion of their development budget in growing their network of knowledge value chain, whereas others ignore the importance. However, any Knowledge map should provide for obtaining and filtering the basic resources. Subsequent evaluation and categorisation by a panel of experts could reduce the size of the haystacks in which future needles have to be found again. Industry and domain specific taxonomies and ontologies could assist in logically structuring the objects and adding appropriate meta data to the content. A subdivided set of information content can be indexed to provide a matrix of words and documents so that active search engines can later retrieve appropriate knowledge objects. A very important key success factor for implementing KM successfully is the KM culture that requires all involved to participate in structuring the Knowledge Map of a company. Thus agreeing on and understanding of the structure is of paramount importance.

4 FUTURE NEEDS

Implementing enterprise-wide KM system will have an important impact on the organisation and the way of working (individually and collectively) as well as a change from sequential to parallel working. This will lead to the reorganisation of some human tasks and responsibilities and at the same time could cause a need for new reward systems. The decentralised way of working requires that the KM system provides integrated services supporting the mobile workers. The more an individual - usually in a multidisciplinary team - works in an autonomous and co-operative way, the more activity support via an integrated, globally consistent framework is needed.

Furthermore, an effective, scalable KM system allows decentralising decision-making. Consequently traditional hierarchical organisation tends to change to network organisation where autonomous and smaller production units co-operate. In this context, new project management techniques, new work management methods, new model/KM approaches as well as the control of systems engineering team are some research issues, for which the methodologies are missing today. Emphasis on multi-medial language complementary to traditional more coded and formalised languages will facilitate not only the representation (modelling) of knowledge but also it's understanding (interpretation) by end users, and thus create more interactions between actors.

A KM system will lead to the definition of new work processes. Model based business process monitoring, control and engineering can only become a reality if an appropriate enterprise wide infrastructure and the repository technology to support portable and interoperable KM systems is implemented as well. This infrastructure is an important condition to develop traceable, self-adapting, evolving solutions, which are features of KM systems. In particular, software packages supporting these services must be interoperable regardless of the type of computing platform used. Adequate human/machine and human/human interfaces are to be developed. Standardisation could be an important contributor to achieve interoperability.

Last but not least the human acceptance will always be determining to make any new project successful. More learning, training, education are required, not only to use new information technology but also to transform the implicit/tacit knowledge into explicit knowledge, so that it can be exploited by information technology. In addition, knowledge dissemination, end user help and consulting should not be neglected. This is particular important for SMEs. More learning, training, education (life cycle support) are required at least under three aspects: (1) learn and train how to extract individual and collective knowledge and put it in an appropriate form, (2) how to involve everybody in the company to use appropriate infrastructure-based computer services, (3) learn and train how to use the knowledge to better perform the daily business and/or manufacturing activities by interacting more quickly and appropriately.

5 CONCLUSIONS

Knowledge Management can be realised and implemented with business process modelling. Employing model based decision support has large potentials. However, it requires a common user oriented modelling language, with a common presentation, visualisation and standardised constructs, as well as common understanding of the construct semantics and of the modelling process. Current industrial applications show a great variety of approaches, motivations and results. A common understanding of scope and content of KM leading to the elaboration of a global framework, will facilitate not only the integration of various necessary viewpoints and methodologies, but also the clarification and dissemination of the KM concept itself.

6 REFERENCES

Davenport, T. and Probst, G. (2000), *Knowledge Management Case Book*, Wiley, London and Erlangen.

ISO 14258, (1998), *'Concepts and Rules for Enterprise Models'*, TC 184/SC5/WG1.

ISO 15705, (1998), *'Requirements for Enterprise Reference Architecture and Methodologies'*, TC 184/SC5/WG1.

Lillehagen, F. (2002), *'Active Knowledge Models and Enterprise Knowledge Management'* these proceedings.

Pre EN ISO 19439, (2002), *'CIM System Architecture - Framework for Enterprise Modelling'* (formerly ENV 40 003), CEN TC 310/WG1.

Vernadat, F.B. (1996), *'Enterprise Modelling and Integration: Principles and Applications'*, Chapman & Hall

Ontologies and their Role in Knowledge Management and E-Business Modelling

Hans Akkermans
Free University Amsterdam VUA, The Netherlands,contact: elly@cs.vu.nl

Abstract: Ontologies are reference conceptual models that formally describe the consensus about a domain and that are both human-understandable and machine processable. Ontologies are a key technology to realising the next, smarter, generation of the World Wide Web, known as the Semantic Web. We give an overview of recent developments, issues, and experiences in Semantic Web research, and especially discuss the role of ontologies in innovative intelligent e-applications. This paper discusses as a particular example the On-To-Knowledge project for ontology-based knowledge management. It aims to speed up knowledge management, dealing with large numbers of heterogeneous, distributed, and semi-structured documents typically found in large company intranets and the World Wide Web, by: (1) a toolset for semantic information processing and user access; (2) OIL, an ontology-based inference layer on top of the World Wide Web; (3) validation by industrial case studies in knowledge management.

1 INTRODUCTION

The World Wide Web (WWW) has drastically changed the availability of electronically available information. Currently, there are around one billion documents in the WWW, which are used by more than 300 million users internationally. And that number is growing fast. However, this success and exponential growth makes it increasingly difficult to find, to access, to present, and to maintain the information required by a wide variety of users. The competitiveness of many companies depends heavily on how they exploit their corporate knowledge and memory. Most information in modern electronic media is mixed media and rather weakly structured. This is not only true of the Internet but also of large company intranets. But as volumes of information continue to increase rapidly, the task of turning them into useful knowledge has become a major problem. Tim Berners-Lee envisioned a *Semantic Web* (cf. Berners-Lee et al., 2001) that provides automated information access based on machine-processable semantics of data and heuristics that use these meta-data. The explicit representation of the semantics of data, accompanied with domain theories (i.e., ontologies), will enable a web with various specialised smart information services that will become as necessary to us as access to electric power.

Ontologies (cf. Staab et al., 2001, Fensel, 2001) are a key enabling technology for the semantic web. They aim to interweave human understanding of symbols with their machine processability. Ontologies were developed in artificial intelligence to facilitate knowledge sharing and re-use. Since the early nineties, ontologies have become a popular research topic. They have been studied by several artificial intelligence research communities, including knowledge engineering, natural language processing and knowledge representation. More recently, the concept of ontology is also gaining tremendous ground in fields, such as intelligent information integration, co-operative information systems, information retrieval, electronic commerce, and knowledge management. The reason ontologies are becoming so popular is largely due to the fact that they cater for an important general need: *a shared and common understanding of a domain that can be communicated between people and application systems*.

Other applications and case studies on the use of ontologies in e-business modelling have been published elsewhere (Akkermans, 2001, Gordijn and Akkermans, 2001, Schulten et al. 2001).

2 TOOL ENVIRONMENT FOR ONTOLOGY-BASED KNOWLEDGE MANAGEMENT

A major objective of the On-To-Knowledge project is to create intelligent software to support users in both accessing information and in the maintenance, conversion, and acquisition of information sources. These tools are based on a three-layered architecture. Most of the tools presented here in Fig. 1 are described below.

RDF*ferret* combines full text searching with RDF querying. It can be used like a conventional Internet search engine by entering a set of search terms or a natural language query and produces a list of links to relevant Web pages in the usual way. However, RDF*ferret*'s indexing and retrieval technique is also designed to use domain knowledge that is made available in the form of ontologies specified as RDF Schemas. The information items processed by RDF*ferret* are RDF resources, which may be Web pages or parts thereof and such pages or segments are effectively ontological instances. During indexing RDF*ferret* assigns content descriptors to RDF resources: terms (words and phrases) that RDF*ferret* obtains from a full text analysis of the resource content and from processing all literal values that are directly related by a property. They also retain structural information about the ontology. In RDF*ferret* the user can select from a list of all the resource types stored in the index. When searching by selecting a resource type, RDF*ferret* adjusts its result list to show only resources of the selected type. The user is also presented with a search and navigation area. The search area shows the attributes of the selected resource type. For each attribute the user can input a search criterion. RDF*ferret* combines the search criteria entered and matches the resulting query against its ontology-based index. In addition, resource types (ontological classes) related by some property to the currently selected type are displayed as hyperlinks. Clicking on such a type then selects that type and in turn displays those types that are related to it. Thus, the user can browse the ontology in a natural and intuitive way. Fig. 2 shows a typical initial query by a user. The user has entered a query for information about an employee called "George Miller". The search engine has returned a ranked list of 73 documents mentioning the terms "George" and/or "Miller". At the top of the screenshot can be seen a drop-down list containing the selection "any...". When returning the 73 results documents, RDF*ferret* has also compiled a list of the classes to which each document belongs. This class list is made available to the user via the drop-down list.

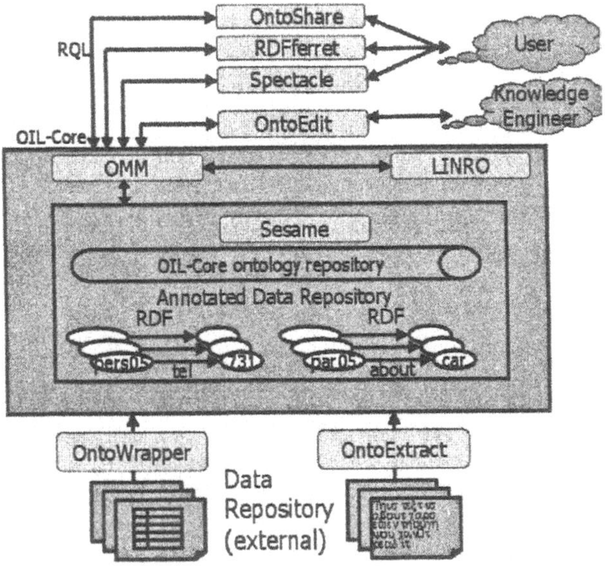

Figure 1: The technical architecture of On-To-Knowledge.

OntoShare enables the storage of best practice information in an ontology and the automatic dissemination of new best practice information to relevant co-workers. It also allows users to browse or search the ontology in order to find the most relevant information to the problem that they are dealing with at any given time. The ontology helps to orientate new users and acts as a store for key learning and best practices accumulated through experience. In addition, the ontology helps users to become familiar with new domains. It provides a sharable structure for the knowledge base, and a common language for communication between user groups.

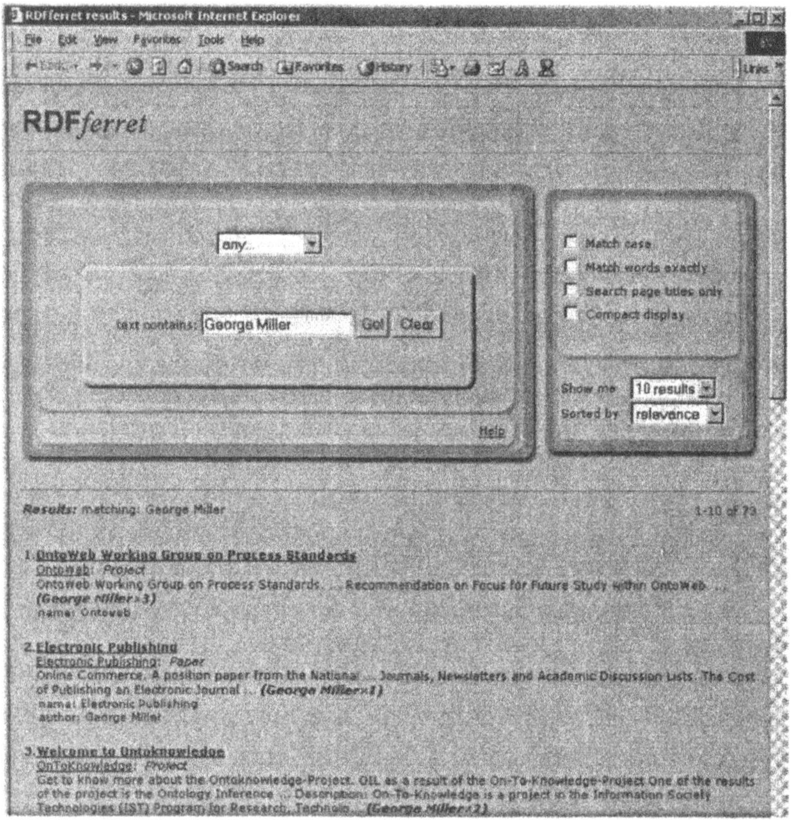

Spectacle organises the presentation of information. This presentation is ontology driven. Ontological information, such as classes or specific attributes of information, is used to generate domain exploration contexts for users. The context is related to certain tasks, such as finding information or buying products. The context consists of three modules: (1) Content: specific content needed to perform a task; (2) Navigation: suitable navigation disclosing the information; (3) Design: applicable design displaying the selected content. The modules are independent. Spectacle consists of the following parts:

- Spectacle server, which handles all interaction between users and exploration contexts;
- Libraries for creating large scale exploration contexts in this server;
- Graphical user interface for building small-scale exploration contexts.

OntoEdit (Sure et al., 2002) makes it possible to inspect, browse, codify and modify ontologies, and thus serves to support the ontology development and maintenance task. Modelling ontologies using OntoEdit involves modelling at a conceptual level, viz. (i) as independently of a concrete representation language as possible, and (ii) using GUI's representing views on conceptual structures (concepts, concept hierarchy, relations, axioms) rather than codifying conceptual structures in ASCII.

Ontology Middleware Module (OMM) can be seen as the key integration component in the OTK technical solution architecture. It supports well-defined application programming interfaces (OMAPI) used for access to knowledge and deals with such matters as:

- Ontology versioning, including branching.
- Security – user profiles and groups are used to control the rights for access, modifications, and publishing.
- Meta-information and ontology lookup – support for meta-properties (such as Status, Last-Updated-By, Responsible, Comments, etc.) for whole ontologies, as well as for separate concepts and properties.
- Access via several protocols: HTTP, RMI, EJB, CORBA, and SOAP.

Sesame (Broekstra et al, to appear) is a system that allows persistent storage of RDF data and schema information and subsequent online querying of that information. Sesame has been implemented in Java, which makes it portable to almost any platform. It also abstracts from the actual repository used by means of a standardised API. This API makes Sesame portable to any repository (DBMS or otherwise) that is able to store RDF triples. At the same time, this API enables swift addition of new modules that operate on RDF and RDF Schema data. One of the most prominent modules of Sesame is its query engine. It supports an OQL-style query language called RQL. RQL supports querying of both RDF data (e.g. instances) and schema information (e.g. class hierarchies, domains and ranges of properties). RQL also supports path-expressions through RDF graphs, and can combine data and schema information in one query. The streaming approach used in Sesame (data is processed as soon as available) makes for a minimal memory footprint. This streaming approach also makes it possible for Sesame to scale to huge amounts of data. Sesame can scale from devices as small as palm-top computers to powerful enterprise servers.

The CORPORUM toolset *(OntoExtract and OntoWrapper)* (Engels & Bremdal, 2000) has two related tasks: interpretation of *natural language* texts and ontology extraction of *specific information* from free text. The latter task requires a user who defines business rules for extracting information from tables, (phone) directories, home pages, etc. The former task involves natural language interpretation on a syntactic and lexical level, as well as interpretation of the results of that level (discourse analysis, co-reference and collocation analysis, etc.). CORPORUM outputs a variety of (symbolic) knowledge representations, including semantic (network) structures and visualisations thereof, lightweight ontologies, text summaries, automatically generated thesauri (related words/concepts), etc. Extracted information is represented in RDF(S)/DAML+OIL, augmented with Dublin Core Meta Data wherever possible, and submitted to the Sesame Data Repository. CORPORUM does not incorporate background knowledge itself, but relies on any knowledge available in the Sesame repository.

3 *OIL*: INFERENCE LAYER FOR THE SEMANTIC WORLD WIDE WEB

The tools discussed in section 2 all exploit ontologies as their common operating ground. All of this requires the existence of a language to express such ontologies. Some basic requirements for such a language are:

- Sufficient expressivity for the applications and tasks (sketched elsewhere in this paper);
- Sufficiently formalised to allow machine processing;
- Integrated with existing Web technologies and standards.

Although much work has been done on ontology languages in the AI community (see e.g. (Corcho & Gomez Perez, 2000) for a recent overview), it is particularly the third requirement that motivated us to design a new language (baptised OIL) for our purposes. In this section, we will briefly describe the constructions in the OIL language, and then discuss its most important features and design decisions.

Combining Description Logics with Frame Languages. The OIL language (Harmelen & Horrocks, 2000, Fensel et al., 2000) is designed to combine frame-like modelling primitives with the increased (in some respects) expressive power, formal rigor and automated reasoning services of an expressive description logic. OIL also comes "web enabled" by having both XML and RDFS based serialisations (as well as a formally specified "human readable" form, see OIL, http://). Classes (concepts) are described by frames, which consist of a list of super-classes and a list of slot-filler pairs. A slot corresponds to a role in a DL, and a slot-filler pair corresponds to either a universal value restriction or an existential quantification. OIL extends this basic frame syntax so that it can capture the full power of an expressive description logic. These extensions include:

- Arbitrary Boolean combinations of classes (called class expressions) can be formed, and used anywhere a class name can be used. In particular, class expressions can be used as slot fillers, whereas in typical frame languages slot fillers are restricted to being class (or individual) names.
- A slot-filler pair (called a slot constraint) can itself be treated as a class: it can be used anywhere that a class name can be used, and can be combined with other classes in class expressions.
- Class definitions (frames) have an (optional) additional field that specifies whether the class definition is primitive (a subsumption axiom) or non-primitive (an equivalence axiom). The default is primitive.
- Different types of slot constraints are provided for universal value restrictions, existential quantification, various cardinality constraints.
- Global slot definitions allow for the specification of superslots (subsuming slots) and of properties such as transitivity and symmetry.
- Unlike frame languages, no restriction exists on the ordering of class and slot definitions, so classes and slots can be used before they are defined.
- OIL also provides axioms for asserting disjointness, equivalence and coverings with respect to class expressions.

Many of these points are standard for a description logic, but are novel for a frame language.

Web Interface. As part of the Semantic Web activity of the W3C, a very simple web-based ontology language had already been defined, namely RDF Schema. This language only provides facilities to define class- and property-names, inclusion axioms for both classes and properties (subclasses and subproperties), and to define domain and range constraints on properties. Instances of such classes and properties are defined in RDF. OIL has been designed to be a superset of the constructions in RDF Schema: all valid RDF Schema expressions are also valid OIL expressions. Furthermore, the syntax of OIL has been designed such that any valid OIL document is also a valid RDF(S) document when all the elements from the OIL-namespace are ignored. The RDF Schema interpretation of the resulting subdocument is guaranteed to be sound (but of course incomplete) with respect to the interpretation of the full OIL document. This guarantees that any RDF Schema agent can correctly process arbitrary OIL documents, and still correctly capture some of the intended meaning. The full details of how this has been achieved, and the trade-offs involved in this can be found in (Broekstra, *et al.*, 2000).

Layering. For many of the applications from section 1, it is unlikely that a single language will be ideally suited for all uses and all users. In order to allow users to choose the expressive power appropriate to their application, and to allow for future extensions, a layered family of OIL languages has been described. The sublanguage OIL Core has been defined to be exactly the part of OIL that coincides with RDF(S). This amounts to full RDF(S), without some of RDF's more dubious constructions: containers and reification. The standard language, is called "Standard OIL", and when extended with the ability to assert that individuals and tuples are, respectively, instances of classes and slots), is called "Instance OIL". Finally, "Heavy OIL" is the name given to a further layer that will include as yet unspecified language extensions. This layering is depicted in Fig. 3.

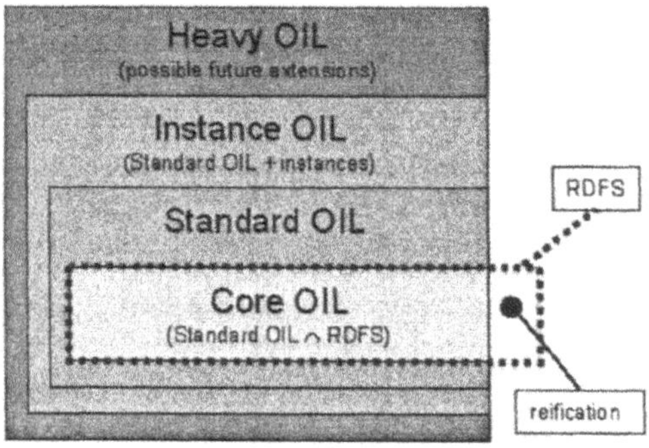

Figure 3: The layered language model of OIL.

Current status. Meanwhile, OIL has been adopted by a joined EU/US initiative that developed a language called DAML+OIL (http://), which has now been submitted to the Web Ontology Group of the W3C (http://), the standardisation committee of the WWW. We can soon expect a recommendation for a web ontology language; it features many of the elements on which OIL is based.

Future developments: OWL. In November 2001, the W3C started a Working Group for defining a Web Ontology language. This WG is chartered to take DAML+OIL as its starting point. Over 40 of the W3C members from academia and industry are currently participating in this effort. It is most likely that such a Web Ontology language will range in power somewhere between the rather simple RDF Schema and the rather rich Standard OIL language. Other efforts are underway to define extensions for this web ontology language - that has been named OWL - such as an ontology-query language, or an extension with rules (which would allow for example role chaining, as done in Horn logic).

4 BUSINESS APPLICATIONS IN SEMANTIC INFORMATION ACCESS

Accounting Information Search. Swiss Life carried out two case studies to evaluate the developed Semantic Web tools and methods. One of these approached the problem of finding relevant passages in a very large document about the International Accounting Standard (IAS) on the extranet (over 1000 pages). Accountants who need to know certain aspects of the IAS accounting rules use this document. As the IAS standard uses very strict terminology, it is only possible to find relevant text passages when the correct terms are used in the query. Very often, this leads to poor search results. With the help of the ontology extraction tool OntoExtract, an ontology was automatically learned from the document. The ontology consists of 1,500 concepts linked by 47,000 weighted semantic associations. It supports users in reformulating their initial queries when the results fall short of expectations, by offering terms from the ontology that are strongly associated with (one of) the query terms used in the initial query. An evaluation of user behaviour showed that 70% of the queries involved a reformulation step. On average, 1.5 refinements were made. Thus, although the ontology is structurally quite simple, it greatly improves search results. Another advantage to using a simple ontology is that it requires no manual effort to build.

Skills Management. Swiss Life's second case study is a skills management application that uses manually constructed ontologies about skills, job functions, and education. These consist of 800 concepts with several attributes, arranged into a hierarchy of specialisations. There are also semantic associations between these concepts. The skills management system makes it easy for employees to create in a personal home page on the company's intranet that includes information about personal skills, job functions, and education. The ontology allows a comparison of skills descriptions among employees, and ensures the use of uniform terminology in skills descriptions and queries for employees with certain skills. Moreover, the ontology can automatically extend queries with more general, more specialised, or semantically associated concepts. This enables controlled extension of search results, where necessary.

Exchanging knowledge in a virtual Organization. The case study done by EnerSearch AB focuses on satisfying the information dissemination needs of a virtual organisation. The goal of the case study is to improve knowledge transfer between EnerSearch's in-house researchers and outside specialists via the existing web site. The study also aims to help the partners from shareholding companies to obtain up-to-date information about research and development results. The main problem with the current web site is that its search engine supports free text searches rather than content-based information retrieval, which makes it fairly difficult to find information on certain topics. To remedy this, the entire web site was annotated by concepts from an ontology developed using semi-automatic extraction from documents on the EnerSearch's current web site. The RDF*ferret* search engine is used to extend free text searches to searches of annotations. Alternatively, the Spectacle tool enables users to obtain search results arranged into topic hierarchies, which can then be browsed. This offers users a more explorative route to finding the information they need (see Fig. 4). Three groups with different interests and needs are involved in the evaluation: (1) researchers from different fields, (2) specialists from the shareholders organisation and (3) outsiders from different fields.

5 CONCLUSION

The Web and company intranets have boosted the potential for electronic knowledge acquisition and sharing. Given the sheer size of these information resources, there is a strategic need to move up in the data – information – knowledge chain. *On-To-Knowledge* takes a necessary step in this process by providing innovative tools for semantic information processing and thus for much more selective, faster, and meaningful user access.

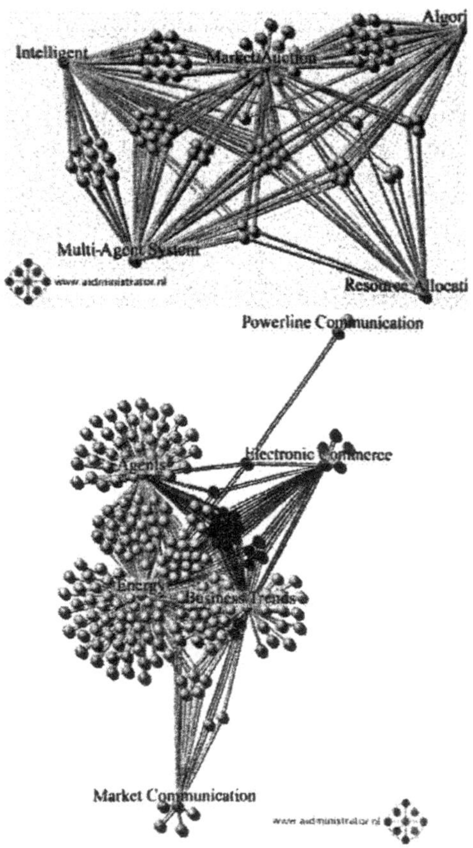

Figure 4: Automatically generated semantic structure maps of the EnerSearch website.

We also encountered a number of shortcomings in our current approach. Building ontologies that are a pre-requisite for - and result of - the common understanding of large user groups is no trivial task. A model or "protocol" that maintains the process of *evolving* ontologies is the real challenge for making the *semantic* web a reality. Most work on ontologies views them in terms of an isolated theory containing a potentially large number of concepts, relationships, and rules. In practice, ontologies must be structured as interwoven *networks* that make it possible to deal with heterogeneous needs in the communication processes that they are supposed to mediate. Moreover, these ontologies change over time because the processes they mediate are based on consensual representation of meaning. It is the network of ontologies and their dynamic nature that make future research progress necessary. Actual research challenges on ontologies are what glue keeps ontology networks together in space and time. Instead of a central, top-down process, we require a distributed process of emerging and aligned ontologies. Most existing technology focuses on building ontologies as graphs based on concepts and relationships. Our current understanding is insufficient when it comes to proper methodological and tool support for building up networks, where the nodes represent small and specialised ontologies. This is especially true of the noisy and dynamically changing environment that the web is and will continue to be.

6 ACKNOWLEDGEMENTS

This paper and the research work it describes is based on contributions from many people, in particular Dieter Fensel, Frank van Harmelen, Peter Mika, Michel Klein (Free University Amsterdam VUA), Jeen Broekstra, Arjohn Kampman, Jos van der Meer (Administrator, The Netherlands), York Sure, Rudi Studer (University of Karlsruhe, Germany), John Davies, Alistair Duke (BT, Ipswich, UK), Robert Engels (CognIT, Oslo, Norway), Victor Iosif (Enersearch AB, Malmö, Sweden), Atanas Kiryakov (OntoText, Sofia, Bulgaria), Thorsten Lau, Ulrich Reimer (Swiss Life, Zürich, Switzerland), and Ian Horrocks (University of Manchester, UK). It has been partially supported by the European Commission through the EU-IST project On-To-Knowledge (IST-1999-10132).

7 REFERENCES

Akkermans, J.M. (2001), *Intelligent E-Business – From Technology to Value*, IEEE Intelligent Systems Vol. 16, No. 4, pages 8-10. Special issue on Intelligent E-Business. Also available from http://computer.org/intelligent.

Berners-Lee, T. Hendler, J. Lassila, O. (2001), *The Semantic Web*, Scientific American, May.

Broekstra, J. Klein, M. Decker, S. Fensel, D. van Harmelen, F. Horrocks. I. (2001), *Enabling knowledge representation on the web by extending RDF schema*. In Proceedings of the Tenth International World Wide Web Conference (WWW10), Hong Kong, May.

Broekstra, J. Kampman, A. van Harmelen, F. (to appear 2002), *Sesame: An Architecture for Storing and Querying RDF Data and Schema Information*. In Fensel, D. Hendler, J. Lieberman, H. Wahlster, W. (Eds.): Semantic Web Technology, MIT Press, Cambridge, MA, to appear.

Corcho, O. Gomez Perez, A. (2000), *A roadmap to ontology specification languages*. In R. Dieng and O. Corby (Eds.), Proceedings of the 12th International Conference on Knowledge Engineering and Knowledge Management (EKAW'00), volume 1937 of LNAI, 80–96. Springer-Verlag.

DAML+OIL, http://www.daml.org

Engels, R. Bremdal, B.A. (2001), *CORPORUM: A Workbench for the Semantic Web*. Semantic Web Mining workshop. PKDD/ECML-01. Freiburg, Germany.

Fensel, D. Horrocks, I. Van Harmelen, F. Decker, S. Erdmann, M. Klein, M. (2000), *OIL in a nutshell*. In R. Dieng and O. Corby (Eds.), Knowledge Engineering and Knowledge Management – Methods, Models and Tools, pages 1-16, Lecture Notes in Artificial Intelligence, LNAI 1937, Springer-Verlag,

Fensel, D. (2001), *Ontologies: Silver Bullet for Knowledge Management and Electronic Commerce*. Springer-Verlag.

Gordijn, J. Akkermans, J.M. (2001), *Designing and Evaluating E-Business Models*, IEEE Intelligent Systems Vol. 16, No. 4), pages 11-17. See http://computer.org/intelligent. Further related work: http://www.cs.vu.nl/~gordijn.

Harmelen, F. van, Horrocks, I. (2000), *Questions and answers about OIL*. IEEE Intelligent Systems, 15(6): 69–72.

OIL, http://www.ontoknowledge.org/oil/syntax/

Schulten, E. Akkermans, J.M. Botquin, G. Dörr, M. Guarino, N. Lopes, N. Sadeh, N. (2001), *The E-Commerce Product Classification Challenge*, IEEE Intelligent Systems Vol. 16, No. 4 (July-August), pages 86-89. (http://computer.org/intelligent).

Staab, S. Schnurr, H.-P. Studer, R. Sure, Y. (2001), *Knowledge Processes and Ontologies*, IEEE Intelligent Systems, Vol. 16, No. 1, pages 26-34.

Sure, Y. Erdmann, M. Angele, J. Staab, S. Studer R. Wenke, D. (2002), *OntoEdit: Collaborative Ontology Engineering for the Semantic Web*. In: Proceedings 1st International Semantic Web Conference 2002 (ISWC 2002), June, Sardinia, Italia.

W3C, http://www.w3c.org

Semantic Bridging of Independent Enterprise Ontologies

Michael N. Huhns, and Larry M. Stephens
University of South Carolina, USA, huhns@sc.edu

Abstract: Organizational knowledge typically comes from many independent sources, each with its own semantics. This paper describes a methodology by which information from large numbers of such sources can be associated, organized, and merged. The hypothesis is that a multiplicity of ontology fragments, representing the semantics of the independent sources, can be related to each other automatically *without* the use of a global ontology. That is, any pair of ontologies can be related indirectly through a *semantic bridge* consisting of many other previously unrelated ontologies, even when there is no way to determine a direct relationship between them. The relationships among the ontology fragments indicate the relationships among the sources, enabling the source information to be categorized and organized. A preliminary evaluation of the methodology has been conducted by relating 53 small, independently developed ontologies for a single domain. A nice feature of the methodology is that common parts of the ontologies reinforce each other, while unique parts are de-emphasized. The result is a *consensus* ontology.

1 INTRODUCTION

Corporate information searches can involve data and documents both internal and external to the organization. The research reported herein targets the following basic problem: a search will typically uncover a large number of independently developed information sources—some relevant and some irrelevant; the sources might be ranked, but they are otherwise unorganized, and there are too many for a user to investigate manually. The problem is familiar and many solutions have been proposed, ranging from requiring the user to be more precise in specifying search criteria, to constructing more intelligent search engines, or to requiring sources to be more precise in describing their contents. A common theme for all of the approaches is the use of ontologies for describing both requirements and sources. Unfortunately, ontologies are not a panacea unless everyone adheres to the same one, and no one has yet constructed an ontology that is comprehensive enough (in spite of determined attempts to create one, such as the CYC Project (CYC, http://), underway since 1984). Moreover, even if one did exist, it probably would not be adhered to, considering the dynamic and eclectic nature of the Web and other information sources.

There are three approaches for relating information from large numbers of independently managed sites: (1) all sites will use the same terminology with agreed-upon semantics (improbable), (2) each site will use its own terminology, but provide translations to a global ontology (difficult, and thus unlikely), and (3) each site will have a small, local ontology that will be related to those from other sites (described herein). We hypothesize that the small ontologies can be related to each other automatically *without* the use of a global ontology. That is, any pair of ontologies can be related indirectly through a *semantic bridge* consisting of many other previously unrelated ontologies, even when there is no way to determine a direct relationship between them. Our methodology relies on sites that have been annotated with ontologies (Pierre, 2000); such annotation is consistent with several visions for the semantic Web (Heflin, Hendler, 2000; Berners-Lee, et al. 2001). The domains of the sites must be similar—else there would be no interesting relationships among them—but they will undoubtedly have dissimilar ontologies, because they will have been annotated independently.

Other researchers have attempted to merge a pair of ontologies in isolation, or merge a domain-specific ontology into a global, more general ontology (Wiederhold, 1994). To our knowledge, no one has previously tried to reconcile a large number of domain-specific ontologies. We have evaluated our methodology by applying it to a large number of independently constructed ontologies.

2 RECONCILING INDEPENDENT ONTOLOGIES

In agent-assisted information retrieval, a user will describe a need to his agent, which will translate the description into a set of requests, using terms from the user's local ontology. The agent will contact on-line brokers and request their help in locating sources that can satisfy the requests. The agents must reconcile their semantics in order to communicate about the request. This will be seemingly impossible if their ontologies share no concepts. However, if their ontologies share concepts with a third ontology, then the third ontology might provide a "semantic bridge" to relate all three. Note that the agents do not have to relate their entire ontologies, only the portions needed to respond to the request.

The difficulty in establishing a bridge will depend on the semantic distance between the concepts, and on the number of ontologies that comprise the bridge. Our methodology is appropriate when there are large numbers of small ontologies—the situation we expect to occur in large and complex information environments. Our metaphor is that a small ontology is like a piece of a jigsaw puzzle, as depicted in Fig. 1. It is difficult to relate two random pieces of a jigsaw puzzle until they are constrained by other puzzle pieces. We expect the same to be true for ontologies.

Ontologies can be made to relate to each other like pieces of a jigsaw puzzle. (Top) Two ontology fragments with no obvious relationships between them. (Bottom) The introduction of a third ontology reveals equivalences between components of the two original ontology fragments

Two concepts can have the following seven mutually exclusive relationships between them: *subclass*, *superclass*, *equivalence*, *partOf*, *hasPart*, *sibling*, or *other*. If a request contains three concepts, for example, and the request must be related to an ontology containing 10 concepts, then there are 7x3x10 = 210 possible relationships among them. Only 30 of the 210 will be correct, because each of the three concepts in the request will have one relationship with each of the 10 concepts in the source's ontology.

The correct ones can be determined by applying constraints among the concepts within an ontology, and among multiple ontologies. Once the correct relationships have been determined, we make use of *equivalence* and *sibling* or, where those do not exist, the most specific *superclass* or *partOf*.

In Fig. 1, the ontology fragment on the left would be represented as *partOf(Wheel, Truck)*, while the one on the right would be represented as *partOf(Tire, APC)*. There are no obvious equivalences between these two fragments. The concept *Truck* in the first ontology could be related to *APC* in the second by *equivalence*, *partOf*, *hasPart*, *subclass*, *superclass*, or *other*. There is no way to decide which is correct. When the middle ontology fragment *partOf(Wheel, APC)* is added, there is evidence that the concepts *Truck* and *APC*, and *Wheel* and *Tire* could be *equivalent*.

This example exploits the existence of the relation *partOf*, which is common to all three ontologies. Other domain-independent relations, such as *subclassOf*, *instanceOf*, and *subrelationOf*, will be necessary for the reconciliation process. Moreover, the reflexivity, symmetry, asymmetry, transitivity, irreflexivity, and antisymmetry properties are needed for relating occurrences of the relations to each other (Stephens, Chen, 1996). Domain concepts and relations can be related to each other by converse/inverse, composition, (exhaustive) partition, part-whole (with 6 subtypes), and temporal attitude. There must be some minimum set of these fundamental relations that are understood and used by all local ontologies and information system components.

In attempting to relate two ontologies, a system might be unable to find correspondences between concepts because of insufficient constraints and similarity among their terms. However, trying to find correspondences with other ontologies might yield enough constraints to relate the original two ontologies. As more ontologies are related, there will be more constraints among the terms of any pair, which is an advantage. It is also a disadvantage in that some of the constraints might be in conflict. We make use of the preponderance of evidence to resolve these statistically.

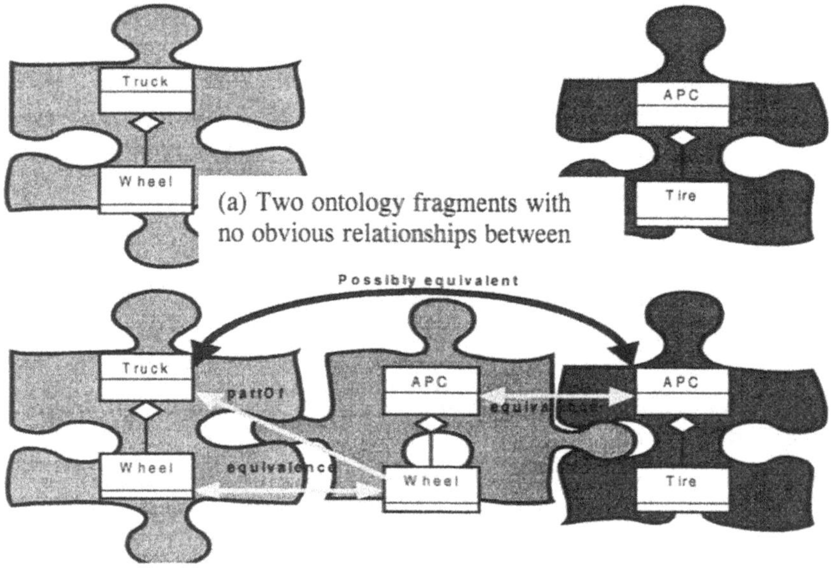

(b) The introduction of a third ontology reveals equivalences between components of the original two ontology fragments

Figure 1: Ontologies can be made to relate to each other like pieces of a jigsaw puzzle

3 EXPERIMENTAL METHODOLOGY

We asked each of 53 graduate students in computer science, who were novices in constructing ontologies, to construct a small ontology for the Humans/People/Persons domain. The ontologies were required to be written in DAML and to contain at least 8 classes with at least 4 levels of subclasses; a sample ontology is shown in Fig. 2.

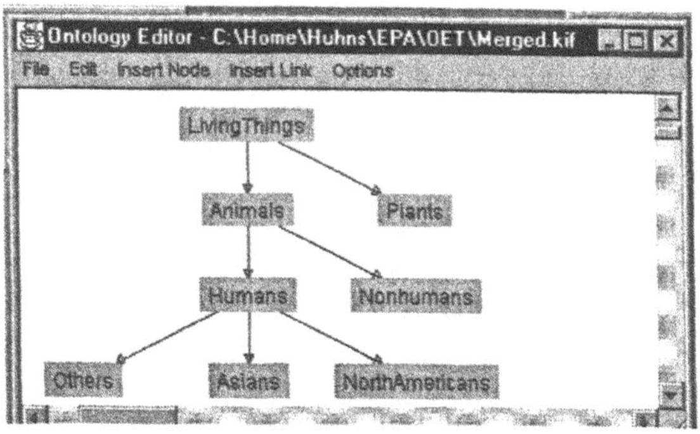

Figure 2: A typical small ontology used to characterize an information source about people (all links denote subclasses)

Using string-matching and other heuristics, we merged the 53 component ontologies. The component ontologies described 864 classes, while the merged ontology contained 281 classes in a single graph with a root node of the DAML concept #Thing. All of the concepts were related, i.e., there was some relationship (path) between any pair of the 281 concepts (see Fig. 3).

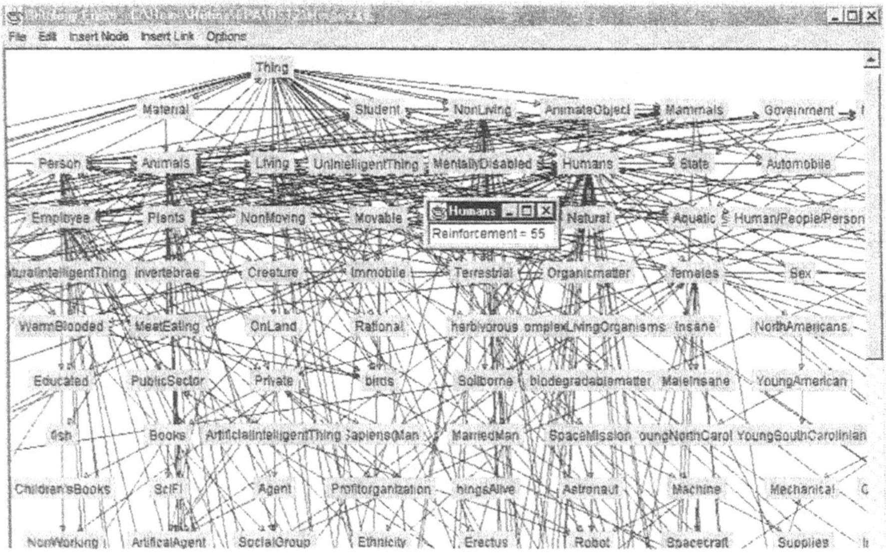

Figure 3: A portion of the ontology formed by merging 53 independently constructed ontologies for the domain Humans/People/Persons. The entire ontology has 281 concepts related by 554 subclass links

Next, we constructed a *consensus ontology* by counting the number of times classes and subclass links appeared in the component ontologies when we performed the merging operation. For example, the class Person and its matching classes appeared 14 times. The subclass link from Mammals (and its matches) to Humans (and its matches) appeared 9 times. We termed these numbers the "reinforcement" of a concept.

Redundant subclass links were removed and the corresponding transitive closure links were reinforced. That is, if C has subclass A with reinforcement 2, C has subclass B reinforced *m* times, and B has subclass A reinforced *n* times, then the link from C directly to A was removed and the remaining link reinforcements were increased by 2. We then removed from the merged ontology any classes or links that were not reinforced.

Finally, we applied an *equivalence heuristic* for collapsing classes that have common reinforced superclasses and subclasses. The equivalence heuristic found that all reinforced subclasses of Person are also reinforced subclasses of Humans, and all reinforced superclasses of Person are also reinforced superclasses of Humans. It thus deems that Humans and Person are the same concept. This heuristic is similar to an inexact graph matching technique such as (Manocha, et al., 2001). Fig. 4 shows the collapsed consensus ontology, now containing 36 classes related by 62 subclass links.

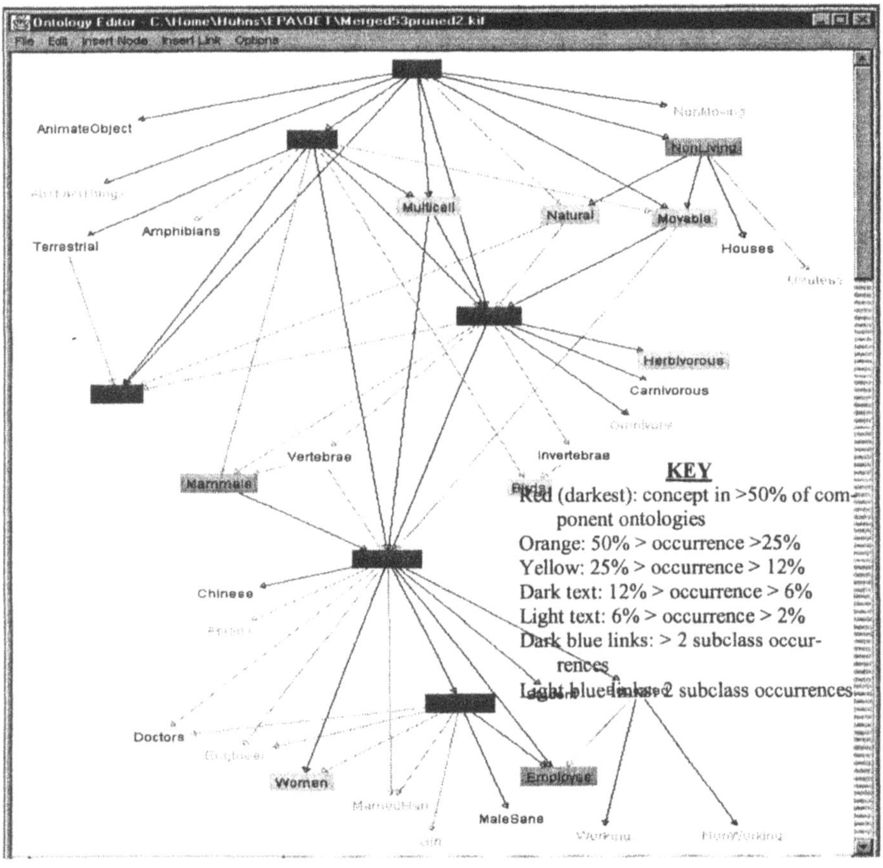

Figure 4: The final consensus ontology formed by merging concepts with common subclasses and superclasses. The resultant ontology contains 36 concepts related by 62 subclass links.

4 DISCUSSION OF RESULTS

A consensus ontology is perhaps the most useful for information retrieval by humans, because it represents the way most people view the world and its information. For example, if most people wrongly believe that crocodiles are a kind of mammal, then most people would find it easier to locate information about crocodiles if it were located in a mammals grouping, rather than where it factually belonged.

The information retrieval measures of precision and recall are based on some degree of match between a request and a response. The length of a semantic bridge between two concepts can provide an alternative measure of conceptual distance and an improved notion for relevance of information. Previous measures relied on the number of properties shared by two concepts within the same ontology, or the number of links separating two concepts within the same ontology (Delugach, 1993). These measures not only require a common ontology, but also do not take into account the density or paucity of information about a concept. Our measure does not require a common ontology and is sensitive to the information available.

Although promising, our experiments and analysis so far are very preliminary. We used the following simplifications:

- We did not use synonym information, such as is available from WordNet, and so did not for example merge "meat eating" and "carnivorous."
- We did not make use of class properties, as in subsumption.
- Our string-matching algorithm did not use morphological analysis to separate the root word from its prefixes and suffixes, and did not identify negated concepts, such as "uneducated" versus "educated."
- We used only subclass-superclass information, and have not yet made use of other important relationships, notably part-of.

Our hypothesis, that a multiplicity of ontology fragments can be related automatically without the use of a global ontology, appears correct, but our investigation is continuing according to the following plan:

- Improve the algorithm for relating ontologies, based on methods for partial and inexact matching, making extensive use of common ontological primitives, such as subclass and partOf. The algorithm will take as input ontology fragments and produce mappings among the concepts represented in the fragments. It will use constraints among known ontological primitives to control computational complexity.
- Develop metrics for successful relations among ontologies, based on the number of concepts correctly related, as well as the number incorrectly matched. The quality of a match will be based on semantic distance, as measured by the number of intervening semantic bridges.

5 CONCLUSION

Imagine that in response to a request for information about a particular topic, a user receives pointers to more than 1000 documents, which might or might not be relevant. The technology developed by our research would yield an organization of the received information, with the semantics of each document reconciled. This is a key enabling technology for knowledge-management systems.

Our premise is that it is easier to develop small ontologies, whether or not a global one is available, and that these can be automatically and *ex post facto* related. We are determining the efficacy of local annotation for Web sources, as well as the ability to perform reconciliation qualified by measures of semantic distance. The results of our effort will be (1) software components for semantic reconciliation, and (2) a scientific understanding of automated semantic reconciliation among disparate information sources.

6 REFERENCES

Berners-Lee, T. Hendler, J. Lassila, O. (2001), *The Semantic Web*, Scientific American, May.

CYC, http://www.cyc.com/publications.html

Delugach, H. S. (1993), *An Exploration Into Semantic Distance*, Lecture Notes in Artificial Intelligence, No.754, Springer-Verlag.

Heflin, J. Hendler, J. (2000), *Dynamic Ontologies on the Web*, Proc. 17th National Conference on AI (AAAI-2000), AAAI Press.

Mahalingam, K. Huhns, M.N. (1997), *An Ontology Tool for Distributed Information Environments*, *IEEE Computer,* 30(6).

Manocha, N., Cook, D. Holder, L. (2001), *Structural Web Search Using a Graph-Based Discovery System*, *ACM Intelligence*, 12(1).

Pierre, J. M. (2000), "Practical Issues for Automated Categorization of Web Sites," *Electronic Proc. ECDL Workshop on the Semantic Web*, Lisbon, Portugal. http://www.ics.forth.gr/proj/isst/SemWeb/program.html

Stephens L. M. and Chen, Y. F. (1996), "Principles for Organizing Semantic Relations in Large Knowledge Bases," *IEEE Transactions on Knowledge and Data Engineering*, 8(3).

Wiederhold, G. (1994), "An Algebra for Ontology Composition," *Proc. Monterey Workshop on Formal Methods*, U.S. Naval Postgraduate School.

Active Knowledge Models and Enterprise Knowledge Management

Frank Lillehagen[1] and John Krogstie[2]
[1]*Computas AS, Norway,* [2]*SINTEF, Norway, fli@computas.com*

Abstract: We present in this paper a novel approach for integrating enterprise modelling and knowledge management in dynamic networked organisations. The approach is based on the notion of active knowledge models (AKM™). An AKM is a visual model of enterprise aspects that can be viewed, traversed, analysed, simulated, adapted and executed by industrial users.
To integrate particular process technologies from the enterprise perspective of generic business process types to the individual work tasks at the instance level, our work is based on our *process modelling reference model*. It identifies 4 layers of process knowledge representation, from general process logic to actual, situated work performance. Process modelling occurs at several levels concurrently, and may start at any level. Learning within and between levels is supported using a framework for process knowledge management.

1 INTRODUCTION

The business environment is getting increasingly dynamic. Co-operation across traditional organizational boundaries is increasing, as outsourcing and electronic business is enabled by the Internet and IS in general. When such co-operation moves beyond the buying and selling of goods and well-defined services, there is a need for a flexible infrastructure that supports not only information exchange, but also knowledge sharing, creation, utilisation and management within and across the traditional organizational borders. To address these challenges, new organizational forms, such as different types of virtual Organizations and extended enterprises flourish. This demands a new approach to enterprise integration and system engineering. Our approach to this area is the use of Active Knowledge Models (AKM).

An Active Knowledge Model is a visual externalisation of knowledge of *enterprise aspects* that can be operated on (viewed, traversed, analysed, simulated, adapted and executed) by industrial users. What does it mean that the model is active? First of all, the visual model must be *available* to the users of the information system at runtime. Second, the model must *influence the behaviour* of the computerised support system. Third, the model must be *dynamic*, users must be supported in changing the model to fit their local situation, enabling tailoring of the system's behaviour.

2 THE AKM APPROACH

AKMs of Enterprises imply that the enterprise is extended by distributed team working on layers of knowledge, and that simultaneous modelling, meta-modelling and work can be performed.

AKM implementation is dependent on a rich generic knowledge base and powerful development and extension capabilities of the infrastructure. Being able to support collaborative work and managing knowledge will decide the quality of the solution, of the methodology, and of the knowledge and solutions created. The usage and value of the solution is mainly decided by the infrastructure, but also by the competence and knowledge of the teams involved.

2.1 Enterprise modelling and knowledge management

The concept of knowledge management has been used in different disciplines, previously mostly in knowledge management and engineering (Skyrme, Amidon, 1997, Schreiber, et al. 2000). Knowledge management is mainly understood by practitioners from manufacturing and the service industry as part of corporate culture and a business-oriented method as "The sum of procedures to generate, store, distribute and apply knowledge to achieve organisational goals".

All main approaches to knowledge management emphasise the process character with inter-linked tasks or activities. Business process modelling is usually done for very specific goals, which partly explains the great diversity of approaches found in literature (Vernadat, 1996) and practice. The main reasons for doing BPM are:

a) To improve human understanding and communication
b) To guide system development

c) To provide computer-assisted analysis through simulation or deduction
d) To enable model deployment and activation for decision making and operation monitoring and control

There are four major knowledge dimensions in any enterprise:

- Products and Services, the results of work and the deliverables of projects
- Organization and People, competence and skills, and resources for work performance.
- Processes and Tasks, including work breakdown structures for different purposes.
- Systems and Tools, technical infrastructure with architectures, interfaces and tools.

The AKM, irrespective of purpose and scope, will always take one or more views from all the four main dimensions into consideration. Which aspects and views to model also depend on the audience and the intended use of the model? The AKM approach is also a holistic approach leaving it to the developers and the users to decide which views, aspects of structures and flows, and which operational solutions should constitute the model to meet expectations and satisfy users and audience.

To integrate in particular process technologies from the enterprise perspective of generic business process types to the individual work tasks at the instance level, our work is based on extending our *process modelling reference model* (Jørgensen, Carlsen, 1999) shown in Fig. 1. It identifies 4 layers of process knowledge representation, from general process logic to actual, situated work performance. Process modelling occurs at several levels concurrently, and may start at any level.

Layer 1 – Describe Process Logic: At this layer, we identify the constituent activities of generic, repetitive processes and the logical dependencies between these activities. A process model at this layer should be transferable across time and space to a mixture of execution environments. Examples of process logic are conceptual value chains and best practice-models of "ways of working" for particular types of organisations.

Layer 2 – Engineer Activities: Here process models are expanded and elaborated to facilitate business solutions. Elaboration includes concretisation, decomposition, and specialisation. Integration with local execution environment is achieved e.g. by describing resources required for actual performance.

Layer 3 – Manage Work: The more abstract layers of process logic and of activity description provide constraints but also useful resources (in the form of process templates) to the planning and performance of each extended enterprise process. At layer 3, more detailed decisions are taken regarding the performance of work in the *actual work environment* with its organizational, information, and tool resources; the scope is narrowed down to an actual process instance. *Concrete* resources increasingly are intertwined in the model, leading to the introduction of more dependencies. Management of activities may be said to consist of detailed planning, coordination and preparation for resource allocation.

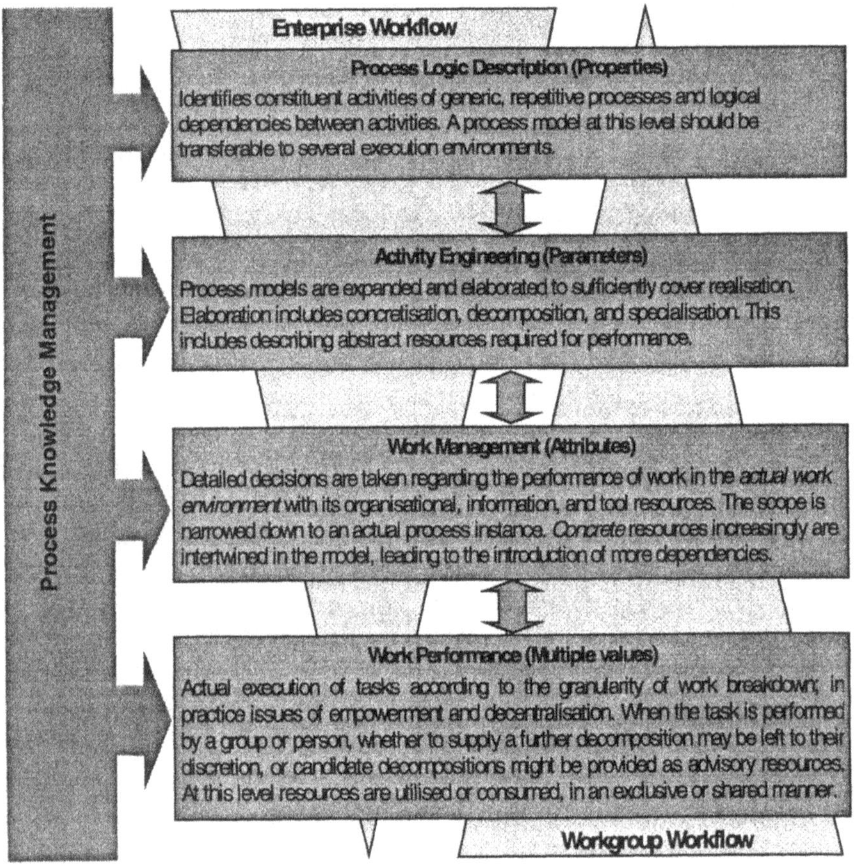

Figure 1: Process modelling reference model

Layer 4 – Perform Work: This lowest layer of the model covers the actual execution of tasks according to the determined granularity of work breakdown, which in practice is coupled to issues of empowerment and decentralisation. When a group or person performs the task, whether to supply a further decomposition may be left to their discretion, or alternative candidate decompositions might be provided as advisory resources. At this layer resources are utilised or consumed, in an exclusive or shared manner.

Process knowledge management can be defined as the collection of processes necessary for innovation, dissemination, and exploitation of knowledge in a co-operating ensemble where knowledge seekers are linked to knowledge sources and a shared knowledge base is cultivated. Process knowledge management is active at all layers of the model, which will be described in more detail below based on (Jørgensen, 2000). Here, our main concern is to understand the mechanisms that enable us to *integrate* process models at various levels of abstraction, so we need a framework that show the activities involved in converting between general (layer 1 and 2) and particular (layer 3 and 4) models. Fig. 2 shows the reference model we have chosen.

Applying a general process model to a particular situation is a case of *reuse* (Reuse may also refer to *copy and paste* of a previously developed particular model into a new process, i.e. reuse must not always occur via a general model. Copy and paste reuse is important to minimise the effort of model building, but less useful for organizational process improvement and knowledge management). Reuse involves selecting a process type (general model) and using it to generate a particular model for process enactment.

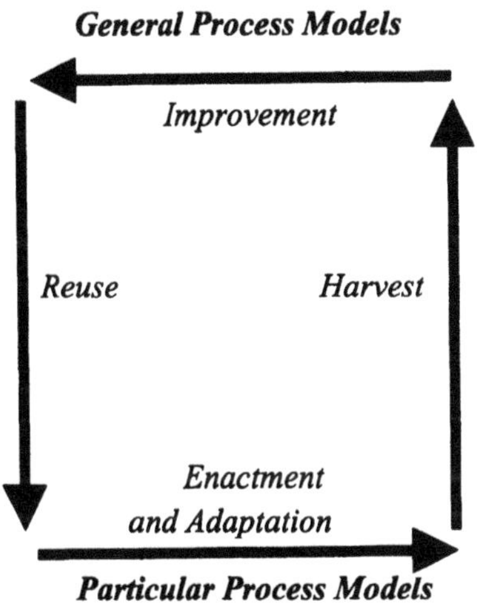

Figure 2: Lifecycle of process model evolution

In some enterprise modelling and process *improvement* initiatives, particular models are seldom used. For such initiatives to be cost effective, they must target general models that are used in several actual processes. The process of transforming one or more particular models into a general one is called *harvesting*. The goal of harvesting is to provide and update templates that may be reused in the future, and to utilise practical experience as an input to assessment and improvement of the general models. Templates include personal, group, and organizational fragments, process examples and patterns, in addition to complete definitions of routine procedures. Following traditional terminology within software process modelling, the activity where people assess and update general models is called *process improvement*. The use and dynamic adaptation of particular models during performance of work, is called *process enactment*.

The activities of process enactment, harvesting, improvement and reuse form a complete learning cycle. If one activity is not performed, the others will not be as effective. This does not imply that all activities need to be explicit or encoded in software. A user may for instance improve a template based on lessons learned in a project, even without software support for harvesting from the particular project model. Similarly, a project model may act as a passive plan and influence practice although automated enactment support is not available.

3 EXTERNAL INFRASTRUCTURE AND APPROACH

In the EXTERNAL project IST 1999-10091 (EXTERNAL, 2000) we are working further to develop a technical and conceptual infrastructure to support the AKM approach as a basis for enterprise knowledge management through process knowledge management.

The most innovative contributions from the EXTERNAL can be summarised as:

- Implementing an Extended Enterprise (EE) based on new capabilities from AKM technology, exploiting meta-models as enterprise integrators and technology convergence enablers.
- Implementing the multiple views of active objects, exploiting the reflective, recursive, repetitive and replicable nature of (situated) work process knowledge. Software methods are defined and linked as properties of visually engineered and managed objects.
- Applying the model evolution and management processes that are enabled by parallel and commercially developed solutions based on the same core concepts and common meta-models.
- Implementing a four-layered infrastructure with open enterprise formation and operation capabilities and architectures for dynamic IT component inclusion, knowledge representation, work and model management and dynamic user environment generation.
- Implementing an integrated methodology supported by the layered infrastructure.

The infrastructure, methodology, case-study solutions, and the EXTERNAL project itself are developed in parallel. The layered infrastructure (Lillehagen, 2002a) will support and implement the methodology, provide project management services, and implement work process driven solutions from re-composable knowledge and software.

Version 1.0 of the infrastructure is an integration of the enterprise and process modelling applications brought into the EXTERNAL project by the partners and further extended there. The following tools provide the core software services of the technical layer.

- METIS, a general purpose enterprise modelling and visualisation tool,

- XChips, a co-operative hypermedia tool integrated with process support and synchronous collaboration,
- SimVision, a project simulator used to analyse resource allocation, highlighting potentials for delays and backlogs.
- Workware, an emergent workflow management system with to-do-lists, document sharing, process enactment and awareness mechanisms.

Together these tools offer varied functionality for creating, maintaining, and utilising shared active knowledge models of the extended enterprise. The models are managed through a shared repository residing on a web server. For the representation and interchange of models, an XML DTD is defined.

As mentioned above, the infrastructure is best described as consisting of four layers. These layers are identified as:

Layer 1, the ICT layer: – defining and describing the ICT platform, the software architectures, tools, software components and capabilities, connectivity and communications. The ICT layer supports multi-user access control and repository management. The architecture has 3-tiers, clients, application servers, and data servers (web services), i.e. server applications communicating with its clients solely through standard web protocols such as HTTP and exchanging data in XML over SOAP.

Layer 2, the Knowledge Representation layer: - defining and describing constructs for knowledge model representation, developing, sharing and managing the contents of model and meta-model repositories. The Knowledge Representation layer defines how models, meta-models and meta-data are represented, used and managed. METIS is used to manage models, modelling languages and meta-data. The model content can be persistently stored in the shared model repository. Future versions will support project, team and work administrative processes and an administrative database. Model contents, meta-model versions, revisions and variants, and meta-data hierarchies that are local, project specific or global will be separately managed. The architecture involves work processes that manage the project administration database (organisation, roles, users) and the meta-model repository, and that save accumulated experiences and life histories for change and configuration management and situated learning purposes.

Layer 3, the Model and Work Management layer; - modelling the customer solution, adapting engineering processes, and implementing work processes, executing and managing models. Model and Work Management will model and implement work processes for the engineering processes, and provides services to support the EE teams. In versions 1.5 and 2.0 we will model and implement work processes as active, reflective objects. Model and work management will therefore be implemented as immersed, rule driven and reflective work processes. The architecture of this layer is the management rules embedded in use case work processes, the model engineering work processes, and the life-cycle management model automatically creating life-history, when teams are executing work processes.

Layer 4, the Work Performance layer; - implementing customer solutions, generating work environments as personalised and context-sensitive views and GUI's being worktops accessed through portal-based user environments, and performing work with life-cycle management control.

4 RELATED WORK

With respect to supporting dynamically networked organisations, most B2B E-business frameworks (Shim, 2000) focus on information exchange and business transactions. This is also the case with newer frameworks such as ebXML and the perceived uses of Web Services. These approaches lack support for the dynamic, collaborative, and knowledge-intensive parts of inter-organisational processes, and knowledge management in this setting.

The major application area of BPM is still Business-Process Reengineering (BPR) and Business-Process Optimisation. The real potential of BPM - real time decision support – is barely exploited.

Enterprise ontologies have been proposed as a way of solving the communication problems arising from different interpretative frameworks in different organisations. This approach is based on conventional notions of model interpretation, i.e. the Turing paradigm, where the technical actor interpretation is fully automated and no interaction is allowed to aid interpretation, and not the more powerful interaction machine paradigm (Jørgensen, 2001; Wegner, 1999). The main characteristic of an interaction machine is that it can pose questions to human actors (users) during its computation. The problem solving process is no longer just a user providing input to the machine, which then processes the request and provides an answer (output); it is a multi-step conversation between the user and the machine, each being able to take the initiative.

Workflow management systems have also been proposed as a solution for inter-organisational collaboration (van der Aalst, Weske, 2001). Knowledge intensive processes are found to require a degree of flexibility not enabled by conventional production workflow systems. Alternative such as Service-flow (Wetzel, 2002) is appearing, but these new approaches are not linked to explicit process modelling.

5 CONCLUSION AND FURTHER WORK

The next version of the infrastructure will be released towards in the beginning of 2002, and we are currently collecting experiences from the case studies as input to further developments. First experiences are reported in (Lillehagen, 2002b), where parts of a quasi-experimental investigation are reported. This paper is focusing specifically on the results reported on communication, learning and trust in an extended enterprise being supported by our model-based infrastructure. Positive trends have been identified within all these areas, making us convinced of the great potential of active knowledge models in this area. Version 2.0 of the 4-layer infrastructure is planned to be available in September 2002. Focus is on implementing EE capabilities as repeatable and reusable work processes and services at layers 2, 3 and 4 of the infrastructure.

6 REFERENCES

Aalst, W. v. d., Desel, J. Oberweis, A. (2000), *Business Process Management*. LNCS 1806, Springer-Verlag.

EXTERNAL (2000-2002), *EXTERNAL - Extended Enterprise Resources, Networks And Learning*, EU Project, IST-1999-10091.

Jørgensen, H. D. Carlsen, S. (1999), *Emergent Workflow: Integrated Planning and Performance of Process Instances*, Workflow Management '99, Münster, Germany.

Jørgensen, H. D. (2000), *Software Process Model Reuse and Learning*, in Proceedings of Process Support for Distributed Team-based Software Development (PDTSD'00), Orlando, Florida. IIIS - International Institute of Informatics and Systemics.

Lillehagen, F. Dehli, E. Fjeld, L. Krogstie, J. Jørgensen, H. D. (2002a), *Utilizing active knowledge models in an infrastructure for virtual enterprises*, Proc. PROVE'02 IFIP Conference on infrastructures for virtual enterprises, Portugal, May, Kluwer.

Lillehagen, F., Krogstie, J., Jørgensen, H. D., Hildrum, J. (2002b), *Active Knowledge Models for supporting eWork and eBusiness*. Accepted at ICE'2002, Rome, June.

Schreiber A. Th., Hoog, R., Akkermans, H., Anjewierden, A., Shadbolt, N., Velde W. (2000), *Knowledge Engineering and Management*. The CommonKADS Methodology. The MIT Press, Cambridge, London.

Shim, S. S. Y., Pendyala, V. S., Sundaram, M. and Gao, J. Z. (2000), *Business-to-Business E-Commerce Frameworks*, IEEE Computer, vol. 33, no. 10.

Skyrme, D.J., Amidon, D.M. (1997), *Creating the Knowledge-Based Business. Business Intelligence*, London, New York.

Vernadat, F. (1996) *Enterprise Modelling and Integration*. Chapman and Hall.

Wegner, P. Goldin, D. (1999), *Interaction as a Framework for Modeling*, in Conceptual Modeling. Current Issues and Future Directions, Lecture Notes in Computer Science 1565, P. P. Chen, J. Akoka, H. Kangassalo, and B. Thalheim, (Eds.), Springer-Verlag.

Wetzel, I. Klischewski, R. (2002), *Serviceflow beyond Workflow? Concepts and Architectures for Supporting Interorganizational Service Processes*. In Pidduck, A. B., Mylopoulos, J. Woo, C. C. and Ozsu, M. T. (Eds.), Proceeding s from CaiSE'14, Toronto, Canada.

Synthesising an Industrial Strength Enterprise Ontology

Chris Partridge[1], and Milena Stefanova[2]
[1]The BORO Program, [2]LADSEB CNR, Italy, partridge@BOROProgram.org

Abstract: This paper presents a report on work in progress of a Synthesis of (selected) State of the Art Enterprise Ontologies (SSAEO) – which aims to produce a Base Enterprise Ontology to be used as the foundation for the construction of an 'industrial strength' Core Enterprise Ontology (CEO). The synthesis is intended to harvest the insights from the selected ontologies, building upon their strengths and eliminating – as far as possible – their weaknesses. One of the main achievements of this work is the development of the notion of a person (entities that can acquire rights and obligations) enabling the integration of a number of lower level concepts. In addition, we have already been able to identify some of the common 'mistakes' in current enterprise ontologies – and propose solutions.

1 INTRODUCTION

This paper results from a collaboration between two projects: the BRont (Business Reference Ontologies) (BORO, http://) and European IKF (Intelligent Knowledge Fusion) (EUREKA, http://) projects.

The BRont project is part of the BORO Program, which aims to build 'industrial strength' ontologies, that are intended to be suitable as a basis for facilitating, among other things, the semantic interoperability of enterprises' operational systems.

This European IKF project has as an ultimate goal the development of a Distributed Infrastructure and Services System (IKF Framework) with appropriate toolkits and techniques for supporting knowledge management activities. The following countries participate in the IKF project, Italy, UK,

Portugal, Spain, Hungary and Rumania. The project will last 3.5 years, and started in April 2000.

There are a couple of vertical applications whose domain is the financial sector. One of these, IKF/LEX – a part of the Italian IKF project – has been selected to undertake a pilot project. IKF/IF-LEX is lead by ELSAG Bank-Lab SpA and its goal is to provide semi-automatic support for the comparison of banking supervision regulations.

There will be two kinds of ontologies developed within the IKF project:

- A *Reference Ontology* composed of a *Top Level Ontology* and several *Core Ontologies* (Breuker, et al, 1997). The top-level ontology contains primitive general concepts to be extended by lower-level ontologies. The core ontologies span the gap between various application domains and the top-level ontology. The IKF/IF-LEX and the BRont projects are collaborating on developing a *Core Enterprise Ontology* (CEO) that IKF will use on this and its other applications in the enterprise domain.
- *Domain Ontologies*. The vertical applications will build ontologies for their specific domains. For example, the IKF/IF-LEX project is building an ontology for bank supervision regulations, focusing on money laundering.

2 SSAEO WORK PLAN

The scope of the synthesis work is large – and so the work has been divided into more manageable chunks.

As Breuker, et al, (1997) states, a core ontology contains "the categories that define what a field is about." A first rough intuitive guess of what these categories might be has proved a useful tool in:

- Helping clarify the scope focus on the important aspects for the CEO, and
- Acting as a basis for segmenting the work.

The selected categories are:

- *Parties* (persons) which may enter in
- *Transactions* (composed of agreements and their associated activities), involving
- *Assets*.

The ontologies to be analysed were selected according to:

- The relevance of their content to the Core Enterprise categories, and
- The clarity of the characterisation of the intended interpretations of this content (Guarino, 1997, Gruber, 1993, Partridge, 1996).

This gave us the following list:

- TOronto Virtual Enterprise - TOVE (Fox, et al, 1993 & 1996, TOVE, http://),
- AIAI's Enterprise Ontology - EO (EO http://, Uschold, 1997 & 1998),
- Cycorp's Cyc® Knowledge Base - CYC (CYC, http://),
- W.H. Inmon's Data Model Resource Book - DMRB (Inmon, 1997, Hay 1997).

The work proceeds by analysing one category in one ontology at a time, and then re-interpreting the previous results in the light of any new insights. Initially, the work focuses on individual ontologies but as it proceeds there is enough information to start undertaking comparisons between ontologies. The final analysis will encompass analyses of both the individual ontologies and comparisons between them.

In each of the ontologies, the concepts and relations relating to the category being considered are examined for the clearness and uniformity of their descriptions and formalisations. Further, each concept is analysed for its coverage and extendibility in cases where the coverage is not complete. Relations between concepts that are not explicitly described, but clearly exist, are identified as well. In addition, for the sake of a clear interpretation, we have found it necessary to consider the top concepts (whether or not they are explicitly described).

An important part of the analysis is testing each concept and its relations against a number of standard examples and more specialised concepts. Further, a check is made against a number of standard difficult cases. Both these checks help to identify weaknesses in the coverage of the ontologies.

A key concern in the analysis is to understand how the various concepts interlink with one another, to better understand the unifying structure of the Enterprise ontology.

At various stages during the analysis an interim ontology is synthesised from the strengths found in the analysis, in such a way as to eliminate the known weaknesses – and itself analysed. In the final synthesis, all the categories in all the ontologies are combined into a base CEO ontology.

At this time, the SSAEO work is concluding the analysis of the Parties (Persons) category for the EO and TOVE ontologies – and early drafts of synthesised ontologies are being reviewed. There is still substantial work that needs to be done in determining the precise relations between concepts, such as *LEGAL ENTITY* and *OWNERSHIP* within the EO.

3 INITIAL FINDINGS

Though both the ontologies have many important insights and provide much useful material – our most general findings, at this stage, are that none of the ontologies:

- Adequately meet our criteria of clear characterisation, or
- Really share a common view of what an organisation is.

Taken together, these findings mean that the creation of the synthesised base CEO ontology cannot just be a simple merging of the common elements of the selected ontologies.

We now illustrate these findings with examples. We also show how we synthesised a resolution to some of these problems - for the two ontologies we have analysed.

3.1 Clear Characterisation

With an unclear characterisation it can be difficult to work out the intended interpretation – in the worst case, impossible to decide between competing interpretations. There are many different ways in which the characterisation can be unclear – as we show below.

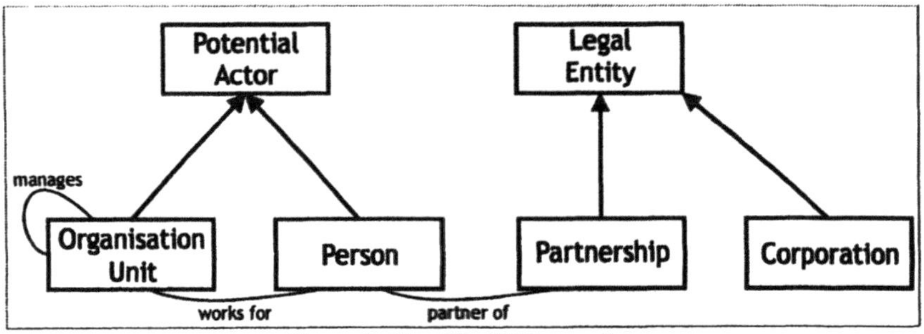

Figure 1: Simplified EO overview

In both TOVE and EO we found no clear overview of the structure – so we developed graphical representations based upon ER diagrams to help us understand it. Fig.s 1 & 2 provide simplified versions of these.

Both TOVE and EO make use of a number of top concepts. A top ontology – or top concepts – can provide a useful structure for defining and using domain concepts and relations – segmenting the enterprise and other domains into general categories. However, if this is not done properly it can have the opposite effect.

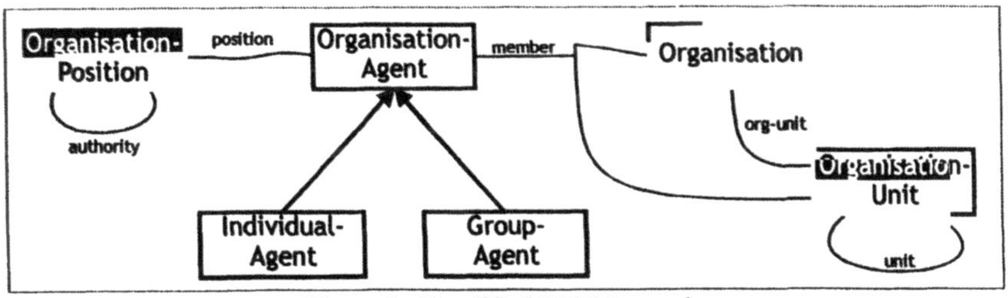

Figure 2: Simplified TOVE overview

Some of the problems we encountered with the top concepts and the domain analysis are:

- Insufficient characterisation of the disjointness of top concepts. For example, in the informal EO the relationship between the top concepts *ENTITY*, and *ROLE* is not clear – in particular, whether *ROLES* can be *ENTITIES* or not, and so whether they can enter into *RELATIONSHIPS*.
- The same lack of care in characterising disjointness (and overlapping) exists at the domain level in both TOVE and EO. We found this can make it impossible to definitely determine the intended interpretation. For example, in TOVE the formalisation allows an *ORGANISATION-UNIT* to be an *ORGANISATION* – though this seems counter-intuitive, and probably not what the authors intended.
- *Not applying top concepts*. TOVE states that a fluent is "a [type of] predicate or function whose value may change with time". But it does not identify which predicates in its ontology are fluents – leaving this to the readers, who have to make their own judgements. Supplying such information would have helped not only the users of the ontology but also its creators and designers. For example, the TOVE's creators end up (probably unintentionally) having to regard *ORGANISATION* as a fluent – when in the normal (common-sense) use of the concept it is not.
- *Messy formalisation trajectories*. EO formalises its concepts in logical systems (Ontolingua and KIF), which rely on their own (different) top concepts. An attempt for a clear formalisation trajectory has been made (Uschold, et al 1997), but unfortunately this does not match very well with the informal specification. For example, in the informal EO it is stated that each *RELATIONSHIP* is also an *ENTITY*, but is not defined as such in the formalisation. Furthermore some *RELATIONSHIPS* are defined in the formalisation as classes and others are defined as relations without explaining what the motivations for these choices are (e.g., *SALE* is a *RELATIONSHIP* formalised as a class, *HAVE- CAPABILITY* is a *RELATIONSHIP* formalised as a rela-

tion). This becomes a more serious problem if the formalisation is meant to be taken as the more accurate version.

- *Failing to use general concepts to achieve uniformity*. Both TOVE and EO fail to use top concepts to describe in a uniform way core relations and concepts. This hampers understanding. Typical examples are the *part-of* relation, used in describing the decomposition of organisations into smaller units, and the relation, which shows the different ways for *participation in organisations*. For example, TOVE introduces two kinds of *part-of* relations: *org-unit* (between *ORGANISATION* and *ORGANISATION-UNIT*), and *unit* (between two *ORGANISATION-UNITs*). These relations express *ORGANISATION* and *ORGANISATION-UNIT* decompositions, but are not explicitly unified under a common relation. In the EO several ways of participating in a company are considered, as a partner (*partner_of* relation between *PERSON* and *PARTNERSHIP*), as an employee (*works_for* relation between *PERSON* and *OU*), as a shareholder in a corporation (only in the informal EO specification, Uschold, et al, 1997). These ways of participation are not unified in the EO.
- *Insufficient analysis*. As an example consider the EO concepts of *OWNERSHIP* and *SHAREHOLDING* (Uschold, et al, 1997), which are formally unrelated, while *SHAREHOLDING* as evident from its informal and formal, definitions represents the ownership relation between a *CORPORATION* and its owners.

3.2 Common view of an organisation

Fig.s 1 & 2 give a broad picture of the concepts included in the analysis of TOVE and EO. As even a cursory glance can tell there are significant differences.

There are many examples in both TOVE and EO of how a better analysis would have led to more similar views:

- *Insufficient analysis*. In TOVE, for example, it seems that an *ORGANISATION* is not an *AGENT*, but has *AGENTS* as members. Yet there are many examples of organisations (such as the EU or NATO), which have other organisations as members.
- *Missing Links*. In the EO, the relation between the concepts *OU* and *LEGAL ENTITY* is unclear. All that we are told is that a *LEGAL ENTITY* "may correspond to a single *OU*" (Uschold, et al, 1997). No further analysis (informal or formal) of the link between these two concepts is given.

- *Implicit context dependencies*. In the EO, the concept *LEGAL ENTITY*, is not well thought out – having several (informally inconsistent) descriptions. It seems that the intended meaning actually depends on a particular jurisdiction (in this case on the current UK jurisdiction) – though it is not clear that the authors recognise this. This dependence is inappropriate in the modern global economy – and it raises potential problems should the UK jurisdiction change. For example, the *LEGAL ENTITY* concept would no longer be the "union of *PERSON, CORPORATION*, and *PARTNERSHIP*".

3.3 Unifying the Core Concepts: Person

Part of the synthesis work is to analyse the ontologies in preparation for a synthesised common view. A vital missing element from both the ontologies is a unifying core category.

To resolve this, we have introduced the concept *PERSON (PARTY)*, which can be a *NATURAL PERSON* or *SOCIALLY CONSTRUCTED PERSON* (*SOCIAL PERSON* in short). This acts as the catalyst for transforming the ontologies into ones with similar characteristics. The next step (which we will undertake soon) is to merge them into a single synthesised ontology.

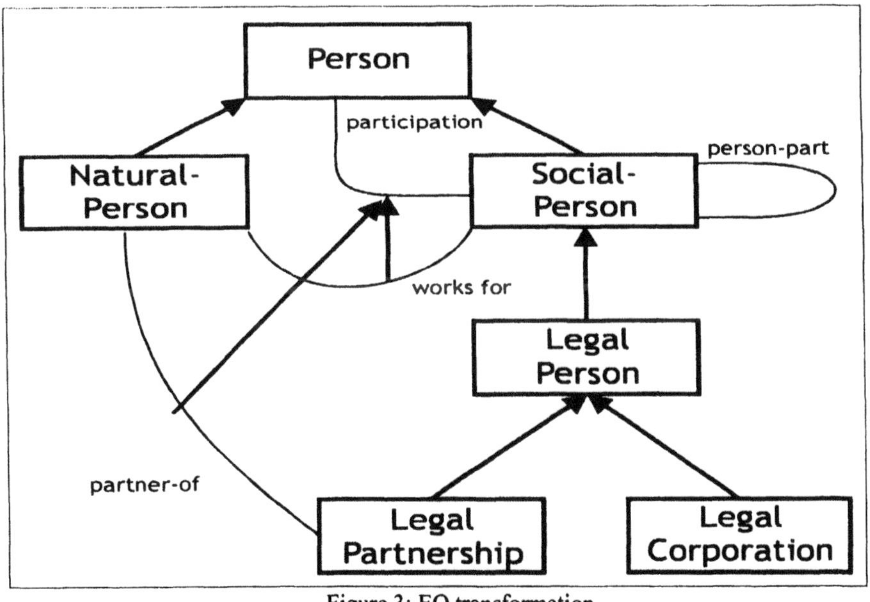

Figure 3: EO transformation

The result of introducing *PERSON* into the EO ontology is shown in Fig. 3. A comparison of this with Fig. 1 shows how *PERSON* has unified the taxonomy.

To give the reader some idea of how the transformation was effected, we describe the steps we went through. The EO concepts *LEGAL ENTITY* and *OU* are generalised into the concept *PERSON*. The EO concept *PERSON* (human being) is renamed into *NATURAL PERSON*. *OU* becomes *SOCIAL PERSON*, while *LEGAL ENTITY* is taken completely out and substituted with the context independent notion of *LEGALLY CONSTRUCTED PERSON* (*LEGAL PERSON* in short).

Note that *LEGAL PERSON* is not the same concept as the EO *LEGAL ENTITY*, since it is intended to represent parties, which are constructed according to a legal jurisdiction, but not necessarily recognised by it as legal persons (in EO terms, *LEGAL ENTITY*s). For example, in UK a partnership is not legally recognised as a person (it cannot sign contracts in its name) but it is a *LEGALLY CONSTRUCTED PERSON*, because there are legal constitution rules for partnerships. Finally the two participation relations, *partner_of* and *works_for* are consolidated under a general *participation* relation, and the relation *manages* is renamed into *person-part* (which is a particular kind of *part_of* relation).

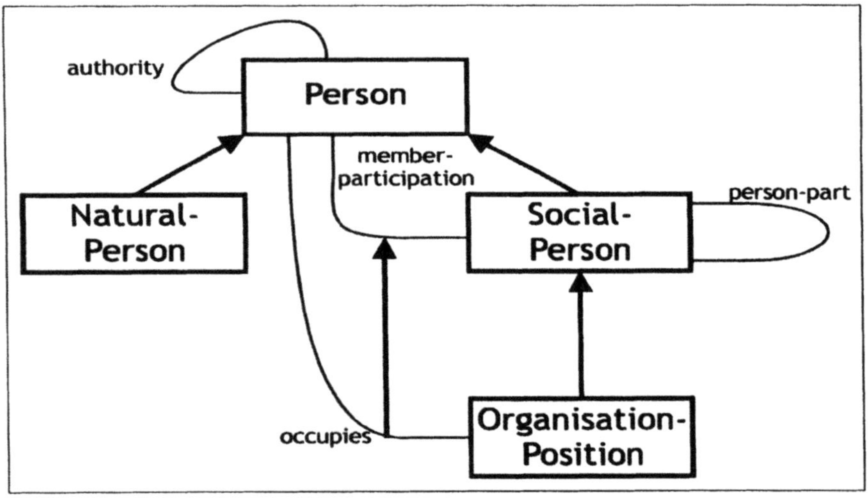

Figure 4: TOVE transformation

The result of introducing *PERSON* into the TOVE ontology is shown in Fig. 4. As before, a comparison of this with Fig. 1 shows how *PERSON* has unified the taxonomy. The transformation steps between Fig. 2 and Fig. 4 are similar in many respects to those between Fig.s 1 and 2.

4 CONCLUSION

Even at this early stage our work has revealed the need for a substantial improvement in enterprise ontologies to bring them up to 'industrial strength'. Hopefully, our work will go some way towards realising this.

5 ACKNOWLEDGEMENTS

We would like to thank to the IKF Project in general and to ELSAG SpA in particular for making this research possible. Furthermore we would like to thank to Allesandro Oltramale, Claudio Masolo, and Nicola Guarino for the numerous fruitful discussions we had on topics related to ontologies and organisations.

6 REFERENCES

BORO, http://www.BOROProgram.org

Breuker, J, Valente, A., Winkels, R. (1997), *Legal Ontologies: A Functional View* in P.R.S. Visser, and R.G.F. Winkels, Proceedings of the First International Workshop on Legal Ontologies.

CYC, http://www.cyc.com/publications.html

EO, http:// www.aiai.ed.ac.uk /project/ enterprise/enterprise/ ontology.html

EUREKA, http://www3.eureka.be/Home/projectdb/PrjFormFrame.asp?pr_id=2235

Fox, M.S. Chionglo, J., Fadel, F. (1993), *A Common-Sense Model of the Enterprise*, Proceedings of the Industrial Engineering Research Conference.

Fox, M.S. Barbuceanu, M. Gruninger, M. (1996), *An Organisation Ontology for Enterprise Modelling: Preliminary Concepts for Linking Structure and Behaviour*, Computers in Industry, Vol. 29, pp. 123-134.

Gruber, T. (1993), *Toward Principles for the Design of Ontologies Used for Knowledge Sharing*, in Nicola Guarino and Roberto Poli, (Eds.) *Formal Ontology in Conceptual Analysis and Knowledge Representation.*

Guarino, N. (1997), *Semantic Matching: Formal Ontological Distinctions for Information Organization, Extraction, and Integration.* In M.T. Pazienza (Ed.) Information Extraction: A Multidisciplinary Approach to an Emerging Information Technology.

Hay, David C. (1997), *Data Model Patterns: Conventions of Thought*, Dorset House.

Inmon, W.H. (1997), *The Data Model Resource Book: A Library of Logical Data and Data Warehouse Models*, John Wiley and Sons.

Partridge, C. (1996), *Business Objects: Re-Engineering for Re-Use*, Butterworth-Heinemann.

TOVE, http://www.eil.utoronto.ca/tove/

Uschold, M. King, M. Moralee, S. and Zorgios, Y. (1997), *The Enterprise Ontology*, AIAI, The University of Edinburgh.

Uschold, M. King, M. Moralee, S. Zorgios, Y. (1998), *The Enterprise Ontology*, in M. Uschold and A. Tate (Eds.) The Knowledge Engineering Review, Vol. 13.

PART 3. ENTERPRISE INTER- AND INTRA-ORGANIZATIONAL ENGINEERING AND INTEGRATION

Virtual enterprises are a new way for SMEs to unite forces, increase their competitiveness and meet today's market needs and jointly behave as one producer towards the customer. But collaboration is not only a technical issue, but also a social and organisational one, as well as a matter of trust.

This section addresses these topics discussing methodologies and reference models for building virtual enterprises as well as their organisational and human aspects. It closes with industrial examples of collaborations.

Two special issues of enterprise engineering and integration are addressed in the workgroup reports. The first group proposes the exploitation of agent technology to obtain solutions applicable for advanced virtual enterprises (Goranson). It includes the use of agent-model pairs applying ontologies and thereby addressing model semantics and its impact on model costs. The second report by Weston is on planning of virtual enterprises and identifies a set of common VE business planning activities and the degree of concurrency between planning processes at different planning levels.

The paper by Bernus describes the need for high quality reference models for virtual enterprises that will speed up the creation of different types of virtual enterprises. The need to develop a set of design principles is identified and demonstrated by some examples.

Focusing on the idea of *process organisation* Levi in his paper reports on a process framework deployed recently in project at a leading energy generation and trading enterprise. The integration of the process framework into the management structure introduces clear focus on consistent and collaborative ways that result in a direct impact on the bottom line.

An approach to the analysis, design and specification of agile and efficient enterprises is presented (Webb). The method enables clear justification of design, definition of interfaces and derivation of validated requirements. Comparisons are drawn to Zachman, ISO 15704 and pre EN ISO 19439.

Cieminski describes a framework for manufacturing systems engineering that is based on the concept of industrial engineering, but uses the life cycle concept as described in the Generalised Enterprise Reference Architecture and Methodologies, GERAM. A generic engineering process is described.

Five papers are concerned with human aspects in enterprise engineering and integration. Starting with the problem of awareness and acceptance, Mendez reports on his efforts in introducing process modelling in Mexico. A concept is described for identifying business process modelling as a solution to a problem in the management decision-making process.

A classification is made based upon properties of teams described in the human factors literature (Byer). A reusable understanding of these characteristic properties should (1) inform on the 'initial design and formulation of enterprise teams', and (2) help focus on 'continuing task development carried out by teams' through their useful lifetime.

Aguilar Savèn addresses human aspects as seen at different levels of an organisation. It describes the perception of the concept of integration by the people involved in the actual enterprise operation. Distinct differences of perception exist between management and the operational staff.

Tolone reports on lessons learned that reflect the human side of enterprise integration, which concerned with the human role, with security and privacy, and the re-examination/definition of traditional business processes.

Focusing on SMEs, the paper by Poller describes a project on knowledge management in the textile industry, evaluating different human related aspects in terms of barriers and potential solutions.

The last two papers present particular application of enterprise engineering. Weston in his contribution explains how 'process aware machine components' have been developed as re-useable building blocks for 'in production' assembly and transfer machine elements. 'Change capable' systems and the role of enterprise modelling in producing 'pro-active systems' are discussed.

Jaekels paper is concerned with simulation of supply chains integrating local models into a complete supply chain process model. The approach enables local maintenance of partial models, and furthermore provides encapsulation according to the needs of chain partners.

The Editors
Kurt Kosanke
CIMOSA Association,
Böblingen, Germany

Roland Jochem
Fraunhofer Institute for Production Systems
and Design Technology (IPK),
Berlin, Germany

Agents and Advanced Virtual Enterprises: Needs and an Approach

Report Workshop 2/Workgroup 1

H. Ted Goranson[1], (Ed.), Guillermina Tormo Carbó[2], Yoshiro Fukuda[3], Lee Eng Wah[4], James G. Nell[5], and Martin Zelm[6]
[1]Old Dominion University, USA; [2]Universidad Politecnica De Valencia, Spain; [3]Hosei University, Japan; [4]Gintic, Singapore; [5]National Institute of Standards and Technology, USA; [6]CIMOSA Association, Germany, tedg@sirius-beta.com

Abstract: see Quad Chart on page 2

1 INTRODUCTION

The following Quad-Chart (Table 1) summarizes the work of the group. It identifies the approach taken to address the issues of infrastructures for virtual enterprises exploiting agent technology and proposes future work on agent technologies and modeling languages.

1.1 Background

The working group decided to make an aggressive re-examination of infrastructure approaches for advanced virtual enterprises. The initial impetus for a significant reappraisal came from the report on new enterprise modeling challenges given in “ICEIMT: History and Challenges” in this volume. That report noted that the enterprise modeling problem set is significantly different now than it was for the first ICEIMT ten years ago. Several major constraints and enabling technologies have changed since then. Some early solutions now seem to present barriers. And in any case, the requirements of advanced virtual enterprises are significantly different than for centralized enterprises. Clearly a fresh look is required.

Table 1: Working Group Quad-Chart

EI3-IC Workshop 2 ***Enterprise inter- and intra-organizational engineering and integration***	**Workgroup 1** ***Agents and advanced virtual enterprises: needs and an approach***	**2002-January 23/25** **Gintic, Republic of Singapore**

Abstract: Virtual enterprises, especially advanced types, have shown promise for some time but have not yet become common. Some techniques that should facilitate progress are knowledge management, agent systems and enterprise modeling. This workgroup examined how these techniques might be applied in concert for information infrastructure for such advanced virtual enterprises.	**Major problems and issues:** – Need a theory of agents that is model centric – Need to migrate from enterprise models that merely represent the process to models that have actors that affect and control process work – Models should be formalized using ontologies – Determine if special modeling techniques are required to support enterprises driven by agents, actors and their models
Approach: – Re-examine infrastructure approaches for virtual enterprises – Assume traditional processes will remain, same outcome metrics will apply (e.g. profitability), financial markets will regard virtual enterprises as normal enterprises, and it is apropos to integrate at the process-model level – Use the levels of agent capability defined in ICEIMT'97 workshop – The information aspect of enterprise components may comprise agents, subagents, and actor-model pairs. Agents consist of subagents, models, and actors, where each actor has a single purpose	**Results and further work needed:** – Agents using enterprise models are the triggers that enable model-driven enterprises to work – Enterprise and process models are used for both reasoning about and controlling the processes – The group introduced an ordered way of bringing the notions of distributed-model integration to the virtual enterprise through the mechanisms of agents and exploiting the benefits of knowledge management – Traditional modeling techniques, if done properly, probably are sufficient to represent model-driven enterprises **Future work:** – Extend Process Specification Language to be more agent friendly and to include state mechanics to allow models to drive the processes – Assure that the Unified enterprise modeling language under development includes requirements inherited from the PSL extension – Research is needed to develop index systems for existing self-organizing model frameworks

But in order to scope the effort, the workgroup decided to accept certain existing assumptions in order to better focus on more important and leverageable matter. The group assumed:

- That whatever forms advanced virtual enterprises take, they are likely to continue a component/responsibility breakdown along well established functional lines such as marketing, financial, human resources and so on.
- That the same outcome metrics will apply to the combined enterprises and major components that apply to old-fashioned enterprises: profitability and deferred profitability in the form of such things as goodwill, market share, knowledgeable workforce and so on. A corollary of this is that financial markets will evaluate virtual enterprises in much the same way in the future as the regular type they might replace.
- That integration at the process/resource model level is the most promising approach for improvement, for example as opposed to integrating applications, services or product data flows. This mirrors the implicit common denominator of the enterprise modeling community. Moreover, the group asserted that for practical advance in virtual enterprises, existing model and model integration paradigms must apply. This means models and methods covered by the unified enterprise modeling and process specification language efforts, and integrating frameworks along the lines of CIMOSA and GERAM.

In the context of these assumptions, the group focused on the leverage of enterprise modeling and knowledge management in the context of agent-supported virtual enterprises.

Luckily a workgroup of an ICEIMT workshop just a month earlier (sharing some members with this group) had devised a complementary strategy between approaches to knowledge management and enterprise integration. This group adopted all the results of that prior work. (See "A Merged Future for Knowledge Management and Enterprise Modeling" in this volume.) Some key issues of that examination were:

- Modeling of uncertainty, for instance the beginning of organizing a virtual enterprise, while the product or opportunity is still being defined
- Modeling of unknowns, for instance managing placeholders for implicit tacit knowledge
- Managing distributed knowledge in terms of "situated" models. The notion behind this is that agents and enterprise components share information and that information only becomes knowledge when registered in context. That registration, the previous workgroup concluded,

can be largely satisfied by normalizing the information in a model and integrating that model fragment system-wide.

- Accommodating non-deterministic outcomes

Concerning the requirements of virtual enterprises, the group adopted the capability model approach from ICEIMT '97. It defines certain levels of capability:

- The lowest is where agents are not discriminated in the enterprise
- Then add the modeling of the effect of the agent. (This is where most virtual enterprise infrastructure is today.)
- Then add the modeling of the agent and integration in a central "location"
- Then add the distribution and autonomy of the agents
- Then add the ability of an agent to change itself to enhance the system. A key behavior here is when the agent acts in a way that is apparently detrimental to itself. In all this, an agent is equivalent to a virtual enterprise component. So an example of this behavior might be a partner, which sacrifices work (and local profit) in such a way that the whole enterprise becomes more profitable.
- Then add the ability of an agent to negotiate and change others in concert with itself

The focus of the workgroup was on virtual enterprises at the last two levels. This is what is meant by "advanced" virtual enterprises. From that ICEIMT '97 workshop, this workgroup was able to begin with a partial, speculative list of functionally aligned agent types of virtual enterprise components.

These types are shown in Table 2 below. The first column denotes the function of the virtual enterprise component (drawn from ordinary enterprises). The second column has a few examples of agents of that type. For instance the first agent, "Opportunity Agent," may be a new type of company that only identifies and defines opportunities as the kernel for virtual enterprise formation. The third column captures some of the modeling/knowledge management issues associated with those agents or components. The top row for example shows that the opportunity agent needs to model information that is "soft," for example product features that are "cool," customer needs that are dynamic and somewhat unpredictable, customer desires that have a certain measure of fickleness, customer values that reflect familiarity or safety in a brand. The rows associated with Distribution/Logistics/Service and Design/Manufacturing are not very populated because they are well known functions with less difficult and understood modeling needs.

Table 2: Types of agents and their relations to functionality

Marketing	Opportunity Agent	Soft Models (Cool, Safe, Dynamic, Fickle)
	Brand Management Agent	
Legal	Liability Agent	Intent Uncertainty
	Risk Mitigation Agent	
	Performance Trigger Agent	
Human Resources	Certification Agent	Knowledge Representation Knowledge Value Trust Metrics Learning Costs Knowledge Distance
	Knowledge Management Agent	
	Collaboration Catalyst	
	Trust Manager	
Financial	Capital Swarmer	Reverse Activity Costing Trust Aggregation Metrics
	Strategic Metric Harmonizer	
	Dispersal and Loan Manager	
Distribution/Logistics/Service	(Ordinary Agents Omitted)	
Design/Manufacturing	(Ordinary Agents Omitted)	Algorithm Fit Metrics State Maps
	Process Reuse Broker	
	Message Registration Agent	
	State Monitor	
General	Role Manager	Role Models Effect Controls
	Deconstructor	Reverse State Propagation
	Dating Service	Speculative Profiling Layered Virtual Exercising

(As an aside, the workshop was held in Singapore, whose economy is centered in shipping and functions related to expediting and scheduling. Advanced virtual enterprises may have several of the new agents noted above, but it is more likely that the expert operation of any one of them can be the focus for such advanced enterprises. It was speculated that Singapore concentrate on the legal function since many types of virtual enterprises are highly distributed. Shared risk implies fairly complex issues associated with who is responsible for events when material and subassemblies are in transit. A successful component focused on this — and concurrent maritime law with arbitration — could form a reusable component for advanced global virtual enterprises "hosted" in Singapore.)

The group decided to focus on the "coordination catalyst" as an example to focus discussion. That agent represents operations that cover all the interesting challenges. In general, the agent monitors the goals of the entire enterprise together with capabilities (including knowledge) of the various components and likely future and alternative components. It manages the optimization of component knowledge and resources in collaboration to optimize the enterprise system. Specific challenges are:

- The product and market of the enterprise system are likely to be partially vague with a set of unknowns and uncertainties. The environment will be dynamic and some dynamism will be unexpected.
- Some of the strategic goals concerning this market will involve indirect benefit such as customer goodwill, brand awareness and improved market share. These "soft" benefits need to accrue in some way to the virtual enterprise components.
- The components must collaborate in a way that individual processes (which contain some tacit knowledge) need to be continuously mapped to global benefit (which is partially soft). Individual and collaborative adjustments in processes must be made to optimize against the global need and or to adjust to changing global need.
- Part of the above includes "learning," and the learning is of several types as outlined by the previous workshop.
- A risk/reward strategy must be maintained, trust incubated, and controversies arbitrated, all in a context which will be non-linear and likely non-deterministic.
- The potential opportunities of the virtual enterprise are drawn partly by external conditions and partly from enterprise capabilities. But since the composition of the virtual enterprise is effectively unlimited, the potential opportunities are not bounded by the relatively simple constraints of old-fashioned enterprises.

2 THE APPROACH

Following general convention, the group assumed an architecture consisting of virtual enterprise components that may be companies or relatively independent operations of a company. Each component is represented as an agent. Each component is likely to have subcomponents, usually defined by process groups, which can likewise be represented as an agent. Each of these agents communicates by information flows. Some of the agents are automated. Some of the information exchanges are explicit, well-formed messages, perhaps from machine to machine. Models capture the mechanics of the process within each agent. And because models also capture the "process" of agent-to-agent interaction, the structure of the information exchanged is defined by those models. An advantage of agents is that some of them can act autonomously and negotiate solutions from the "bottom up."

Here is where the real work of the group began. The problem is that at present there is not a well-founded theory of agents that is model-centric in a way that can exploit the relatively mature mechanisms of model integration by frameworks. This is combined with the problem noted in the previous workshop that there is no well-founded collection of practices that relates knowledge in the enterprise in a leverageable way to models of what goes on in the enterprise. The group set out to make the first step toward such foundations, with the goal of defining a new research agenda. In the near term, the group intended to "pass the torch" to the next ICEIMT workshop for further exploration of foundations and research issues.

The participants of the group were talented thinkers, familiar with the nature of representation and abstraction. In general, when faced with a situation where current abstractions are inadequate, the logical solution is to introduce a new layer of abstraction. That way, the resulting demands on the abstraction primitives are reduced, but at the cost of greater complexity in the mechanics of the representation system. The way around that problem is to if at all possible design the new layer so that it uses as much of the existing representation mechanism as possible.

After some experimentation and debate, the group introduced the notion of actor-model pairs as the key components of an agent. Any agent has as its "active" components this pair. An agent can have many such pairs: where the level of granularity of agents varies, and agents can have subagents. The same is not true of actor-model pairs. Actors are extremely simple, capable of one action only. The associated model would be considered a model fragment in the standard lexicon of process modeling. Many such fragments (and associated actors) would be needed to do some meaningful work. For the remainder of this paper, "model" is used for such model fragments.

Note that the model is not just a notational record divorced from the process. Early modeling was indeed that: a way of describing the work without directly affecting or controlling the work. The model in this scheme is used for both reasoning about the process and controlling the process. So the model actually "stands for" lots of stuff within the agent: resources, data stores, sensors, people, perhaps material and so on. The model is both the representation of that part of the world of such stuff that matters, and the surrogate for that stuff.

Consider the example case of two simple agents, each of which consists of one actor-model pair. Typically, each agent would have many such pairs. The primary information flow is between actors: one a "sender" and the other a receiver" actor. Each model is serviced by resources, data stores and so on. The group simplified the exchanges by making each actor capable of only one action. This was at the cost of greater numbers of components, but they all are of the same basic type. The group presumes that model integration via frameworks can be accomplished feasibly, so further decomposition of models is not unmanageable.

The granularity of the actors is defined by the granularity of the messages between agents. Fortunately, there is a well-founded theory of such messages in both the agent and virtual enterprise worlds. It is based on the same simplification used for the actors: interactions between agents are of a very few simple types. In the agent world, these are called "speech acts," for which there is a robust formal understanding. In the virtual enterprise domain, these are sometimes called "transactions" to emphasize the collaborative nature and the sometimes explicit contractual nature of these messages.

The U. S. National Institute of Standards and Technology has a program to codify virtual enterprise transactions in the context of speech act theory. Typical types are: query, affirm, solicit, request, refuse and ship. There is some debate on what are the optimum few types. A standard would be desirable, though not necessary. The workgroup assumed that this NIST work, or a similar standard that results from it would be the driver in determining the size, type and number of actors and accompanying models.

The introduction of the actor is one of two new layers of abstraction introduced by the group. The actor is suggested, even required by the formalism of transactions, so it goes a long way toward harmonizing the virtual enterprise and agents. But it introduces one new problem on the modeling side. Modeling is sufficiently mature now that there are projects underway to thoroughly formalize the approach. This centers on the notion of an "ontology," which is a formal specification of the laws used by a particular modeling method.

Ontologies are useful in two ways. The first is that they bring a mathematical formalism to modeling; this makes possible such things as automated correctness and systems for reasoning. In fact, the formalism of ontologies is what allowed the previous workgroup to assert that a simple bridge could be made between models and knowledge representation.

Knowledge is "situated" information. The message passed between two agents is simple information. The receiving actor "situates" that information, turning it into knowledge. The actual process is that the actor simply registers the information in the model. But behind the scenes, that model is linked to all other models in the enterprise, so that the act of placing in the model fragment actually places the information, or situates it, in a global context. The formalism of ontologies makes this possible.

Process ontologies have another benefit. Suppose that the originating agent used a different modeling method or lexicon than the receiving one. In this case, the message would consist of the message itself, and the ontological information about that message so that the actor on the right could perform the necessary translations to register it in its differently conceived model. A simple example message might be: "it is 2:00," which might be all you need if both models are part of precisely the same world. If not, the ontological wrapper might say: "2:00 is a measure of time; for us, time is an irreversible sequence of half-seconds; for us, 0-12 o'clock always means AM; for us, the number refers to GMT; for us, GMT is..." and on and on until everything is "explained" in simple standard terms that both sides understand.

The standard ontology for process models is the PSL, process specification language. Process ontologies are the trickiest of ontologies; where most ontologies focus on noun-like things, process ontologies are concerned with verb-like things. That means the ontology has to capture the notion of state, because it is a different thing that a process can happen, compared to it has happened, or even that it is happening. The mechanics that PSL uses is the formal mechanism of situation calculus, which can simply define, relate and reason about those three states.

But the group introduced some new states. The process between the model and actor represents a new state: in the originating agent it is a "pre-state" state. A process can now be in the state of having its actor get it ready for happening, or beginning to happen. In sequence, this new state goes in between "can happen" and "is happening." This is a new complication. The current use of situation calculus can be extended to account for these new state situations, but it is likely not a simple extension.

The introduction of the actor is at some cost, as noted above. But it solves a key problem: the explicit mechanism for "situating" information to become knowledge in a global context. However, it alone does not solve the other problems of concern having to do with soft, uncertain and dynamic knowledge. For that, the group introduced a second new layer of abstraction.

3 A META PROCESS: MODEL AND ACTOR

The new notion is a "meta-actor and meta-model." The basic role of this pair is to monitor the functioning of the agent and modify it under certain conditions. In this way, the agent can "learn" and adapt its performance. Such mechanisms are common in the agent community, but there is a special, novel constraint proposed by the group.

The clarity of process modeling is the core concept being leveraged. There is no reason that the processes that govern learning be considered — or modeled — any differently than the processes, which actually do the work of the enterprise. The novelty suggested by the group is to have all these actors/meta-actors and models/meta-models use the same conceptual infrastructure — the same speech act performatives as well. The state mechanics are apparently simpler, not more complex. (This presumption is revisited below.)

In fact, the group believed that except in two cases, there need be no special accommodation. Treat them all the same.

The first special case addresses the integrating strategy for the meta-models apart from the enterprise models. A non-automated example will clarify this. Consider that a function, perhaps a manufacturing cell, is part of an enterprise. It has processes that contribute to the work of the function, and thereby the work of the enterprise. Now suppose that the enterprise hired a management consultant to do process re-engineering. The consultant would introduce actors into the manufacturing cell to observe and change. But the algorithms used to do the process re-engineering would be part of the corporate knowledge of the consultant.

In this scenario, there are two integrated enterprise models: that of the manufacturing enterprise, and that of the consultant. They touch at the level of individual processes, but are integrated separately. If the manufacturing enterprise chooses to become distributed and virtual, with a distributed and virtual model integration strategy, there is no coincident requirement for the consultant to do so as well. In fact process re-engineering processes are likely to be better served by central, CIMOSA-like integration. Those algorithms are probably more static, and the decisions to "trigger" them likely to require more centralized oversight.

More generally, meta-models can refer to, or be shadowed from a central meta-model repository. This is seen as a simplifying constraint, designed to manage the complexity of introducing a new level of abstraction. It also reflects a reality.

But there is another special case that is trickier. The group wanted to explore the ability to have processes that were explicitly adaptive. This would at least require that the meta-model is wholly within the processes of the enterprise. This also reflects a reality: for instance manufacturing managers that are smart enough to improve their processes without the wisdom of an external consultant.

But the additional mechanics of having unconstrained meta-meta-models may not be so friendly. The problem is not additional burdens on modeling, control, or integration because those are all handled as before, in the simple case. It is instead the burden of preventing circular linkages and the state control problem of initiating a change at a high level of a process that "is in the middle" of something at a lower level. Apparently, the circular problem is manageable by restricting what actors can do, and by "hardwiring" them. In other words, the composition of an agent's actor-model composition cannot change from its birth, only the model. So if sufficient checks are done at birth, the risk of circular processes is eliminated. (Actors can only act on one model.)

The other problem is the problem of state mechanics. Earlier it was mentioned that the state mechanics for the internal actors is simpler than those who act outside the agent. But now it seems that the complexity is the same. In extending the state ontology, the system at all levels needs the state of "something is getting ready to happen" which can alert downstream processes. As noted before, this seems to be a manageable problem.

All in all, the group introduced an ordered way of bringing the notions of distributed model integration to the virtual enterprise through the mechanism of agents and exploiting the benefits of knowledge management. The group believes this preserves and leverages the considerable tools and advantages of CIMOSA-like modeling and integration. While it introduces new levels of abstraction, these appear to be manageable within the current tool framework and by extending existing standards rather than proposing new ones.

4 SOFT: TACIT, UNCERTAIN, DYNAMIC AND NONDETERMINISTIC KNOWLEDGE

The question remains. How many of the "grand challenge" problems identified in the previous workgroup does this resolve, and what additional work needs to be done?

Some problems are solved in a straightforward way: namely those associated with "tacit." knowledge. Tacit knowledge is all that stuff that you know that you never think about. In a normal interaction there is lots of tacit knowledge. So when one looks at the content of a message, the message has to be considered in context, or "situated," to use the term from above. For example, if someone says "hand me the small spanner," the speaker assumes certain tacit knowledge. Such knowledge is the target of much in knowledge management. It is also the bane of virtual enterprise components since there is a high level of tacit knowledge in normal process transactions, and the components don't have the shared past of built implicit understandings.

The group's notion of using model integration as a situating knowledge framework provides a formal basis for the identification, explication and management of tacit knowledge. There don't seem to be any significant technical barriers to exploiting this approach; they all are probably cultural barriers between industrial engineers and social scientists, which is a tacit knowledge problem in itself.

"Dynamic knowledge" is the obvious problem of using models (which are relatively static) to characterize and drive an intrinsically organic, evolving enterprise. The group feels that this problem has been deftly solved by introducing the meta-actors and meta-models. The reason for this confidence is that the processes of learning are relatively static. Separating the models of work from those of learning about work is a particularly elegant approach to leveraging the two communities without trying to synthesize them.

"Non-deterministic results" are those, which cannot be precisely seen by examination of the populated models involved and their inputs. Typically, non-deterministic results can come from many agents interacting with each other to optimize a result. Often that optimization is unintuitive and possibly would never have been found by predictive methods. The underlying power of virtual enterprises is some measure of non-determinism: such enterprises are expected to improve themselves in ways and at speeds unattainable from centrally managed enterprises.

The architecture described above is a conventional agent architecture that supports any measure of emergent behavior. That is not the problem. The problem is that many external metrics require the enterprise to appear deterministic in important ways. For instance, financers want to know where their money will be going and how it will map to the working of the enterprise. Some balance of distributed model independence (for non-deterministic behavior) and model integration (for whole system analyses) must be supported. This is left as an open issue for the next workshop.

The final issue is the problem of modeling uncertainty. Uncertainty in this context goes beyond the non-determinism described above. The major need is modeling uncertainty in the external environment, for example to understand completely unexpected, singular threats or opportunities. Often, these can be explained after the fact, but rarely predicted. But there are all sorts of uncertainties with internal processes as well.

What is needed is a model entity for a suspected but unknown fact, or a collection (or "situation") of them. All of these four issues may be helped by expanding the current use of the situation calculus in process ontologies to a more full-blown situation theory. In particular, situation theory shows significant promise for modeling uncertain and tact situations.

These outstanding issues are passed to the next ICEIMT workshops. The next workshop (number 3) was on "Enterprise Inter- and Intra-organizational Engineering and Integration," an appropriate topic for these matters.

5 CONCLUSIONS

The workgroup had two sets of results: suggestions for extending existing standards and proposals for new research.

The mapping of agents to virtual enterprise components depends on a standard set of speech act related transaction primitives. A standard set of transaction primitives should be adopted. This should be relatively easy because small differences among the definitions in common use convey no clear special advantage. The work, which has been started at the U. S. NIST, is a promising start toward such a standard.

The process specification language is mature, workable and well on its way through the standards process. But it was designed to support the translation between model methods, which is the second use described above. It is not agent-friendly, and does not have the additional state mechanics described. The standard should be so extended.

By definition, the workgroup assumed that exploitation of existing modeling methods was necessary. So it is likely that the work on a unified enterprise modeling language does not inherit any new requirements from agent-based virtual enterprises. But this should be checked, with particular attention to the implications of the PSL extensions, and the introduction of the "meta-actor" which can change models.

There is a standard model integration framework, coming from a CIMOSA legacy. This framework presumes a top down, relatively static organization, where all relevant processes are accessible. Again, it was the assumption of the group that the principles in this standard be exploited. There probably is a significant research project required to determine how best to distribute an index that allows distributed, diverse model fragments to self-organize against such an integrated registration framework.

The workgroup believes this to be a highly leverageable approach to evolving an infrastructure for virtual enterprises from existing foundations.

Research projects are needed.

Work needs to be done on the theory of situations to allow PSL's state mechanics to be extended in a formal manner as noted above:

- Research also needs to be focused on developing index systems for self-organizing model frameworks of the familiar type; and.
- Research must explore the issue of enterprise features that accommodate non-determinism but support apparent determinism.

Virtual Enterprise Planning Methods and Concepts

Report Workshop 2/Workgroup 2

Richard H. Weston[1], (Ed.), Cheng Leong Ang[2], Peter Bernus[3], Roland Jochem[4], Kurt Kosanke[5], and Henry Ming[2]

[1]Loughborough University, UK, [2]Gintic, Singapore, [3]Griffith University, Australia, [4]FhG-IPK, Germany, [5]CIMOSA Association, Germany, r.h.weston@lboro.ac.uk

Abstract: see Quad Chart on page 2

1 THE NEED FOR VE PLANNING METHODS AND CONCEPTS

Global working can enable an enterprise to: (i) gain access to overseas customers; (ii) improve utilisation of idle capacity in a falling industry sector; (iii) search for new business (e.g. to offset effects of trading cycles); (iv) satisfy a need to develop new products with high margins (Brooke, 1986).

Global working requires: (a) a broad base of relevant skills; (b) a wealth of experience and practice in a number of local markets (Ohmae, 1995).

Bleeke and Ernst, (1995) describe a common way of meeting criteria and to form "partnerships" between existing businesses to provide new skills, experience and practice. However, establishing and exploiting such partnership places very significant requirements on the planning of such collaborations (Berry, 1999).

The following Quad-Chart (Table 1) summarises the work of the group that addressed those requirements. It identifies the approach taken to resolve the issues in this domain and proposes a concept for planning such collaborations. In addition it states some ideas for future work for testing and enhancing the proposed solutions.

Table 1: Working Group Quad-Chart

<table>
<tr>
<td>EI3-IC Workshop 2
Enterprise Inter- and Intra-organisational Eng. and Integration</td>
<td>Workgroup 2:
Virtual Enterprise Planning Methods and Concepts</td>
<td>2002-January-23/25
Gintic, Singapore</td>
</tr>
<tr>
<td colspan="2">Abstract:
With the move to global markets and the emphasis on core competencies the need for inter-organisational collaboration increases. Such collaborations usually try to exploit business opportunities usually at short notice as well. Support of this type of business is not yet well established (Schweiger, Very, 2001).
This working group explored planning methods that will increase the efficiency of enterprise collaborations, which increasingly will deploy virtual environments (NIIP, 1998).</td>
<td>Major problems and issues:
– How to achieve more efficiency in the identification, establishment and exploitation of collaborations in virtual environments?
– How to enhance known concepts like the GERAM framework and business process modelling to provide guidance during the life cycle of such inter-organisational relationships?
– How to define languages and methods to describe business strategies and business models in relations to the life cycle phases of the GERAM modelling framework?</td>
</tr>
<tr>
<td colspan="2">Approach:
– Use the life-cycle concept and the GERAM modelling framework to structure the different tasks to be carried out during the life cycle of the collaboration
– Focus on the identification and concept phases of the life cycle and define the relevant tasks both in relations to the envisioned market and the capabilities of the potential collaborators – with emphasis on SMEs
– Identify relations between the different contributors and support the development of scenarios for proposed collaborations
– Propose planning methods/processes to support the development of business models and business plans for the virtual enterprise
– Extend the planning processes to cover strategic, tactical and operational planning for the virtual enterprise</td>
<td>Results:
– Identification of a set of common VE business planning activities: Market/Capability Analysis, Scenario Generation, Business Analysis, Business Plan Generation, Monitoring of BP Implementation
– Identification of degree of concurrency between planning processes at different planning levels
Future work:
– Test concept in real practical applications:
– Local SME environment (Printing enterprise network in Singapore)
– Global industrial environment (in Automotive Industry)
– Global environment (IMS project Globemen)
– Investigate concept relations to concepts in human and management science
– Investigate communication and negotiation needs with emphasis on human relations</td>
</tr>
</table>

2 OBSERVED NATURE OF VE FORMULATION

Virtual enterprises have a greater range of business opportunities to which they might respond than would a single company (Schweiger, Very, 2001). However before a response can be actioned and funded it must be properly justified and planned (Berry, 1999). VE Business Planning is focused on abstract, human enacted analysis and decision-making. Generally it requires at least one team of people with necessary clout and capabilities (Ohmae, 1995). Knowing about the kind of business processes that work in their market segment they must acquire knowledge about the capabilities of potential partners and reason about affiliations between partners with potentially complementary competencies that might operate competitively as a coordinate whole (Brooke, 1986).

In general VE Business Planning Process will be iterative in nature. The planning team(s) will assess alternative distributions of responsibility for product realisation amongst a selection of businesses. Thereby the business planning team will develop a business plan based on scenario analysis, which identifies a viable VE configuration or set of configurations.

The VE planning processes is based on the GERAM life cycle concept. The planning teams(s) must consider product, process and resource issues at a relatively high level of abstraction. They will investigate alternative scenarios trying to understand/identify key aspects of the opportunity. In so doing they will perform Middle-Up-Down analysis and preliminary VE design work at a high level of abstraction but will require enough detailed knowledge to realise financial justification. By such means the VE business planning team(s) will begin to flesh out "concept design" and "requirements definition" aspects of an enterprise configuration.

3 AGREE AIMS OF THE PLANNING TEAM

It was observed that GERAM conformant enterprise engineering architectures and methodologies (such as CIMOSA) can enable many aspects of VE formulation, implementation and evolution. It was assumed that:

a) GERAM concepts and methods are sufficient to structure and support the set of projects generated as an output of the VE business planning process
b) Developed descriptions of VE formation processes and configurations specified using GERAM concepts will prove useful when convincing partners to collaborate on business grounds.

To understand better the implications of (a) and (b) the agreed aim of workgroup discussion was to begin to flesh out common VE business planning activities and to achieve a better understanding of how typical business planning processes should be resourced.

Table 2: Common VE Business Planning activities

VE Business Planning Activities Needed	Typical Inputs	Typical Output
Market Analysis - Assess customer Requirements - Assess market characteristics - Anal. market share opport. etc.	Product & market data - which may be used as input to existing strategic planning meth. & tools	- What can to be made, when, in what sort of quantities & what their likely geographical distribution will be
Capability Analysis - Acquire knowledge of competitor capabilities - Acquire understanding of possible partner capabilities	Estimated capability & capacity data on potential partners - which may be used as input to existing analysis meth. & tools	- SWOT analysis of potential partners & suppliers relationships taking into account their relative location
Scenario Generation and Simulation - Capture descriptions of alternative product realisation processes - Capture descriptions of alternative VE configurations	Domain process models and as required detailed business planning models developed by designated detailed business planning teams	- Dynamic simulation models that allow experimentation regarding alternative operational process flows & VE configurations
Business Analysis - Value analysis of alternative operational process distributions among partners - Cost estimate of forming alternative partnerships	- Experimentation results, simulation models, - Process based cost-calculations related to partnerships	- Spreadsheet results & graphical summaries showing financial pro's & con's of value added processes & their distribution amongst different partners
Business Plan Generation -Specify business plans for staged project engineering - Harmonise activities of tactical & operational planning teams	- Validated processes and their distribution among partners - Plans of tactical and operational teams	- Business Plan specification - Harmonised team plan
Decompose/Release B Plans -Breakdown plans into fundable engineering projects - Release & manage projects within GERAM framework	- Stages of business plans/specification - Cost prerequisites - Project framework (GERAM)	- Project plans - Released projects with defined milestones according to GERAM
Monitor Plan Implementation Oversee & measure projects based on GERAM principles through their life-time	- Feedback from project milestones - Project plans	- Measurement and monitoring result sheets
Ongoing Change to B Plans Based on event value analysis recommend change tactical & operational process distribution - Modify business plans re-decompose into fundable engineering projects & release & monitor project outcomes	- Change requests from 'monitoring' or directly from 'partners' - Project plans - Monitoring result sheets	- Modified plans

4 COMMON BUSINESS PLANNING ACTIVITIES TO BE CARRIED OUT DURING VE PLANNING

Table 2 lists a set of common VE business planning activities identified by the working group. The table also lists typical inputs and outputs from the common activities. The need for multiple levels of planning was observed and the need to harmonise multiple team-based activity. Also time dependencies between activities were observed, as were opportunities to iteratively develop and evaluate alternative scenarios using established management techniques possibly supported by enterprise modelling tools.

For example the business processes of candidate configurations of resources can be represented and their dynamic behaviours modelled using business process simulation tools. This might lead to "what if" analysis focused on alternative distributions of processes amongst candidate partners and their business units. The application of suitable management theories would then facilitate selection amongst viable VE configurations, e.g. based on short, medium and long term financial and lead-time considerations.

5 NEED FOR CLEAR DIFFERENTIATION BETWEEN DIFFERENT TYPES OF PROCESS

With respect to VE business, tactical and operational planning the working group understood the importance of drawing clear distinctions between product, process and resource building business processes, and their planning, as follows:

- Business Planning Processes will be used to develop a business case for one or more VE configurations.
- Tactical Planning Processes will consider how business cases can be realised by considering the use & benefits of alternative product, process & resource building processes.
- Operational Planning Processes will test the technical, economic & practical feasibility of realising specified business cases.

Business, Tactical & Operational Planning Processes will generate product realising process specifications & will assess the use of alternative ways of configuring VE partner capabilities so that product realisation is achieved in the correct quantities, in the right place, on time and at acceptable quality levels. Business, Tactical & Operational Planning Processes will be resourced by suitable teams of agents (normally requiring at least one human agent).

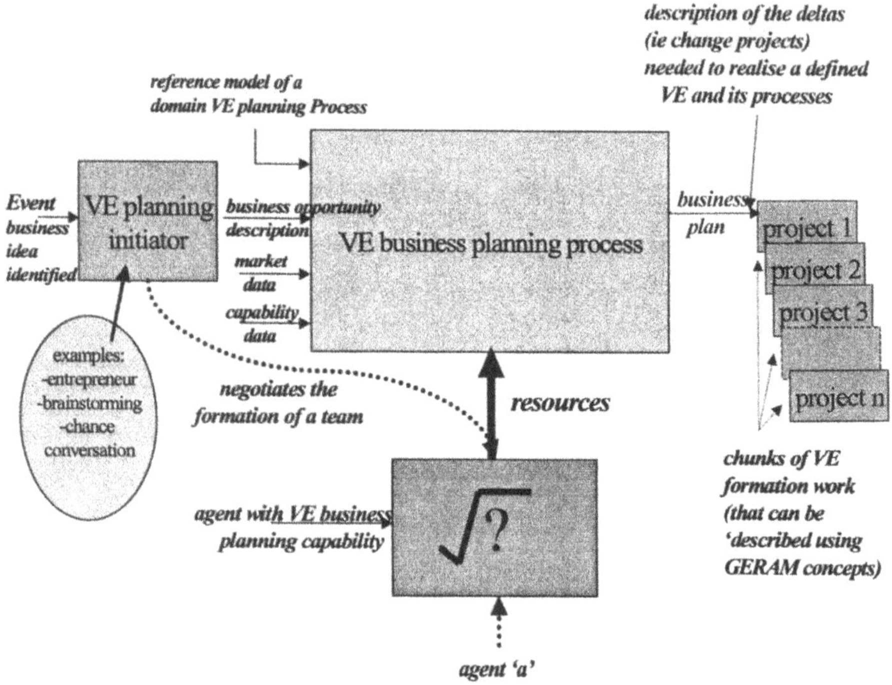

Figure 1: Basic VE Business Planning Process

As illustrated by Fig. 1, invariably the complexity of VE business planning processes requires them to be resourced by appropriate human teams. In the case of a basic VE business planning process, a planning initiator driven by some event (like a business idea) will negotiate the formation of a planning team, which comprises planning agents with necessary business planning capabilities to resource the VE planning process required. Negotiating the formulation of teams will be complex and political but critical to the quality of the business plan generated. The agreed composition and objectives of the team will also depend upon the intended nature of VE partnerships, e.g. whether alliances, mergers and/or acquisitions are the likely outcome.

Fig. 2 illustrates derivatives of the basic VE planning process shown in Fig. 1. This shows that business planning teams will negotiate the formulation and terms of reference of more focused planning teams and will take into account their detailed findings.

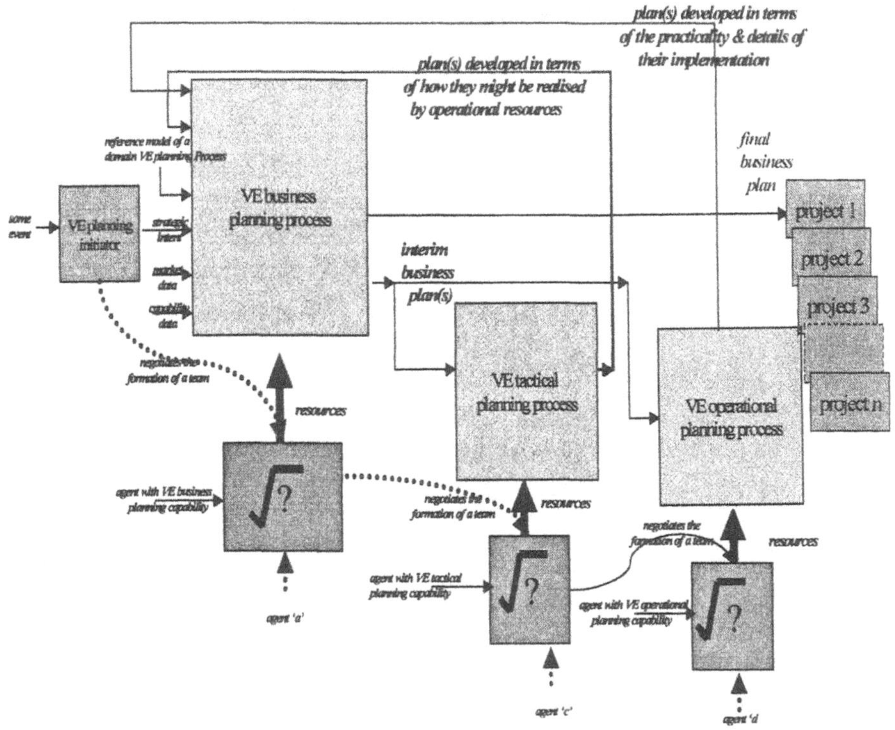

Figure 2: Possible Concurrent VE Planning Processes

6 OPPORTUNITIES TO APPLY AND FURTHER DEVELOP WORKING GROUP FINDINGS

6.1 Case Study Work

The working group identified illustrative case study work that could apply and develop the VE business planning concepts identified in concert with established GERAM concepts and business process modelling tools. Three possible test cases have been identified, which have a global reach, but are located in different environments:

- Printer enterprise network currently established in Singapore
- Global manufacture of car engines (COMPAG, COMPANION)
- Industrial project currently carried out as an IMS project (Globemen)

Action officers for each of the three test cases have been identified.

6.2 Interface with Established GERAM Concepts

The Working Group understood that improved formalisation of its VE planning concepts and their interface to established GERAM concepts would have potential to improve VE planning processes. Bearing in mind that a significant number of VE formation and implementation processes are ongoing around the world and that they will impact significantly on global economies, then enabling reuse of best VE business planning practice would also be of significant importance. A natural extension of this working group activity would therefore be to specify and test modelling constructs that formalise aspects of VE planning processes and their interfaces to enterprise engineering, and thereby to capture and promote examples of best practice.

6.3 Interface to Established Practice in Management and Human Sciences

VE business planning processes might be improved by generating reference models of the use of management theories, concepts and tools in support of VE formulation, implementation and evolution. Linked to this reference models describing team roles and responsibilities and the supporting methods, tools and infrastructure services they require could also be developed. Of fundamental interest might be a study of the way that planning teams adapt and evolve their tasks, behaviours, processes and structures so that co-ordinated planning is achieved amongst teams.

7 REFERENCES

Brooke, M. Z. (1986), *International Management: A Review of Strategies and Operations*, Hutchinson.

Ohmae, K. (1995), *Putting Global Logic Firs"*, Harvard Business Review, Vol. 78(1).

Bleeke, J. Ernst, D. (1995), *Is Your Strategic Alliance Really a Sale?* Harvard Business Review (01/02).

Berry, C. (1999), *Mergers Management: Acquiring Skills*, IEE, Manufacturing Engineer, Vol. 78(2).

Schweiger, D.M and Very, P. (2001), *International Mergers and Acquisitions Special Issue*, Journal of World Business, Vol.36(1).

NIIP, (1998), *Introduction to NIIP Concepts*, NIIP Consortium.

makes them very economical, because VEs do not have any overhead, they only dispose over resources as much as they actually need them to produce or add value.) However, from the customer's point of view when a VE designs and builds a product, this VE no longer exists during the time the product operates or when the product is maintained or decommissioned.

Thus the 'Business Model' behind a VE not only has to describe the ways of providing a market with goods and services of a certain kind, but it also must identify and describe a number of other entity types necessary to support the complete life of the products and services (including operational support and dissolving or recycling these as needed).

In order to create such a 'Business Model', one must define *all* involved enterprise entity types and their roles, interests and relationships within the overall value chain, and guarantee that the overall model operates as if all products and services were produced and supported as necessary for their entire life by one larger enterprise.

In addition to this ephemeral property, what makes VEs different from conventional businesses that use electronic means to integrate information flow is the *extent* and *completeness* of this integration, where this extent is measured in terms of how well this integration hides the fact that the VE achieves its mission through co-operation and co-ordination of many partners rather then through being a single large enterprise. We shall further elaborate on the required characteristics of this co-operation and co-ordination in Section 4.)

A common way to satisfy the above needs is to create Enterprise Networks. An Enterprise Network is an alliance of businesses formed for the exploitation of some kind of business opportunity. Participation in a Network (as well as entry and exit) is guided by contracts and agreed processes, which allow any member of the Network (according to the Network's co-operation contract) to respond to a customer order and to draw upon the resources of other members in the network. Enterprise Networks may be characterised by defining the set of functions that partners share as opposed to the set of functions that they do *not* share.

E.g. a network may be a sole marketing network, where partners only share marketing functions but do not share production and service delivery. Another network may share part of its logistic functions, such as transport and delivery, as well as marketing and quality control and product classification (such as in some agricultural co-operatives) but not share the production processes.

Some networks make a viable or even attractive business proposition (win-win for *all* parties, including the customers) and some do not (win-loose or loose-loose). The 'business model' needs to be investigated from the point of view of value adding and gain distribution, to establish that the interest of all parties is compatible with the overall business interest.

Since a Network is a relatively stable organisation - with or without substantial resources of its own (typically without) - it is possible for the participating businesses to develop a set of contracts and processes that guide all aspects of joint co-operative action. Thus at the time a business need arises (i.e. when a customer enquiry or order is received) network partners can create a virtual enterprise on demand - or immediately.

What makes the VE *different* from a supply chain with integrated information flow is the nature of this information flow. While conventional businesses have been exchanging information using electronic means for decades (such as orders, delivery schedules, payments etc. or as in recently popularised B2B transactions), a VE extends this information flow to all levels of management to achieve complete co-ordination of the joint activity. Partners involved in VEs have co-ordination on the strategic, tactical and operational levels, as well as this co-ordination extends to product related and resource related management decisions.

3 THE NEED FOR REFERENCE MODELS

From the above considerations it is clear that to exploit the idea of the VE, today's businesses need blueprints or 'Reference Models' that describe viable building blocks and combination rules of partner-, network- and VE functions. These building blocks can then be customised and combined for a given type of network and its VEs. Thus we develop reference models of processes for network and VE creation as well as reference models of business processes that are performed by the network and its VEs.

While blueprint-type Reference Models are necessary for businesses to be able to embark on a new way of doing business, it is usually also required that there be a 'step-by-step' methodology that businesses could follow to build networks and networks can follow to build VEs.

Since a methodology in general is a collection of procedures, rules, methods and tools, it is not necessarily possible to attain this goal. However, the more specialised the objective (the more we know the intended business model and therefore the *type* of network we intend to create), the more chances there are that a specialised methodology can indeed be developed in form of a step-by-step procedure.

Who should develop such step-by-step methodologies? - Clearly, the research community in business management and IT/IS can only be made responsible for the development of generic methodologies, those that are *not* step-by-step. This is because the amount of detail and know-how that is needed for a step-by-step methodology is a commercial commodity, thus not likely to be published in the open literature.

Consulting companies are another possible source of such step-by-step methodologies, but the problem is that selling such a methodology cannot be repeated many times over.

End users (i.e. potential partners in Networks) would be ideal candidates to pool resources and develop their in-house methodology, but with the exception of large companies, resources for such planning and development are scarce. Even though the IT resources that are needed for forming simple VEs and networks become cheaper and more accessible, the complexity and the risk involved in designing viable and significantly new business networks and VEs limits the level of innovation that small and medium enterprises can attain.

The result of the above situation is a potential polarisation of the business world, with large companies being the only ones that can tap into the new potentials of VEs, and small companies being only followers. It is argued, that for a healthy business world the innovation capability must be preserved for small and medium enterprises in the same way as for large companies, and a second set of reference models is necessary. This second set of reference models would be ones describing how government, industry associations and small and medium sized companies may be able to exploit the techniques necessary to successfully invent new ways of doing business.

One example of a model with similar aim is the European Community's 6th Framework Programme, which takes an active role in the support of small and medium sized enterprises to enter this new era of competition. While many projects (in the precursor 5th Framework Programme) have been directed at the use of information technology (such as supporting electronic business) - with many interesting results and demonstrations - there has been a lack of strong results in terms of legal and business management, or really innovative ways of doing business. Most projects use information integration technologies to create quite usual value chains that operate faster, more reliably, with better co-ordination, or implement a business involving geographically dispersed business participants.

Results of some pre-competitive research projects - as developed in the Globemen Consortium and other IMS consortia - address this gap, especially in terms of responsibility structures for businesses, networks of businesses and virtual enterprises, as well as in terms of using the agent concept to dynamically build global enterprise entities (Bernus, Nemes, 1999). The Globemen consortium developed management models for Partners, Networks and Virtual Project Enterprises, identifying the interfaces and responsibility structures between their respective decision frameworks. These models - on the high level - describe the decision roles and frameworks for each of these enterprise entity types (the description is given in form of GRAI Grids). On the more detailed level the activities and information flow is described, as it occurs within and among management roles relevant to the Network and VE formation, operation and decommissioning. These more detailed descriptions are produced as IDEF0 models.

A note on the side: we differentiate between activity (functional) models (expressed as GRAI Grids, IDEF0 models, Use Case diagrams, etc), and process models (expressed in IDEF3, CIMOSA / First Step, UML collaboration diagrams, or UML sequence diagrams, etc). It has been observed in the practice of reference model development (Kalpic, Bernus, 2002) that sufficiently generic activity models can always be produced, i.e. for any level of management and for any function. In contrast, process models (behavioural models) can only be produced for such functions, which have a procedural nature, and/or only for a level of detail where generic procedures exist. Even if in many cases a behavioural model can be produced, usually this can only be done at the expense of genericity. Thus the Reference Model needs to be based on activity models, which can later be detailed using process (behavioural) models for more concrete cases. Of course for those functions that have an industry wide accepted step-by-step procedure there is no obstacle to developing behavioural descriptions as well.

The way we express these reference models for a given network type could be an example for other types of networks to describe their own reference models. The present models describe networks formed by a few larger partnering companies that co-operate in the design, procurement and construction of one-of-a-kind products as well as in after sales service. Further work includes the identification of transactions among the management roles for partners, networks and virtual project enterprises (using IDEF1X schemata and where generic procedures exist, IDEF3 process diagrams). Thus while lower level functions do not need to be prescribed procedurally (individual companies may follow different internal procedures), the inter-enterprise transactions will be described by process models and accompanying information models defining the information content of messages in these transactions.

Note that a similar approach is followed in B2B standardisation efforts, such as Rosettanet; however, these B2B models start with process models, where business to business transactions are expressed in form of behavioural models (such as UML collaboration diagrams). This may be suitable for operational level transactions, but at all except the highest level of granularity common procedural standards are not desirable when business-to-business transactions for tactical and strategic levels are designed (such as for joint scheduling, planning and strategy making). Procedural transactions on these high levels may be restricted to 'business protocols' in the sense that transactions need to consist of propositions / counter-propositions and acceptance and delivery reports. This is because procedural standards on the lower levels of granularity (i.e. beyond defining the signatures of interacting entities) contradict the information hiding principle. The violation of this principle either forces total homogeneity on the partners, where every partner uses the same algorithms to develop joint plans and schedules - an unlikely to succeed approach -, or increases the complexity of the joint action - thus making the solution more brittle in the face of change in any of the involved parties.

4 REFERENCE MODELS AND PRINCIPLES FOR THEIR FUTURE DEVELOPMENT

To develop generic and practical transaction protocols between businesses one may consider the applicability of *design principles* developed during the past 25 years for the design of complex systems. Note that Suh (1990) has developed two axioms of design that promise to encompass all design principles in all disciplines, thus it is an interesting research question whether all necessary principles listed above are indeed consequences of these two axioms. If this is indeed the case then the relationship between Suh's axioms and the desired set of design principles as advocated in this article is the same as the relationship of the axioms of logic to the theorems of mathematics while in theory it was not necessary to know more then the axioms, this fact did not eliminate the need for Russel and Whitehead's *Principia Mathematica* (1910,1912,1913). Specifically, principles we wish to consider should apply to enterprise entities that consist of humans and automated systems (machinery and computers). Some principles are presented below.

- Principles for the reduction of complexity in system design and eventually implementation and maintenance (one example for these principles is information hiding, application of which allows independent changes to occur in subsystems / component systems and in general prevent the proliferation of change effects);
- The principle of constructing systems of systems, where a complex system is built from less complex systems so as the apparent complexity of the resulting system does not compound the complexities of the constituents. By this we mean that without knowing the internal structure of the component systems the 'interesting' properties of the higher-level system may be derived without knowing the internal structure of the components. Of course the result of such system construction is dependent on what the interesting properties of the higher-level system are deemed to be. E.g. in constructing audio systems us-

ing operational amplifiers the knowledge of the actual circuitry of the components is unnecessary, because for the derivation of the behaviour of the complete system we only need to know the transfer function of the involved operational amplifiers (and some operational boundaries);

- Construction principles of human / socio-technical systems, as derived by behavioural scientists, ergonomists, sociographers, organisational psychologists and management scientists. E.g. humans need to receive positive reinforcements, motivation, and a sense of progress to perform at the best of their abilities; humans and organisations in general need stretches of 'secure' times to be able to concentrate on the job at hand; conflicts of interest arise when decisional powers are distributed in certain undesirable ways as humans are unable to completely disassociate themselves from playing multiple contradicting roles, etc;
- Good designs are orthogonalised, meaning that the designed entity's function is specified as a combination or interaction of independent (or orthogonal) functions. This ensures that even if not at all possible combined functionality has been specified at the outset, the system is capable of being extended and developed at minimal cost since new combinations do not need to change the constituent functions (and their implementing modules) - only new combinations need to be implemented, or simply configured.

This list is not exhaustive but we tried to list some of the most important principles to be able to demonstrate how they apply to the design of suitable reference models for partner companies, networks and VEs. We should list many more principles, applicable to specific technical domains, such as the layering principle of computer infrastructure services, even though most are in fact specific cases of the above more generic principles.

One such generic reference model that could satisfy the above criteria is the 'multi-agent' model. We refer to agents as defined by the Artificial Intelligence community, more specifically the 'distributed AI' community (Wooldridge, Jennings, 1995; Barbuceanu, Teigen, 1998). Thus an agent for our purposes is an aware agent, a) with objectives, b) autonomously acting using its own resources, c) with the (apparent) ability to plan its actions, d) observing its own progress towards the objective, e) taking remedial action if necessary, and f) interacting with its environment that may contain other agents. In addition to this basic definition, distributed AI agents have the ability to g) reason about and negotiate with other agents to agree on joint objectives and joint action (plans and schedules), h) apply the same ability as (d) and (e) onto the joint action. In distributed AI many additional properties are defined, depending on which the agent may have additional desirable properties, but for the purposes of this discussion we only require the above basic characteristics.

While this model has been developed for the purposes of understanding how artificial agents may be created and aggregated, it is important to note that such properties are the most desirable properties of *any* enterprise entity, be it a company, a network of companies, a project or a complex product (system) comprised out of either humans or machines or both. After all the aim of VE design is to create enterprises (virtual ones and real ones) so that the enterprise acting in a negotiated co-operation with other enterprises could achieve a joint objective, while satisfying its own objectives, and if its actions do not contribute to the progress any more it should be aware of this and take remedy. We call the enterprise that behaves as an agent an *aware enterprise.*

A further characteristic of some distributed AI systems is that agent negotiation for joint action always follows the same negotiation protocol. While this is not *a priori* necessary, some models stipulate such a self-similarity property. For discussion of such models see Warnecke's 'Fractal Factory' (Warnecke, 1993).

General Systems Theory (von Bertalannfy, 1968) describes systems as processes in dynamic equilibriums with their environment, and Holonic Systems (Koestler, 1968; Tharumarajah et al, 1996) describe systems of autonomous entities forming higher-level autonomous entities (Holons), placing further constraints on what desirable designs we must encapsulate in the Virtual Enterprise Reference Model.

The Globemen consortium VE Reference Models - in their state, as of this writing (March, 2002) - define functional requirements of the management structure of partners, networks and Virtual Project Enterprises. The co-ordination links between these entities (and among their constituent decision centres, or management roles) define the *user requirements* of such systems. The Globemen Management model and Functional Model (Olegario, 2000; Tolle, *et al*, 2002] describe the necessary management and control functions and interfaces.

The next step in the development of a high quality reference model would have to be the application of the listed principles to this user requirements model. The resulting model could be called a *system requirements* model in which all desired user requirements are satisfied, but in addition to that the model would display the properties of orthogonality, self-similarity and holonic autonomy, and each interface - as identified in the user requirements model - would be defined on the basis of agent negotiation protocols, including its behaviour in terms of transactional processes (Petri Nets, CIMOSA, FirstSTEP, UML sequence diagrams or IDEF3 process models) and state (IDEF3 Object State Transition diagrams, state-charts, or Coloured

Petri Nets, etc), but trying to hide as much as possible from the internal state and behaviour of the negotiating entities, so as to satisfy the complexity reduction and information hiding principle. Still the *contents* (the messages) in these protocols would have to be defined one by one, using traditional data modelling techniques.

Finally we address an important quality of enterprise models, which have a strong indirect influence on the characteristics of the technical and human systems built on the basis of these models. It is widely believed that completeness and consistency are indispensable qualities of enterprise models. However, it is not elaborated in detail what we mean by these terms, nor is it specifically stated what level of completeness and consistency is required from enterprise models. To investigate this issue it is helpful to consider the pragmatic use of models and *define* completeness and consistency in terms of what is required from these models when people use them or machines process them.

As has been discussed in (Bernus *et al*, 1996) there is a marked difference between mathematical, or formal, completeness and consistency and pragmatic completeness and consistency. The former applies to machine processed models, while the second applies to all models. Thus, for example, formal models alone are *inadequate* if used in isolation, because only a subset of the pragmatic uses is satisfied by their being complete mathematical models. Formal models (where completeness entails the ability to automatically analyse and execute them to derive desired properties of the entity modelled) are desirable in case of Generic Enterprise Models (IFIP-IFAC Task Force, 1999; ISO 15704-2000, 2000) whereby ontological theories allow the precise definition of the semantics of modelling languages use to represent enterprise models. However, formal completeness is (perhaps surprisingly) *not* a sufficient condition for pragmatic completeness. Reference Models to be pragmatically complete, they have to have the quality of understandability and uniform interpretability by humans, and not just humans in general, but by the population of humans who need to use and reuse these models. Bernus *et al* (1996) have discussed what measures one can take to ensure pragmatic completeness. If these are not considered (even in case of the availability of formal models), Reference Models will not be able to be utilised effectively for designing and building Virtual Enterprises.

5 CONCLUSION

The holonic manufacturing community develops technology usable for the detailed design and implementation of the Globemen model, i.e. based on distributed AI principles. The B2B community develops specialised transactions that could populate the high level Globemen models with specialised content, but would have to be enveloped in agent negotiation protocols, otherwise the 'standards' (proliferating at an alarming speed) would create a jungle much as what the programming language community has suffered from for three decades.

Further research is needed to ensure that the application of humanistic principles (third bullet point in the list) get intrinsically built into the architectural (preliminary) design phase when functions get aggregated into 'enterprise modules' ('CIMOSA functional entities' (Vernadat, 1996)) - i.e. humans, groups of humans, machines and complex automated systems, as well as complex human-machine systems.

This article placed emphasis on the need for reference models, but not just *any* reference model: the ones we are looking for must satisfy a number of design principles. The challenge is to bring together the various communities behind these principles; otherwise we shall end up with many competing and still unsatisfactory solutions. Finally, the concept of completeness and consistency of reference models were discussed (both formal and pragmatic) as conditions of their usability.

6 REFERENCES

Barbuceanu, M. Teigen, R. (1998), *System Integration through Agent Coordination*, in P. Bernus, K. Mertins and G. Schmidt (Eds.) Handbook on Architectures of Information Systems, Springer-Verlag.

Bernus, P. Nemes. L. (1999), *Organisational Design: Dynamically Creating and Sustaining Integrated Virtual Enterprises*, Proc IFAC World Congress, Han-Fu Chen, Dia-Zhan Cheng and Ji-Feng Zhang (Eds.) Vol. A, Elsevier.

Bernus, P. Nemes, L. Morris, B. (1996), *The meaning of an enterprise model,* in P. Bernus, L. Nemes (Eds.), Modelling and Methodologies for Enterprise Integration, Chapman and Hall, London pp 183-200.

Bertalanffy, L. v. (1968), *General systems theory: Foundations, development, applications.* Braziller, New York.

Doumeingts, G. Vallespir, B., Chen, D. (1998), *GRAI Grid Decisional Modeling*, in Handbook on Architectures of Information Systems, Springer-Verlag.

Hatvany, J. (1985), *Intelligence and Cooperation in Heterarchic Manufacturing Systems, Robotics & Computer-Integrated Manufacturing*, 2(2).

IFIP-IFAC Task Force (1999), *The Generalised Enterprise Reference Architecture and Methodology* (GERAM) V1.6.3. http://www.cit.gu.edu.au/~bernus

ISO 15704 (2000), *Requirements for Generalised Enterprise Reference Architectures and Methodologies,* TC 184 SC5/WG1

Kalpic, B. Bernus, P. (2002), *Reference Models of a Project Enterprise*, Int. J Technology Mgmt (submitted)

Koestler, A. (1968), *The Ghost in the Machine*, The Macmillan Company.

Olegario, C. (2001), *A Partial Enterprise Model for the Management and Control in an Extended Enterprise Scenario*, Masters Dissertation, School of CIT, Brisbane: Griffith University.

Suh, N. P. (1990), *The Principles of Design*, Oxford University Press.

Tharumarajah, A. Wells, A.J. Nemes, L. (1996), *Comparison of the bionic, fractal and holonic manufacturing system concepts*, in International Journal of CIM, 9(3).

Tølle, M. Bernus, P. Vesterager, J. (2002), *Reference Models for Virtual Enterprises*, Proc. PRO-VE'02, Kluwer.

Vernadat, F.B. (1996), *Enterprise Modelling and Integration: Principles and Applications*, London: Chapman & Hall.

Vesterager, J. Bernus, P. Larsen, L. B. Pedersen, J. D. Tølle, M. (2001), *Use of GERAM as Basis for a Virtual Enterprise Framework Model*, in J. Mo and L. Nemes (Eds.), Global Engineering, Manufacturing and Enterprise Networks, Kluwer.

Whitehead, A.N. and Russell, B. (1910, 1912, 1913) *Principia Mathematica*, 3 Vols, Cambridge University Press.

Warnecke, H. J. (1993), *The Fractal Company: A Revolution in Corporate Culture*, Springer-Verlag.

Wooldridge, M., Jennings, N. (1995), *Intelligent Agents: Theory and Practice*. The Knowledge Engineering Review 10(2).

The Business Process (Quiet) Revolution

Transformation to Process Organization

Meir H. Levi

Interfacing Technologies Corporation, Canada, meir@interfacing.com

Abstract: The competitive global market climate of the new millennium has raised awareness of business processes as the most important management paradigm. The idea of the *process organization* is gaining strong momentum; the process 'option' is now becoming a mandatory requirement. The integration of the Process Framework into the management structure introduces clear focus on consistent and collaborative ways to achieve results that directly impact the bottom line; hence, delighted customers and stakeholders.

This paper addresses the definition of the Enterprise Process Framework, the process of creating it and the benefits it can generate. The paper concludes with an initial report from a recent process framework deployment project at a leading energy generation and trading enterprise. This enterprise is using the open methodology implemented by Interfacing Technologies in its CIMOSA-like FirstSTEP and the EPC (Enterprise Process Center) process management solutions.

1 INTRODUCTION

Although research into business processes was conducted earlier, e.g. IBM, (Engelke, et al, 1984) and CIMOSA (AMICE, 1989), it was Michael Hammer (1990) who first raised the visibility of business processes with the introduction of BPR - Business Process Reengineering - in the early 90's. In subsequent years, BPR has often been associated with drastic change and downsizing initiatives, rather than improving practices. The emergence of Business Process Management (BPM) in the new millennium has given renewed focus to the process promise and has been a solid, yet quiet, business revolution.

To understand why an entire enterprise would begin instituting process structure and transforming functional management into business *process* management, we must understand the primary characteristics of the business process, or the *Process Construct* and the benefits brought about by BPM.

The traditional "Function Enterprise" is the product of the Industrial Revolution in which the guiding principle for organizing enterprises by function is the distribution of work by labor specialization.

In the Process generation, the functional organization of enterprises may not completely disappear, but rather be transformed into the context or grid for performing processes that bring value to customers.

Technological superiority, innovation, or longevity are no longer what makes or breaks companies - it is how well they are organized to respond to and serve their customers.

The only way to achieve such sustainable customer satisfaction and results is to become a process centric organization. Table 1 below highlights the important cultural differences between a functional organization and a process-centric one.

Table 1: Functional vs. Process Enterprise

Enterprise Behaviors	**Functional Enterprise**	**Process-centric Enterprise**
Managers Manage	Resources and Work	Customers and Results
Teams Operate	Independently	Collaboratively
Organization Dynamics	Rigid to adapt – Frequent re-org.	Flexible to new demands and self-reorg.
Resources Focus	Meeting job requirements	Best results, Customers
Knowledge Dissemination	Islands of Information	Integrated across the enterprise
Culture	Closed	Open

2 PROCESS CLASSIFICATION

Quite often, business processes at different levels are seen as synonymous with workflow, application automation and/or application integration. These "automated processes" are a sub-set of the overall "human processes" which make up the process framework of the organization. While selected steps of human processes are traditionally automated using Workflow solutions and/or specifically designed applications, such automation applies to a very specific set of repeatable and frequent processes, sub-processes and activities. Common examples include: 'Call Routing' in the Help Desk process, 'Order Entry and Tracking' in the Order Fulfillment process, Automated Core processes (trading transaction processes, on-line banking, etc.). It is important to note that every automated process is typically triggered by a human activity or sub-process. The 'Call routing' sub-process is triggered by a call or an email to the help-desk and may commence with a support person responding, which then invokes an automated flow of subsequent activities.

To ensure successful process transformation, both automated and human processes must be managed under the same comprehensive framework (Table 2). The basic criteria for a successful business process is that it: a) is visible to all process stakeholders, b) adds value, and c) is streamlined and focuses on contributing to customer satisfaction.

Hundreds (and sometimes thousands) of processes make up the process framework of a given enterprise. Classifying them in a manageable top layer (typically consisting of up to 10 top processes) and distinguishing between 'core' also called 'identity' and 'support' processes enables clarity in the process forest. Identity processes are those that make the enterprise unique in its market space, while support processes are the same from one enterprise to the other (Finance, Admin, HR, and others). Once created, the process hierarchy must be maintained like the enterprise's organizational chart.

Table 2: Business Process Classification

The Enterprise Process Framework:			
Support Processes		**Core/Identity Processes**	
<u>Finance</u> – Credit Authorization – Budgeting – Auditing	<u>Marketing</u> – PR/ Communication – Web Marketing – Lead Generation – Events/Trade Shows	<u>Banking/Finance</u> – Straight Through Processing – Acct Provision-ing – Loan/Credit Processing	<u>Services</u> – Outsourcing – Application Development – Logistics
<u>Operations/Logistics</u> – Purchasing – Contracts – Invoicing – Shipping	<u>Sales</u> – Qualification – CRM – Pre-sale – Negotiation & Closing – Channel Mktg.	<u>Energy</u>: – Short Term Trading – Long Term Trading – Strategic Plan-ning	<u>Pharmaceutical</u> – Clinical testing – Drug Submis-sions – Clinical Re-search
<u>Human Resources</u> – Hiring – HR Develop-ment – Performance Evaluation	<u>Legal</u> – Contracts – Policies – Acquisitions	<u>Manufacturing</u> – Product R&D – Product Engineering – Quality Assur-ance – Production	

An enterprise is a microcosm, with people, behaviors, activities, goals, aspirations and so on, working in a non-isolated environment of market pressures, customers, suppliers, regulatory laws.

ployees and shareholders - but to actually delight them with great experiences.

The way to achieve this dimension of satisfaction is to create an environment in which all parties collaborate towards common goals and results.

The business process framework enables:

a) Alignment and Consistency: Process stakeholders gain a clear understanding of their process and align to execute the process in a consistent manner.
b) Execution: A clear process holds its owner accountable for a high degree of execution and maximization of results.
c) Optimization: Well-defined processes are easier to improve and optimize.

3 PROCESS MANAGEMENT WITH FIRSTSTEP® AND THE EPC

Business Process Management is about availability of critical and up-to-date process information to process managers who are accountable for the process execution and goals. A number of process-modeling and -management solutions supporting several methodologies have been developed over the years. Being a methodology independent enterprise modeling and simulation application, FirstSTEP®, developed by Interfacing Technologies since 1994, has been adopted by industry leaders in such diversified sectors as manufacturing, finance, telecom, energy, healthcare, public, and services. Its concept and approach comply with the CIMOSA framework (1996).

Today, the FirstSTEP family of process modeling products includes two members: FirstSTEP Designer for modeling business processes, performance analysis and simulation, and FirstSTEP Charter for process mapping in Microsoft-Visio®. Both modeling environments use XML as a bridge between each other and to the Enterprise Process Center™ – the EPC.

The EPC is a web-based (J2EE) knowledge and process portal that enables creation, management and access to process information (maps, models, documents, applications, process instances, process data) from every desktop in the enterprise. Optional extranets allow the extension of the EPC to customers and suppliers outside of the enterprise boundaries.

While the FirstSTEP process design environment is evolving to become an integral component of the EPC, the EPC will also support other process meta-models so enterprises can benefit from existing investments in process design work.

3.1 The New Business Process Construct

Perhaps the most concise definition for 'business process' is the one suggested recently by Michael Hammer (2001): "an organized group of related activities that together create a result of value to customers" (Fig. 1).

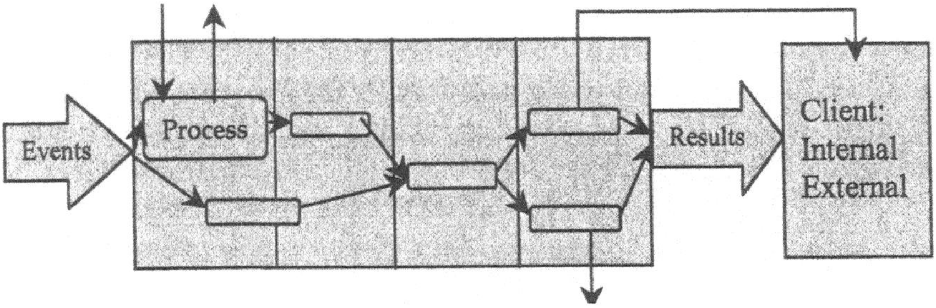

Figure 1: The Business Process Construct

Note that distinct activities, performed by a single or multiple functions within a single or multiple enterprises, join together to deliver results to a client. Operating through a process ensures that focus is kept on everything that makes the process results shine and the process client delighted.

With an enterprise-wide view on processes, a process chart, or 'Process Framework', emerges. Such a framework provides a well-defined structure for the web of processes that traverse the enterprise. All key processes and their inter-relations, links to enterprise objects, human resources, assets, information, knowledge and supporting applications ultimately make up the framework.

Processes can be defined at many different levels and with various boundaries. To derive the important benefits that processes bring to the enterprise management infrastructure, we need to consider a number of important process characteristics:

Process Results: These are clear performance targets linked to the organizational strategic objectives and designed to support the mission and the direction of the enterprise.

Process boundaries: Boundaries define the scope of the process: its beginning and end-points. Furthermore, they determine the '*touch points*' with other processes.

Process Instantiation: Instances of processes can be a) *transactions* in a repeated transactional process, b) *projects* in a project driven business unit, or c) *programs* in a service delivery organization.

Process Client: The ultimate customer who will enjoy the benefits of the process and receive the value generated by all process constituents.

Process Manager/Owner: This is where the responsibility and accountability for the performance of the process lies.

Now, to create a beneficial *Process Framework* for the organization, one must create the 'top-level tier' first to provide a clear vision or structure, and then proceed to subsequent process tiers in a way that can lend itself to a clear and efficient process distribution through the organizational functions.

In constructing the top-level tier, it is important to make a distinction between core processes, which are sometimes, also called 'identity processes' (Keen, 1997) and 'support processes'. Support processes are typically transactional in nature. Identity processes can accommodate all types of process instantiation. The types that are used depend on the nature of the enterprise.

3.2 FirstSTEP Methodology and Process Standards

Charting the enterprise processes can be accomplished in many ways - from the simplest diagramming approach to the creation and management of multi-layered and complex maps coupled with various organizational objects and linked to systems, information, networks and other dimensions. There is a renewed effort to converge Business Process Modeling standards. The OMG and the BPMI are making efforts to create a common language.

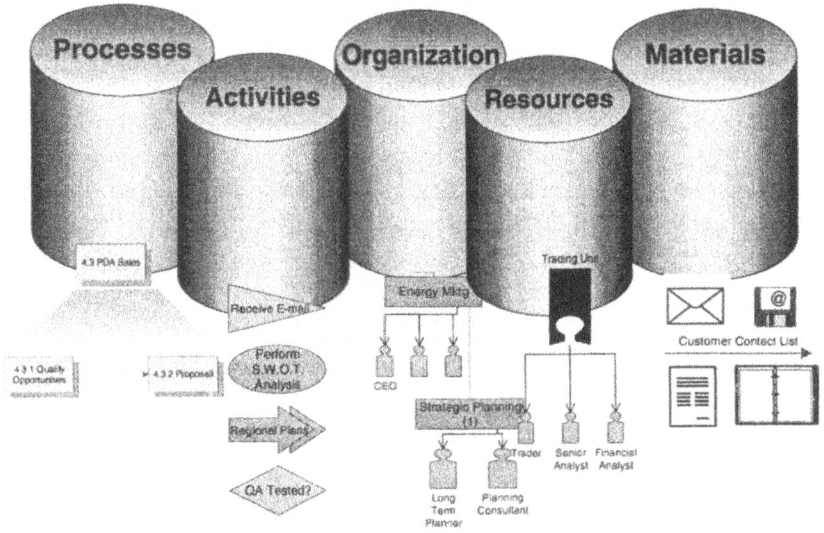

Figure 2. The FirstSTEP® Methodology

The FirstSTEP modeling class definition has been described in *FirstSTEP Process Modeler - a CIMOSA compliant Modeling Tool* (Levi, Klapsis, 1999). It consists of 5 classes of objects: Processes (with hierarchical layers of sub-processes), Activities (the lowest-level work steps, classified in sis, 1999). It consists of 5 classes of objects: Processes (with hierarchical layers of sub-processes), Activities (the lowest-level work steps, classified in 6 generic groups for clarity and standardization), Organizational units (functions, departments), Resources (performers of activities, human and non-human – systems, machinery), and Materials (products, documents, information objects -- all entities flowing through, linked to, or referred to the process). This definition represents a common subset of most evolving standards and thus is designed to adapt to most enterprise needs.

The FirstSTEP methodology (Fig. 2) focuses on the business design needs first and foremost. By being simple and generic, it can be easily understood by business users at all levels, a pre-condition to the acceptance of process framework and getting the buy-in from all users in the enterprise.

The methodology is evolving to support integration to other meta-models such as represented by evolving standards; BPML, IDEF and WfMC.

4 DEPLOYING PROCESS FRAMEWORK WITH THE ENTERPRISE PROCESS CENTER

The transition to Process Enterprise takes a concentrated level of effort. It requires attention to many operational, organizational, human and systems/application details. The process may take several months and considerable effort to shape up, depending on the size and state of the enterprise. But the rewards justify the process and the efforts.

Through working with a number of enterprises, Interfacing Technologies has developed a three-step program, titled PI-3 (Fig. 3), supported by the Enterprise Process Center technology and designed to take an organization through the transition path.

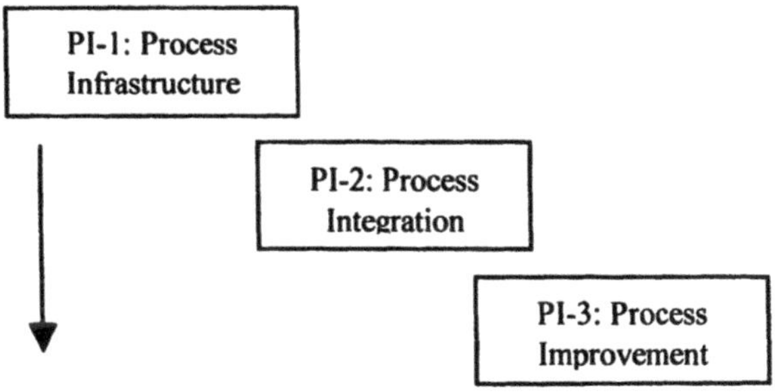

Figure 3: The PI-3 Program

While many enterprises attempt the transition, to ensure success it is necessary to have clarity in the transformation process, especially when it involves everyone in the organization.

Coupled with the FirstSTEP methodology, and supported by both FirstSTEP and the EPC, PI-3 takes advantage of the hierarchical approach to setting the high-level process framework in its first step: the *Process Infrastructure*. The enterprise process framework is created in this step. The framework typically consists of 6 to 10 primary processes at tier 1, with decomposition to multiple levels of sub processes, organized in a methodical fashion. The EPC, with its hierarchical view (process trees) enable easy access to any layer and process.

Processes have numerous links to one another, and it is in those links that errors in executions often occur. The *Process Reconciliation* is a key step required to ensure clarity in what is expected from one process to support the other. Inter-process collaboration is fundamental element for a solid infrastructure.

The second step in the program focuses on integrating the process framework into the Enterprise's existing systems, applications, content and knowledge. This is the *Process Integration*.

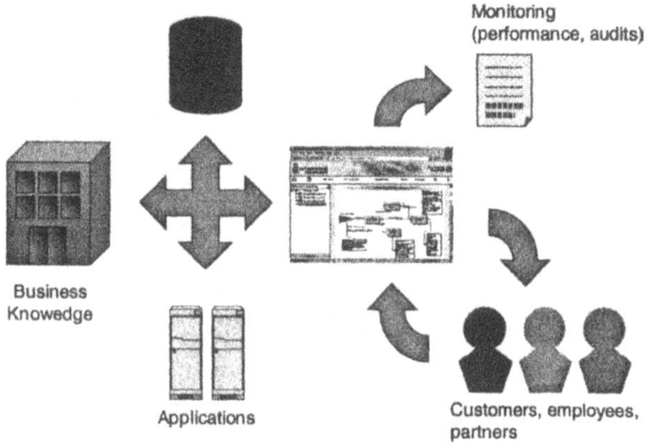

Figure 4: Process Knowledge

Process knowledge captured within the first two steps in either maps or in related content, creates enthusiasm for the promise of process management (Fig. 4). Users, managers, and external partners can view the processes they own or in which they participate, access critical knowledge and applications, and execute with clear view of the impact of their actions. Process knowledge must be live and incorporated into the evolving business practices.

The Process Framework can now become an integral part of the organization, enabling access to critical data and specific applications directly from the corresponding process step. To complete the integration, applications and systems must be able to feed back critical information to assess and audit the processes.

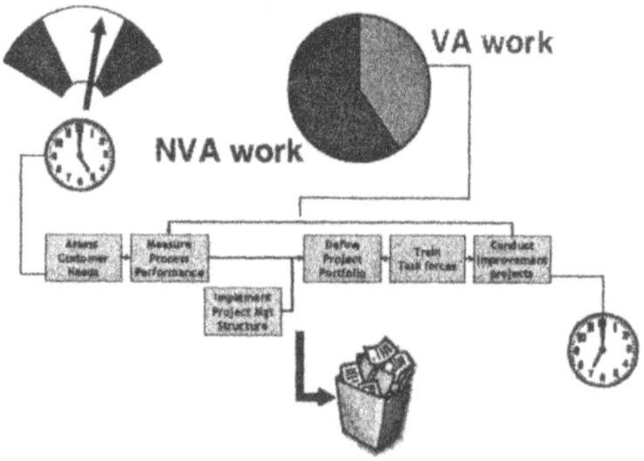

Figure 5: Process Improvement

The third element in ensuring that processes meet and exceed set targets and results is captured in the *Process Improvement* phase (Fig. 5). Goals must be measurable and verifiable. Improvement proposals can then be implemented with clear ability to assess the incremental results.

quality methodologies; the most process oriented one is Six Sigma (Pande, 2002). Others target both organizational and process key measures; the most prevalent being Balanced Score Card (Kaplyn, Norton, 1996). The goal is to ensure that the standardized processes deliver results that meet customers' requirements in the most effective way, and if they do not, to act on the real cause of process waste.

Integrated with the other two dimensions of PI-3, the process improvement dimension will enable the process-led organization to:

- Maximize customer satisfaction
- Eliminate ambiguity, by making critical decisions based on measured facts
- Document priorities for processes improvements
- Eliminate root causes of poor performance
- Maximize, sustainable business results by focusing on added value work.

5 INITIAL CASE REPORT

Since the first release of FirstSTEP in late 1994, Interfacing Technologies has worked with hundreds of enterprises, to support BPR initiatives. Focus has shifted over the last few years to BPM and the transformation to process organization. In this paper we report notable initial results from a significant project with a leading US based energy generation and trading organization. In 2001, this enterprise decided to become a process-driven enterprise. To support them in this mission, they turned to Interfacing's FirstSTEP and EPC technology.

Key expectations for drastic impacts were:

1. Making business process knowledge available to team members across the organization, thus adding value through cross-functional knowledge transfer.
2. Increasing operating efficiency with the following mechanisms:
   - Continuously challenging and improving how they do business
   - Clearly defining accountability
   - Defining IT needs through process workflow
   - Enhancing auditing capabilities
   - Enhancing performance management capabilities
   - Providing flexibility in a growth environment
3. Streamlining the process of integration in a merger or acquisition
4. Self-reorganization through dynamic process changes:
   Changes in process ownership, boundaries and definition automatically imply new roles and responsibilities, hence accomplishing in a

more natural manner what was traditionally done through functional reorganization, often viewed as a drastic, costly and unpleasant process.

5. Improving Human Resource management:
   Skill gaps identified through resource needs of processes - workforce planning, a much quicker process of hiring and reallocation of resources.

The project started by outlining and achieving consensus on the high-level process framework structured around 10 primary core and support processes. Process Leaders were appointed to each of the process areas. This was the start of the Process Infrastructure phase, during which hundreds of processes were mapped through a well-defined procedure involving process owners and process stakeholders.

In addition to a simple but effective process review and validation, the Process Reconciliation phase was a very important step in completing the process framework. In this phase, cross-functional and cross-process teams identified key process touch points where critical interfaces between distinct process areas occur.

The project is supported by the EPC, which enables a single-point access to all of the key processes and their related content.

This includes:

- Business Process Flows, Maps and Diagrams
- Application and System Linkages / Integration
- Process Content Integration; documents, spreadsheets, web sites, presentations, procedures, templates
- Process Ownership and Accountability

While the enterprise is still in the midst of the transformation process, with the ultimate benefits yet to be realized, promising observations and results have already been reported:

- The whole Company is "process aware"
- Process members share information and are enthusiastic about process definitions and executions
- One place has been created for employees to go for process knowledge
- Applications fall into the "Big Picture"
- Operations are gaining higher level of consistency, which is expected to yield direct impact on the bottom-line.

To achieve the initial results major challenges had to be overcome (see Table 3).

The *Infrastructure* phase of the project is near completion at the time of writing this paper. The *Integration* phase is being executed with process content and initial applications being integrated. The process improvement initiative, which will tie organizational and process metrics to the Framework, will be launched next. With the ability to set, measure, monitor and manage process related KPI's; complete process ownership and accountability will be realized. This will enable true management by process results to maximize the organization output and bottom line.

Table 3: Challenges in the project

Typical Challenges	**Steps taken**
Resistance to Process Change	Regular Process Leader and Process Manager meetings, short/frequent education sessions
Learning new ways and practices	Get everyone involved in capturing and reviewing Process information
Assuming responsibility for results	Clear KPI based incentive programs
"No time ..."	Prioritization
Doubts and all of the above	"Drive and vision from the very top – the CEO and his management team"

6 SUMMARY – CONCLUSION

It is only now that the biggest impact of process framework technology has the opportunity for significant realization. While many enterprises have embarked on one level of BPM initiative or another, only those who completely change their culture to become Process Enterprises will gain sustainable rewards. Initial results reported in this paper illustrate pioneering work of one of the first enterprises to embrace process culture, and the initial results are significant. Challenges and threats to such deployments are very real and the deployment teams had to continuously find ways to meet the challenges. Without a doubt, the most critical success factor was the complete and unconditional directive from the very top - the CEO - to see the transformation through to its completion.

Future work, already underway, will see the smooth integration of the high-level Process Framework to the enterprise application and data layers to increase the ability to automate and integrate. The general BPM market is expected to experience the merging of multiple methodologies and metamodels to support similar initiatives.

7 REFERENCES

AMICE (Eds.), (1989), *Open Systems Architecture for CIM*, Springer-Verlag

CIMOSA Association, (1996), *CIMOSA, Open System Architecture for CIM,* Technical Baseline, Version 3.2", Release, private publication.

Engelke, H. Grotrian, J. Scheuing, C. Schmackpfeffer, A. Schwarz, W. Solf, B. Toman, J. (1985), *Integrated Manufacturing Modeling System*, IBM Journ. of Research and Development, 29(4).

Hammer, M. (2001), *The Agenda: What Every Business Must Do to Dominate the Decade*, pp. 51 – 60, Crown Business Publishing Gr.

Hammer, M. (1990), *Re-engineering Work: Don't Automate, Obliterate*, Harvard Business Review, pp. 104-112, July-August.

Kaplan, R.S., Norton, D.P, D.P., (2001), *The Balanced Scorecard, Translating Strategy into Action*, Harvard Business School Press, Boston

Keen, Peter G.W. (1997), *The Process Edge: Creating Value Where It Counts*, Harvard Business School Press,

Levi, Meir H. Klapsis, Marios P. (1999), *FirstSTEP Process Modeller – a CIMOSA Compliant Modeling Tool*, special Issue Computers in Industry, 40(2-3)

Pande, Peter S. Newman, Robert P. Cavanagh, Roland R. (2002), *The Six Sigma Way; An Implementation Guide for Process Improvement Teams*", The McGraw-Hill Companies, Inc.

Enterprise Architecture and Systems Engineering

Peter Webb
BMT Defence Services Limited, United Kingdam, pwebb@bmtdsl.co.uk

Abstract: Factors such as the end of the cold war and increasing globalisation in terms of competition and social responsibility have significantly increased the complexity of events addressed both by governments and by industry. Furthermore, their responses are increasingly more constrained by economic and environmental issues. It is clear that the large complex "socio-techno-economic" systems (i.e. enterprises) that may comprise as well as create and deliver these responses must be both agile and efficient. An approach to the analysis, design and specification of such enterprises is presented which draws on state of the art Enterprise Architecture concepts and Systems Engineering techniques. The resulting EA/SE method enables clear justification of design, definition of interfaces and derivation of validated requirements. Comparisons are drawn to Zachman and ISO 15704 - GERA concepts.

1 INTRODUCTION

This paper describes how enterprise architecture concepts and systems engineering techniques have been combined to provide a new approach to the analysis, design and specification of enterprises. It has been named the Enterprise Architecture / Systems Engineering (EA/SE) method.

Enterprise architecture provides a rich but generic framework that guides the construction of enterprise models. Furthermore, it enables stakeholder needs and expectations to be identified and traced to technological design while facilitating integration efforts.

Systems engineering provides the tools and techniques to create "well engineered" enterprise architectures. In particular, it uses modelling as a means for analysis, improvement and validation of proposed solutions. Systems engineering also supports possible integration between interfacing enterprises or systems by identifying and defining interfaces.

The systems engineering core process first defines desired behaviour and determines structural options. Then the defined behaviour is allocated onto each structural option to create a range of potential solutions. Finally a trade-off is undertaken against pre-defined effectiveness measures in order to identify a near-optimal solution.

In this way an enterprise architecture description is supported by a number of products comprising diagrams, models and logic statements underpinned by a data repository. Where necessary architecture descriptions can be easily translated into requirement sets that are coherent, consistent, complete and validated as correct.

2 ENTERPRISE ARCHITECTURE

EA/SE considers an enterprise to be an organisation of people performing processes and supported by technology that cooperate to create and deliver solutions to customers (Fig.1). These solutions comprise products and/or services that may themselves be whole or part enterprises. Furthermore, an enterprise can interface with others to form a larger integrated enterprise.

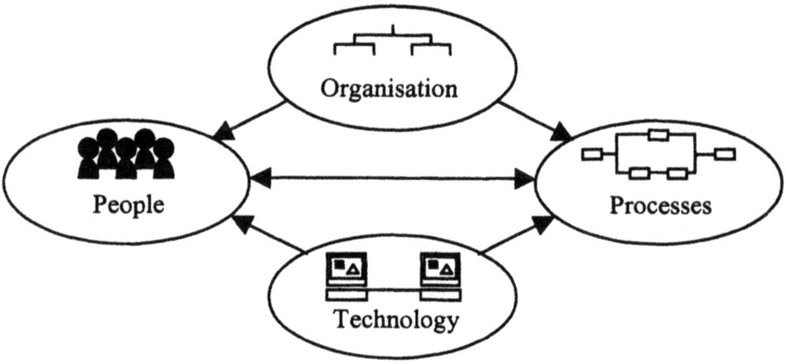

Figure 1: Composition of an Enterprise

Examples of enterprises include procurement, support or design project teams and company organisations (people biased) as well as industrial facilities and ships, aircraft or other transport systems (technology biased).

The behaviour and structure of an enterprise is defined by its architecture. If an enterprise is to create and deliver quality solutions that, where practical, meet all stakeholders' needs and expectations within cost, schedule and risk constraints then it should have an efficient, enabling architecture. Such enterprise architectures may be considered "well engineered".

The EA/SE approach does not analyse the behaviour and structure of a whole enterprise at once. Instead, it progressively analyses the enterprise from successive degrees of abstraction, building the complete enterprise architecture in a structured manner. These degrees of abstraction provide perspectives of the architecture and are analogous to supporting layers where each layer defines the context and provides input data for the layer beneath (Fig. 2). The analysis activity starts at the most abstracted layer and works down through the layers. The systems engineering core process outlined in section 1 above is applied once to each layer.

Figure 2: EA/SE Layers of Abstraction

Table 1:Comparison of EA/SE Abstraction with ZEAF Perspectives and Lifecycle Phases of ISO 15704 (GERA) and Pre EN ISO 19439

EA/SE Abstraction	**ZEAF Perspective**	**ISO 15704 - GERA & EN ISO 19439 Lifecycle Phase**	
Capability / Market	Not Addressed	*Domain* Identification	
Asset / Business Unit	Not Addressed	Concept *Definition*	
System Context	Scope (Contextual) - Planner	Requirements *Definition*	
System Concept	Enterprise Model (Conceptual) - Owner	Preliminary Design	Design *Specification*
Logical Representation	System Model (Logical) - Designer	Detailed Design	
Physical Realisation	Technology Model (Physical) - Builder	Implementation *Description*	

The EA/SE layers of abstraction reflect the "perspectives" of the Zachman Enterprise Architecture Framework (ZEAF) (Zachman, http://). However, EA/SE adds two higher layers while missing the lowest ZEAF perspectives dealing with "Detailed Representation" and "Functioning Enterprise".

Furthermore, the EA/SE layers are analogous to the life cycle phases of the Generic Enterprise Reference Architecture (GERA), (ISO15704, 1999) and (pre EN ISO 19439, 2002). These relationships are shown in Table 1 with typical outputs of EA/SE in Table 2.

Table 2: Typical EA/SE Outputs by Abstraction Layer

EA/SE Abstraction	**Typical EA/SE Outputs**
Capability / Market	– Analysed threats and incidents with desired countermeasures and responses. – Allocation to capability / market areas – Outline capability (user) requirements.
Asset / Business Unit	– Analysed capability behaviour – Detailed user requirements – Allocation to assets / business units – Outline system requirements
System Context	– Analysed asset / business unit behaviour – Detailed system requirements – Allocation to systems – Outline system architecture
System Concept	– Analysed system behaviour – Detailed system architecture – Allocation to sub-systems – Preliminary design
Logical Representation	– Analysed sub-system behaviour – Detail design Allocation to Objects – Outline build
Physical Realisation	– Analysed object behaviour – Allocation to components – Build description

Confusingly, in the ZEAF each "perspective" is divided into "abstractions" that deal with six fundamental interrogatives. Within EA/SE, each layer of abstraction (i.e. perspective) is similarly divided to address the six interrogatives, but they have been renamed as foci and are defined as follows:

a) What entities are being processed? i.e. inputs and outputs;
b) How are entities processed? i.e. functions (tasks, activities);
c) When do processes occur? i.e. events and times (sequence);
d) Where are processes occurring? i.e. locations and environmental conditions;
e) Who is involved in processes? i.e. organisation, jobs, roles, skills (depending on perspective);

f) Why are processes undertaken in the way they are? i.e. principles, standards.

The EA/SE approach aims to answer each of these interrogatives: Analysis of behaviour addresses the interrogatives "what", "how" and "when", while the consideration of structure addresses "where" and "who". Similarly, consideration of motivation and purpose addresses the interrogative "why". The abstraction layers and foci combine to form the EA/SE framework as shown in Table 3.

Table 3: The EA/SE Framework

EA/SE Abstraction	**Behaviour**			**Structure**		
	What?	**How?**	**When?**	**Where?**	**Who?**	**Why?**
Capability						
Asset						
System Context						
System Concept						
Logical Representation						
Physical Realisation						

The output from analysis is a description of the Enterprise Architecture model at the respective layer of abstraction. Combining the descriptions for each layer provides a description of the complete Enterprise Architecture model. In fact, the description of each architecture layer can take the form of a requirement specification for the layer beneath.

3 SYSTEMS ENGINEERING

The Systems Engineering Core Process adopted by EA/SE is adapted from Oliver, et al, (1997). The associated process flow diagram is reproduced in Fig. 3. The overall process description given in their book has been adapted in relation to the EA/SE approach and is included in the following paragraphs.

Step 1 evaluates and categorises available information for input to the modelling steps. It identifies and eliminates information redundancy, inconsistency and obvious incompleteness. In all but the first abstraction layer this is simply a matter of accepting context information output from is simply a matter of accepting context information output from analysis in the preceding layer.

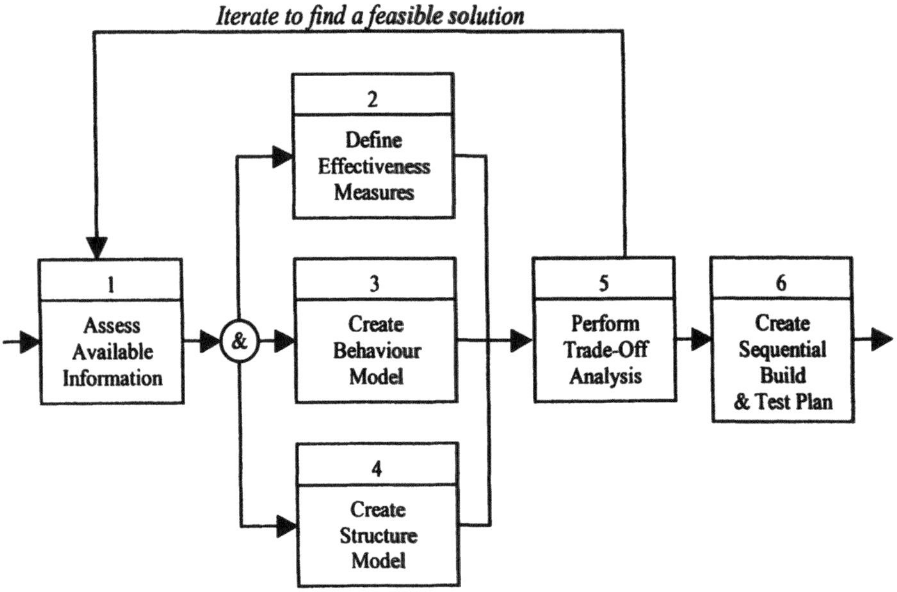

Figure 3: The Six Steps of the Systems Engineering Core Technical Process

Steps 2, 3, and 4 are concurrent activities. They can be ordered; however in practice it is found that they are highly inter-dependant and iteration is required. In other words, as understanding progresses in one of the tasks, it suggests changes in the other two.

Step 2 defines the criteria for optimisation. These are effectiveness measures and are written against the behaviour needs expressed in the input/context data. They can be represented by perhaps three to fifteen key requirements, even for large complex systems. They are the criteria that mean success or failure and against which options are assessed during trade-off analysis (see step 5).

Step 3 defines the behaviour that is desired. Behaviour is a rigorous description of what is to be done. It includes the functions to be performed, the sequencing control of those functions and the inputs and outputs from the functions. Two partial views of behaviour can be created diagrammatically: The first view, that tends to be scenario specific, shows functions and sequence, e.g. Functional Flow Block Diagrams, IDEF3 or Catalyst Process Charts. The second view is not scenario specific; it shows functions, inputs and outputs, e.g. Data Flow Diagrams, N-squared charts, or IDEF0. In addition, Entity-Relation diagrams can be used to simplify or "normalise" the inputs and outputs. These views provide static models of behaviour. Either alternatively, or in addition, an executable model, such as Coloured Petri Nets, can be used to analyse behaviour.

When the desired behaviour is modelled separately from structure, then alternative structures can be readily identified. Step 4 defines these alternate structure models. In either step 3 or step 4, an allocation of modelled behaviour onto structure is made so that each option exhibits the same emergent behaviour. Further options can be derived through partitioning the modelled behaviour in different ways for each structure.

Step 5, trade-off, selects among the alternative architectures. The best feasible design is selected based on the effectiveness measure values defined in step 2. This is a critical, or key, best practice in the engineering of complex systems because it finds a near-optimal solution while guaranteeing the desired emergent behaviour. One possible branch from Step 5 is iteration back to the beginning, made necessary when no alternative architecture meets the effectiveness criteria. When this occurs, either steps 1 to 5 are repeated to find feasible solutions, or effectiveness criteria are relaxed so that a previous non-feasible solution is accepted.

Step 6 creates a plan for further modelling and implementation when a feasible and near optimal architecture has been found. The plan takes into account identified issues, risk reduction, etc. It may also include a description of the architecture at the related layer of abstraction in the form of requirements for the next layer.

The six systems engineering core steps are applied once for each layer of abstraction forming a complete Enterprise Architecture model. As stated previously, this analysis activity should start at the most abstracted layer and work down through the layers since each layer defines the context and provides input data for analysis in the layer beneath.

4 CONCLUSION

A method for analysing, designing and specifying enterprises that can comprise complex technological or business systems has been presented. Known as Enterprise Architecture / Systems Engineering (EASE), it combines enterprise architecture principles with systems engineering techniques.

EA/SE simplifies the modelling of Enterprise Architectures by using a relatively small number of constructs from the systems engineering process and applying these successively to layers of abstraction. The result is a rich but simple model that enables user and other stakeholder needs to be traced to technical design. Furthermore, justifiable and validated requirements can be derived from the model against each layer of abstraction.

5 REFERENCES

ISO 14258, (1999), *Concepts and rules for enterprise models*, available from http://www.mel.nist.gov/sc5wg1/

ISO 15704, (1999), *Industrial automation systems - Requirements for Enterprise Reference Architectures and Methodologies*, http://www.mel.nist.gov/sc5wg1/

Oliver, D. W. Kelliher T. P. Keegan Jr, J. G. (1997), *Engineering Complex Systems with Models and Objects*, downloadable free from http://www.incose.org/newsline.html

Pre EN ISO 19439, (2002), *Enterprise Integration - Framework for Enterprise Modelling,* CEN TC 310, WG1, ISO TC 84 SC5 WG1.

Zachman Enterprise Architecture Framework, http://www.zifa.com

6 ADDITIONAL REFERENCES

Levis, A. H Wagenhals, L.W. (2000), *C4ISR Architectures 1: Developing a Process for C4ISR Architecture Design*, Journal of INCOSE Vol. 3(4), Wiley ISSN 1098-1241, available from http://viking.gmu.edu/http/courses.htm

US Chief Information Officers (CIO) Council, (1999), *Federal enterprise Architecture Framework,* Version 1.1, available from http://www.cio.gov/

US Chief Information Officers (CIO) Council, (2000), *A Practical Guide to Federal Enterprise Architecture*, version 1, available from http://www.cio.gov/ or directly from http://www.itpolicy.gsa.gov/mke/archplus/ea_guide.doc

US Chief Information Officers (CIO) Council, (2000), *Architecture Alignment and Assessment Guide*, available from http://www.cio.gov/

US DOD, (1997), *C4ISR framework* Version 2.0, available from http://viking.gmu.edu/http/c4isrAFI/archfwk2.pdf

Vernadat, F. B. *Enterprise Modelling and Integration: Myth or Reality*, available from http://www.cit.gu.edu.au/~bernus/taskforce/

Vernadat, F. B. (1997), *Enterprise Modelling Languages*, ICEIMT'97 Enterprise Engineering and Integration - International Consensus, Springer-Verlag, http://www.mel.nist.gov/workshop/iceimt97/pap-ver3/pap-ver3.htm

Proposal of a Reference Framework for Manufacturing Systems Engineering

Gregor von Cieminski[1], Marco Macchi[2], Marco Garetti[2], and Hans-Peter Wiendahl[1]
[1]IFA, University of Hannover, Germany [2]DIG, Politecnico di Milano, Italy,
marco.macchi@polimi.it

Abstract The authors propose a "generic" reference framework to integrate modelling methods typically adopted to engineer a manufacturing system under a unifying "umbrella". These methods – from formal descriptive modelling to analytical modelling and simulation modelling methods – originate from different disciplines and serve different purposes.

1 INTRODUCTION

Manufacturing systems engineering is an intrinsically established process: starting from product specifications resulting from product design activities, quality, time and cost constraints imposed by enterprise corporate decisions and, eventually, physical constraints from existing plants, it encompasses process specification, system design and system evaluation to finally accomplish manufacturing system design solutions. Despite the fact that the majority of production engineers essentially agree on the steps laid out for this process model and utilise it accordingly, the authors deem that a "unifying" reference framework – comprising both engineering activities and related modelling methods – is missing. In fact, until now, only "particular" reference frameworks have been established in specific fields and with reference to specific engineering methods. Besides being frameworks with a specific scope, they also often deal with a single step of the manufacturing systems engineering process only – consider, for example, the engineering process typically proposed to carry out manufacturing simulation projects.

A shared reference framework might be obtained, in the authors' opinion, through the complete specification of the two main elements involved in the engineering cycle, the *manufacturing system* itself and the *modelling activities*. In the paper, a reference architecture – in the remainder, the "Morphology" – will determine key characteristics of a generic *industrial manufacturing system* (IMS). A reference engineering methodology – the "Engineering Process" – will define the *modelling activities* to accomplish engineering solutions and the methods – and tools – required. The "Morphology" is inspired by the GERAM framework (IFIP, 1998). It inherits – and specialises in the manufacturing context – GERAM life cycle and view concepts. It extends GERAM with a thorough definition of the manufacturing system entity, subject of the engineering study. The "Engineering Process" is derived from methodologies established in the industrial engineering discipline (simulation (Carrie, 1988) and facility layout planning (Meller, Gau, 1996)). It integrates, as a unique methodology, the utilisation of descriptive modelling methods – to provide formal representations of system properties – and evaluation methods (analytical, algorithmic, heuristic or simulation methods) – for system performance calculation.

Section 2 presents the "Morphology of IMS" in its constituting elements. Section 3 outlines the "Engineering Process of IMS". The overall framework is presented in section 4 discussing its position with respect to GERAM and its functionality in the manufacturing system design process. Section 5 eventually outlines ongoing and future research activities.

2 THE MORPHOLOGY OF INDUSTRIAL MANUFACTURING SYSTEMS

2.1 The Scope of the Morphology

The "Morphology" is defined to guide the modeller to establish the nature of the manufacturing system that he or she would like to model. To this regard, manufacturing systems engineering is comparable to the widely accepted product design process and its adoption of morphological analysis. According to (Zwicky, 1948), the morphological analysis consists of two main steps. Firstly, the definition of the parameters that are the key characteristics of a product (system), and of all possible distinctions of these parameters. This allows a systematic description of the product (system). Secondly, the possible or feasible combinations of different distinctions of different characteristics then represent the possible configurations of the product (system) under scrutiny. Transferred to the context of manufacturing systems engineering, the use of the morphological analysis is intended to facilitate a systematic definition of the characteristics of a manufacturing system that is to be engineered. It is proposed to help the modeller in the identification of those modelling methods that might be suitable for description, engineering or re-engineering activities of the manufacturing system itself.

2.2 The Morphology Modelling Framework

A morphological matrix of manufacturing systems is thus established in order to be able to distinguish between the different types of such systems (Fig. 1).

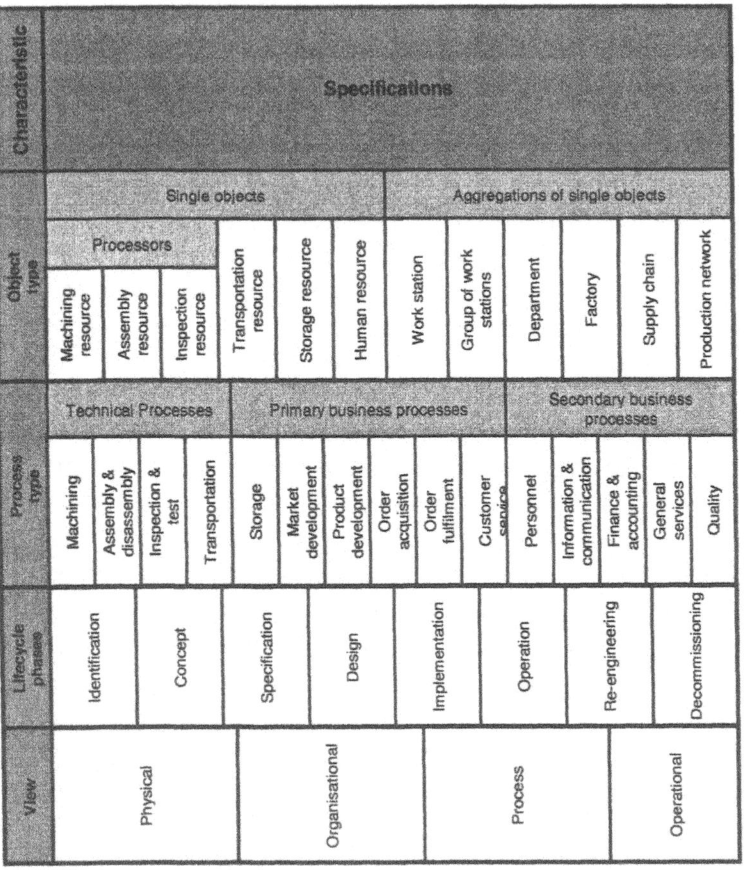

Figure 1: Morphological matrix for manufacturing systems

The rows of the matrix correspond to the unique key characteristics that a manufacturing system possesses. *Object* and *process type* define, respectively, a system and a process taxonomy of a manufacturing system entity, whilst *lifecycle phase* and *view* specialise the GERAM lifecycle model and modelling views for a manufacturing system. For each of the four characteristics, one can distinguish between several specifications. All systems that exist can be described in detail by one combination of specifications of the four core characteristics.

The *object type* characteristic is concerned with the physical entities that represent a manufacturing system. All single objects encompass processing resources (machining, assembly and inspection processors), material handling resources and human resources. The aggregations of single objects are ordered according to accepted hierarchies of manufacturing systems. Starting from a workstation that might combine a machining, a human and a storage resource, the focus is widened until one arrives at the production network.

The *process type* characteristic defines three process types: technical processes, primary and secondary business processes. The definition of process types is inspired by similar models available in literature: compare, e.g., with CIMOSA definitions (CIMOSA, 1993), which group processes into manage processes (to develop business objectives and strategies and manage the overall behaviour of an organisation), operate processes (directly related to transformation of information and material into products) and support processes (providing assistance and maintenance to enterprise resources). The main scope of the "Morphology" are the operate processes (separated in technical processes and primary business processes) and the support processes (secondary business processes).

The *lifecycle phase* definition is inspired by lifecycle models proposed in enterprise engineering literature, in particular the GERAM life cycle model (IFIP, 1998) and comprise all phases of a manufacturing system.

The *view* characteristic specifies the modelling views that can be used to observe certain aspects that typically characterise a manufacturing system. Any manufacturing system is in fact characterised by four aspects: its physical structure, its production processes, the organisation of human resources and their relation with the physical structure and, finally, the co-ordination, control and support of its production processes. Thus, the physical view is concerned with the representation of all levels of physical aspects of a manufacturing system entity, from material flows to physical processes occurring during its operations. The organisational view focuses on the structure of responsibilities between the various human operators within a manufacturing system. The subjects of the process view are the "technological" processes, which set out the way that materials are transformed throughout the plant.

The operational view, finally, refers to procedures to control and support the operation of the production processes.

3 THE ENGINEERING PROCESS OF INDUSTRIAL MANUFACTURING SYSTEMS

3.1 Engineering methodologies in literature

Up to now, "particular" methodologies have been developed for specific scopes. Amongst the methodologies developed around a specific method, the simulation methodology is a significant example. Simulation practitioners seem to agree on a common methodology consisting of a sequence of two main activities. A modelling activity aims at creating a "logical" (descriptive) model of the system (of its properties and behaviours (Vernadat, 1999)). A simulation activity then adopts a simulation model, created from system "logical" views, to evaluate the system behaviour (its expected performances) under experimental conditions. Amongst methodologies developed for specific applications, methodologies for facility layout planning are a significant example: the methodologies are built upon the final objective of the study, the determination of the physical organisation of production systems, and adopt proper methods for its accomplishment. A systematic layout design methodology is presented by Sly, et al (1997). Meller, Gau (1996) provide a comprehensive survey of the facility layout problem and methodologies and algorithms commonly adopted for its solution. Similarly to simulation, a common procedure is shared amongst practitioners consisting of "logical" modelling and performance evaluation activities. "Logical" modelling concerns the technological cycles (operations sequences/material flows), the plant structure and equipment (area requirements/constraints, equipment features) and the plant "design skeletons" (Sly, et al, 1996), where manufacturing "blocks" are assigned to material flows, laid out and graphically connected in a network-like fashion. The evaluation activity then seeks to analyse the performance of the "skeletons": e.g., distance-based evaluation methods are adopted to further detail "skeletons" through inter-"block" distances (Meller, Gau, 1996) and to evaluate performances either by means of algorithmic or heuristic models (Sly, et al, 1996).

The next section presents the proposal of a generic "Engineering Process of IMS", aiming to synthesise, in a "generic" methodology, existing "particular" models, such as those discussed in this section.

3.2 "Generic" engineering methodology

The "Engineering Process" is structured as an IDEF0 model (Fig. 2). Five *modelling activities* are defined: A1 concerns the identification of the production system organisation, automation level and control policies; A2 is concerned with the specification of the required technical processes; A3 concerns the specification of the system design solutions; A4 serves to develop the evaluation model and A5 concerns the evaluation of the system performance of the design solutions against performance objectives.

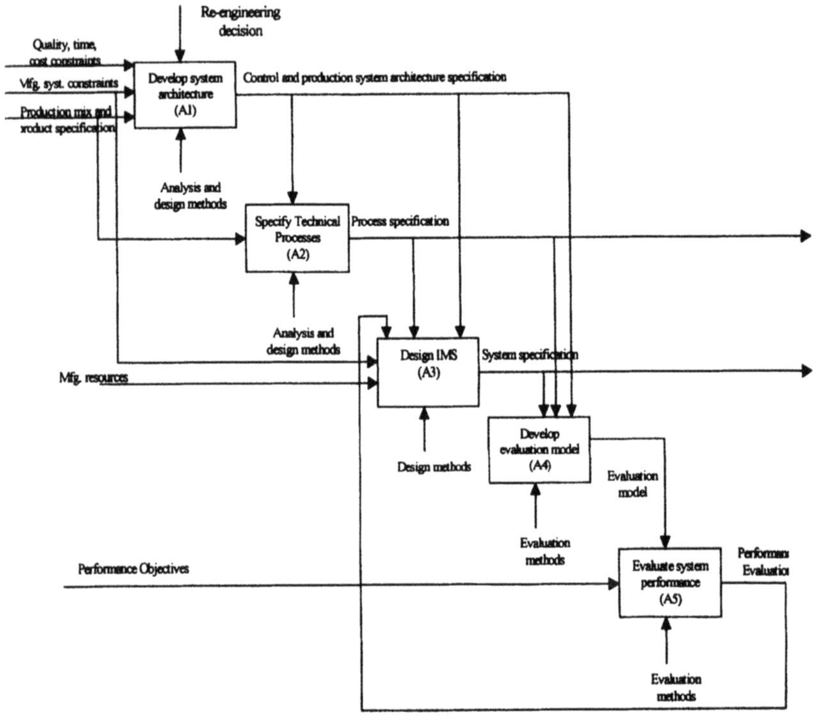

Figure 2: The generic engineering methodology - process model

The "best" solution results from one or more "evaluation" loops until performance objectives are achieved. The "best solution" is then utilised as an input to future implementation activities, both, as a process and system specification.

Having a closer insight, "develop system architecture" (A1) starts from given product designs and production mix (product bill of material, drawings and demand forecast), management decisions (cost, time, quality targets) and physical constraints (for existing plants) and sketches out a first draft of the design solution at architectural level. "Specify technical processes" (A2) is concerned with the specification of the production requirements (the process plans). "Design IMS" (A3) further details architectural drafts and defines a possible design solution capable to operate the "technical processes". It defines the allocation of manufacturing resources (equipment and human resources), their organisation (facility layout and human organisation), the processes and rules to *operate* and *support* the "technical processes". The final phases then seek to evaluate the design specifications. "Develop evaluation model" (A4) consists of translating the specifications into the ontology of the method adopted for evaluation. Hence, "evaluate system performance" (A5) represents the calculation of performance measures and the subsequent comparison with performance objectives by means of the evaluation model developed.

The methodology finally distinguishes amongst 3 types of modelling methods: analysis methods (for system analysis), design methods (for design specification), evaluation methods (for the evaluation of design specifications). The methods adopted by "particular" methodologies should be included in these types of methods. Referring, e.g., to facility layout planning, technological cycle diagrams are methods used to "specify technical processes" with an analytical scope. "Block" diagrams are adopted to set out draft layout designs. Distance-based methods are examples of evaluation methods.

4 THE FRAMEWORK

The framework proposed in this paper is concerned with the requirements and preliminary design phase of manufacturing system life cycle (Fig. 3). It defines a "generic" methodology to produce a manufacturing system design starting from its system concept. The methodology integrates analysis, design and performance evaluation methods. Being actually concerned with a generic definition, specific methods are not addressed. The authors foster integration of formal descriptive methods, derived from enterprise engineering and industrial engineering disciplines, and evaluation methods, typically adopted in industrial engineering (Garetti, Macchi, 2000). Whilst descriptive methods serve to produce formal views of the physical, operational, organisational and process aspects of manufacturing systems – thus serving analysis and design purposes –, evaluation methods integrate system views in a unique model for the sake of performance evaluation.

The engineering methodology is thus proposed as a unique basis for engineering method selection amongst all available methods. It should guide the modeller in selection either of suitable descriptive or evaluation methods competent to express models of all areas defined in the "Morphology". The

"Morphology", in particular, is intended to guide the modeller in the selection of methods based on the *life cycle phase* and *view* concepts. Beside, the *system* and *process taxonomy*, facilitating the description of the "specific" configuration of the system under study, may further help in the selection whenever methods are "specific" to the system configuration itself.

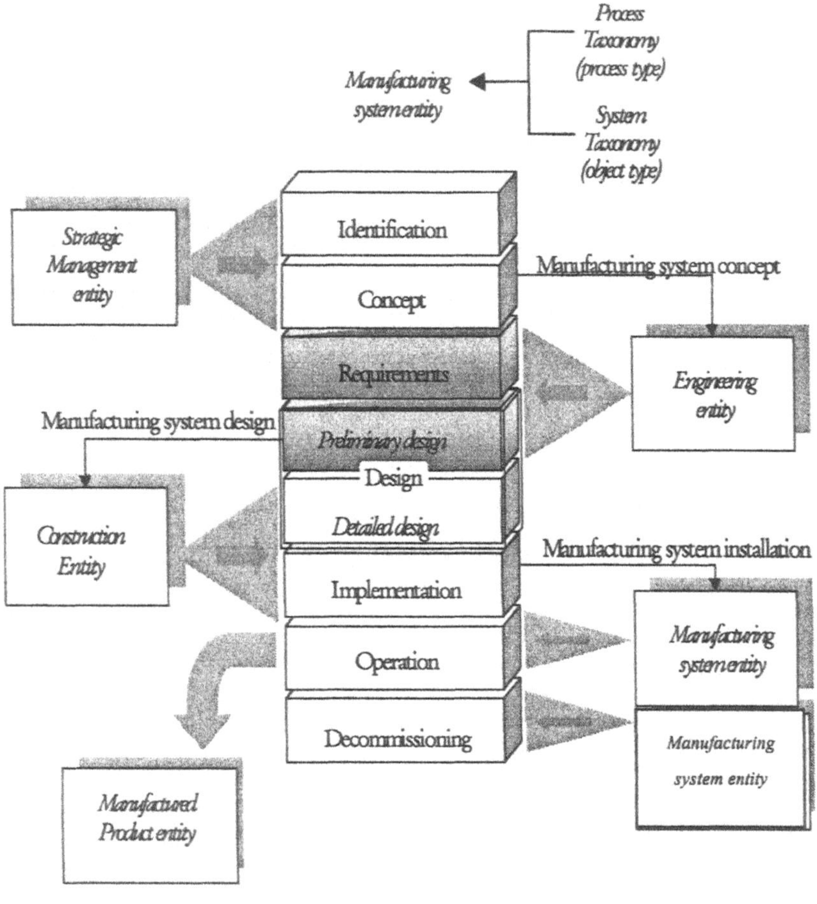

Figure 3: The Mfg. Systems Engineering framework and its relations with GERAM life cvcle

As an example of utilisation, the integration of methods for facility layout planning in the "Morphology" is shortly discussed. Their integration can be "generically" based on the *life cycle phase* and *view* concepts. Thus, e.g., technological cycle diagrams pertain to the requirements phase and process view. Block diagrams concern the design phase and physical view. Thereafter, referring to methods defined for "specific" system configurations, analytical models specifically built to calculate the throughput time of production lines, being either assembly, disassembly or machining lines, can be mentioned (Dallery et al., 1989, Di Mascolo et al., 1991). In the "Morphology", these methods should be classified as evaluation methods for specific *process types* - assembly, disassembly and machining processes - and *object types* - departments organised as lines - and subsequently selected.

5 CONCLUSIONS

In this paper, the authors' propose a framework that constitutes a unifying "umbrella" to be shared amongst both researchers and industrial practitioners. The framework represents a shared reference to integrate existing engineering methodologies. The proposal presented in this paper, however, is an initial proposition that is predominantly based on an abstract view of industrial engineering. Industrial case studies are envisaged that should provide a basis for comparison with more practical approaches to manufacturing systems engineering. In this way, the framework could be validated, especially in terms of a more "pragmatic" view, and extended by elements that are so far missing from an industrial point of view.

Future development work has then to validate the applicability of the manufacturing systems engineering approach proposed to the practical context in industry. The facility planning cycle has already been used as a basis for comparison in this paper. Wiendahl (1997) defines a systematic procedure that chiefly fulfils the practical requirements of real implementations. As is common practice in the industrial context, formal manufacturing modelling methods are not part of this practical development process.

Thus it has to be analysed what benefits a combination of this practical approach with the more theoretical approach presented in this paper holds. One unanswered question is whether this combination can help industry to effectively apply the formal modelling methods available in order to improve the design of manufacturing systems. Conversely, one ought to examine to what extent the confrontation with rather more "hands-on" approaches can lead academia to propose modelling methodologies that improve the planning quality and ease the planning efforts for practitioners.

The framework, and, in particular, the "Engineering Process of IMS", will be further developed to include specific modelling methods. In fact, methods, currently being reviewed within the VIMIMS project (see acknowledgements), will be classified in the "Engineering Process " – i.e. formal descriptive modelling methods such as IDEF0, UML and MES, analytical modelling methods such as Markov Chains, Fluid Markov Chains, Queuing networks, Petri-Nets and the Theory of Logistic Operating Curves, and discrete-event simulation methods. In VIMIMS, the framework will be used as a means to classify methods provided as educational contents.

6 ACKNOWLEDGEMENTS

The results presented in this paper have been achieved as a side-activity of the VIMIMS project, an ongoing project funded by the European Union under action "MINERVA - Trans-national Projects in the Field of Open and Distance Learning (ODL) and Information and Communication Technology (ICT) in Education", whose aim is the establishment of a Virtual Institute for the Modelling of Industrial Manufacturing Systems (VIMIMS) for delivery of on-line courses to undergraduate students in production engineering disciplines. The authors thank other VIMIMS partners for collaboration and support: METID centre of Politecnico di Milano, Italy; LAG-INPG, France; SZTAKI, Hungary; ENFAPI Briantea, Italy.

7 REFERENCES

Carrie, A. (1988), *Simulation of manufacturing systems*. John Wiley & Sons.

CIMOSA (1993), *CIMOSA Open System Architecture for CIM,* ESPRIT Consortium AMICE (Eds.), Springer-Verlag.

Dallery, Y., David, R. Xiao-Lan, X. (1989), *Approximate analysis of transfer lines with unreliable machine and finite buffers*. IEEE Transactions on Automatic Control, 34(9).

Di Mascolo, M., David, R. Dallery, Y. (1991), *Modelling and analysis of assembly systems with unreliable machines and finite buffers*. IIE Transactions, 23(4).

IFIP-IFAC Task Force, (1998), *GERAM: Generalised Enterprise Reference Architecture and Methodology*. Annex A to ISO 15704, ISO TC184/SC5/WG1 N423.

Garetti M. Macchi. M. (2000), *Manufacturing systems modelling and enterprise modelling: do they need to be integrated?* In: Proceedings of the Information Technology for Business Management ITBM - XXVI IFIP World Computer Congress WCC, Beijing, China.

Meller, R.D. Gau, K.-Y. (1996), *The facility layout problem: recent and emerging trends and perspectives*. Journal of Manufacturing Systems, vol. 15(5).

Sly, D.P., Grajo, E.S. Montreuil, B. (1996), *Layout design and analysis software*. IIE Solutions magazine, part 3 of a 3 part series.

Sly, D.P. (1997), *Before dynamic simulation: systematic layout design from scratch*. Proceedings of the Winter Simulation Conference.

Vernadat F. B. (1999), *Requirements for simulation tools in Enterprise Engineering*. 15th Int. Conf. On CAD/CAM, Robotics and Factories of the Future, Aguas de Lindoia, SP, Brazil.

VIMIMS Consortium, (2000), *Virtual Institute for the Modelling of Industrial Manufacturing Systems (VIMIMS)*. Proposal submitted to MINERVA action of European Commission.

Wiendahl, H.-P. (1997), *Betriebsorganisation für Ingenieure* (Business Management for Engineers). 4th edition, Carl Hanser Verlag, Munich, Vienna.

Zwicky, F. (1948), *The Morphological Method of Analysis and Construction*. Courant, Anniversary Volume, pp. 461-470, Intersciences Publishing, New York.

The Users View of Enterprise Integration and the Enterprise Process Architecture

Juan Carlos Mendez Barreiro
ADN INTERNACIONAL, S.A. DE C.V., Mexico, jcmendez@adn.com.mx

Abstract: The paper describes an architecture and methodology developed for the introduction of enterprise integration into the enterprises in Mexico. Starting from the experience gained as a management consultant a concept has been developed that allows the user to understand, evaluate and implement integration technology.

1 INTRODUCTION

Most of the companies around Mexico searching for adequate technology (including tools and methodologies) to be used in their company have to see (preferable on paper) the structure and the related knowledge of their processes. This information is needed to support a number of decisions to be made in the course of acquiring and implementing such technology. Therefore, any supporting architecture or methodology must provide them with the capability to present the business processes and the related information in order to be used in the improvements of their enterprise operations. Preferably the resulting representations should be reusable eliminating the need of re-mapping the process each time.

During the years in the management consulting business we experienced that all the companies are facing the same problem: none of the people involved really uses their operation manuals as a tool to analyze the operation and evaluate the improvements needed to increase the process efficiency, productivity and/or profitability. When a company needs the process mapping or operation manuals the people do not trust the validity of the manuals because they find them never to be up to date and complete.

Therefore, the companies are looking for new ideas in the field of information technology. Standardization is looked at as a way to simplify the process of documentation and to keep it better under control and ISO 9000 is being seen as a standard solution to their needs.

Another problem today is that nobody in the companies is really responsible to maintain the operational procedures. The reason: companies see the subject of operational procedures as being important, but still feel it does have only a rather low priority. We see that many companies assign the responsibility of maintaining the operational procedures to the lowest management level in the organization. On the other hand the companies want more control on their operation in order to increase their market share and their profits.

The use of enterprise models for documenting operational procedures is not really accepted. The main reason is lack of awareness on enterprise integration technology. But even people who do know do not see it yet as a solution to their problem. The available tools and methodologies do not seem to provide the flexibility needed to easily create and maintain the models of their processes.

In the consulting industry the consultants are using tools, but those tools are not really developed to model the enterprise from a user point of view and to cover all the elements (views) needed to capture the enterprise knowledge and to present it to the user. Even more important, such models cannot be maintained by the user themselves, which makes them of no interest for operational use. So, the companies are not using models, but employ mostly common sense as their main tool to make improvements in their operation.

2 HOW TO INTRODUCE MODELLING TECHNOLOGY

Introducing enterprise technology to companies in Mexico cannot be done by presenting the contents of know architectures and methodologies (Vernadat, 1996) like ARIS, (http://), CIMOSA (1996) or GERAM (Bernus, et al, 1996) or even the ISO and CEN standards (ISO, 1999 and 2001, Pre EN ISO, 2002). The users in the companies do not understand these technologies and cannot decide on the value for their own organization. Are these just fashions or a trend that should be followed? We must tell them that architectures like CIMOSA cover a hole in the management process that presently hinders or at least delays the improvement of their companies. The focus has to be on the hole rather than on the solutions. Why to fill a management process hole? Because there is a need it to see all elements of the enterprise operations prior to make a decision on the improvements. So they need to know that the need a technology that will provide the capability and the How's to capture the knowledge of the operation and will allow its maintaining it as well.

Many of the enterprises do not really understand what it is they are missing, because they do not see the reason nor a convincing application for this technology. Some see it as a methodology to create operational procedures. Also when we tell them that the technology will integrate their company, they do not see what is the different compared with supply chains and some others methodologies identified as integrated methodologies. Many people see ERP as an IT solution to integrate their company.

We have developed a methodology aimed at companies in Mexico to create an understanding for the process modeling and integration technology. Recently, we started to use what we call an "Enterprise Process Architecture" in order to focus the thinking on processes, more than on enterprise integration, because the users understand the term process much better. Thus my clients are now beginning to see what is the reason for this technology. We are telling them that this technology is an Enterprise Process Architecture and it is part of Process Engineering.

3 THE ENTERPRISE PROCESS ARCHITECTURE

The concept was developed to support integration of the enterprise. Why named Enterprise Process Architecture (EPA)? Lets look into public definitions (Cambridge, http://):

- **Enterprise:** an organization, especially a business, or a difficult and important project, one that will earn money. Enterprise is also eagerness to do something new and clever, despite any risks.
- **Process:** a series of actions or events that are part of a system or a continuing development, or a series of actions that are done to achieve a particular result.
- **Architecture:** the art and science of designing and making buildings or constructions.

Combining these definitions means we can speak of an enterprise or business as being constructed as a number of business processes producing the things that enable it to earn money and with the business construction guided by an Enterprise Process Architecture (see Fig.1). The Enterprise Process Architecture itself is a suitable collection of available methods and tools capable of supporting business process modeling from an operational use perspective.

So the first thing we tell the users is that EPA is a development to cover the big hole in the management process, which exists between the Enterprise Real World (ERW) and the use of Enterprise Technique and Methodology (ETM). Only the use of these techniques and methodologies in a continuous cycle (CLC – Fig. 2) of re-engineering will keep the enterprises competitive, keep their place in their respective markets and open up new ones and will achieve the necessary financial and operational results (FOR). The re-engineering has to be guided by QCD – improving product quality, lowering operational costs and shortening delays.

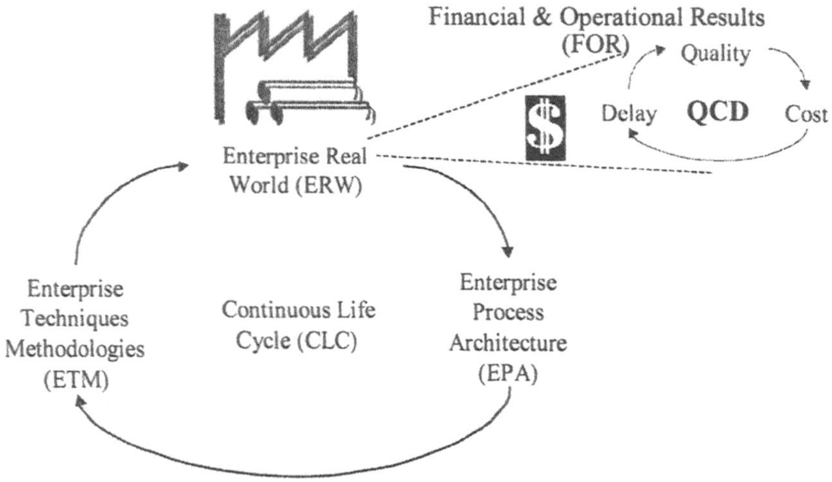

Figure 1: Components and Results of Continuous Re-engineering Process

The final stage in the introduction of EPA is reached when the end user starts to see what the ETM said about their implementation. All of them mention the need to model the Enterprise Processes but none of them really explains how to make an Enterprise Model.

Now if we tell the user that the EPA was made to fill this big management process hole that exist between ERW and ETM they will understand why this technology was developed and will be eager to apply and exploit it.

3.1 BENEFITS OF THE EPA

When the users discover the place of the EPA they now will understand the need of the EPA. Now we are in the position to start to explain them the benefits that will come when they create, use and maintain enterprise process models (EPM) with the EPA.

Some enterprise users instantly start to mention the impact they can have with an EPM. But the concept has to be re-enforced telling them that the three big impact areas to get benefits are: Quality, Cost and Delay. Now we have to document the potential benefits and using these against the old fashioned concepts. In addition, we will create a matrix of the different ETMs needed and how the EPA will support those ETMs.

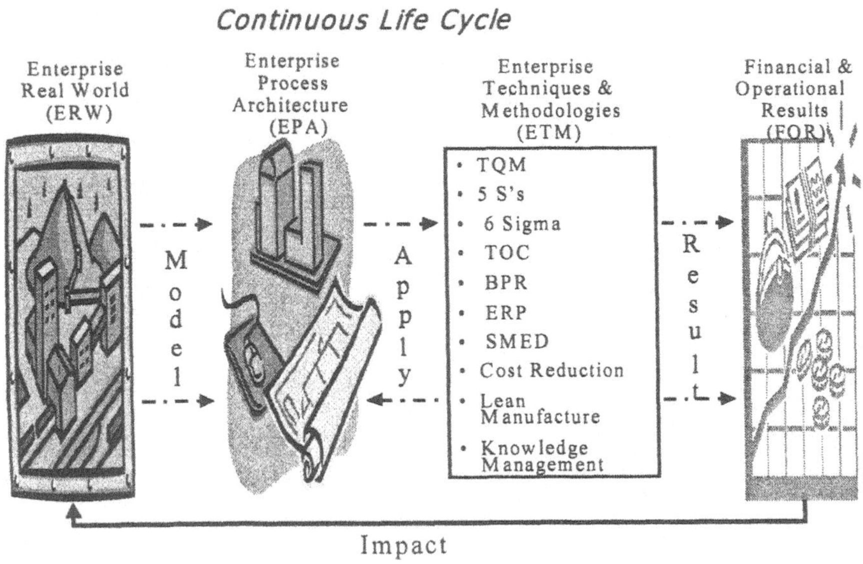

Figure 2: Continuous Life Cycle Re-engineering Process

4 CONCLUSION

In order to create a big response on EPA and EPM from the industry the users have to understand that the EPA has to be positioned between ERW & ETM. We must tell them that the EPA was made to improve the benefits of the ETM. Finally they will see the need to use all the ARIS or CIMOSA model formalism in order to have a complete Enterprise Model that can be reusable.

The users must have the feeling of not capturing sufficient process knowledge; a lack of knowledge will come if they only model workflows. Also we must create the understanding that Business Process Modeling is not a fashion or new management trend, but is a significant step towards information and knowledge models for decision support in the management of inter-organizational collaborations.

5 REFERENCES

ARIS, http://www.ids-scheer.com/ARIS

P. Bernus, L. Nemes, T.J. Williams (Eds.), (1996), Architectures for Enterprise Integration, The findings of the IFAC/IFIP Task Force, Chapman & Hall

CIMOSA Association, (1996), CIMOSA - Open System Architecture for CIM, Technical Baseline; Version 3.2, private publication

Cambridge On-Line Dictionaries, http://dictionary.cambridge.org/

Pr EN ISO 19439, (2002), Enterprise Integration – Framework for Enterprise Modelling, CEN TC'310 WG1 together with TC 184 SC5 WG1

Pr EN ISO 19440, (2002), Language Constructs for Enterprise Modelling. CEN TC'310 WG1 together with TC 184 SC5 WG1

ISO 14258, (1999), Industrial Automation Systems – Concepts and Rules for Enterprise Models, TC 184 SC5 WG1.

ISO 15704, (2001), Requirements for Enterprise Reference Architecture and Methodologies, TC 184 SC5 WG1.

Vernadat, F.B. (1996), Enterprise Modelling and Integration: Principles and Applications, Chapman & Hall, London.

Matching Teams to Business Processes

Nikita Byer, and Richard H. Weston
MSI Research Institute, Loughborough University, United Kingdom, r.h.weston@lboro.ac.uk

Abstract Progress has been towards developing 'reference models of human teams'. A classification is made based upon the purpose of teams by accounting for observed differences in 'structure' and 'composition' properties of teams described in the human factors literature. A reusable understanding of these characteristic properties should (1) inform the 'initial design and formulation of enterprise teams', and thereby match the composition and structure of teams to specified business processing needs and (2) help focus 'continuing task development carried out by teams' through their useful lifetime.

1 TEAMWORKING – BENEFITS AND PROBLEMS

Although industry at large is deploying many types of team a number of problems have emerged (Anantharaman, et al, 1984). Recent reports indicate that 70% of teams fail to produce desired results (Tranfield, et al, 1998). Improperly applied team working can increase lead-times needed to complete tasks. The empowerment of individuals and work teams can generate high-pressure environments with low slack and little buffering. A darker side of team working and empowerment can be strong group norms and powerful individuals stifling individual flair and self-expression (Barker, 1993). A connected downside is the lack of visibility of the individual working of teams, particularly in complex environments where multiple team solutions are needed and hence effective inter-team integration is a requirement.

Certainly with respect to specific instances of enterprise need, it is unclear as to whether or not team working constitutes a more efficient organisation of social system resources when compared to more traditional organisational forms used by industry. Also necessary prior to any completion of team tasks: (1) 'team design requirements' must be specified team tasks: (1) 'team design requirements' must be specified (e.g. the 'composition' and 'structure' of teams and the 'responsibilities' of team members must be determined) and (2) activities needed to satisfactorily realise task completion require formulation and planning.

The longer it takes to design and operationalise a team the longer it will take the team to complete its first task. But the quality of the team design will directly influence the quality with which tasks are completed. Hence matters of team design can be very important and should not be trivialised.

2 STATIC AND DYNAMIC ASPECTS OF TEAMS

Seminal work of Syer and Connolly (1996) considers teams to be systems. Some elements of any team system causally and temporally link processes realised by the team or processes that transform a team. Generally this leads to circular cause and effect relationships that over time effect dynamic behaviours of the team. Inputs to the team system are transformed to outputs via *team processes* in accordance with *team structures*. Syer and Connolly (1996) considered the structure of a team to comprise those relatively static and enduring aspects, while its processes concern patterns in the team's system behaviour in response to relatively dynamic or transient factors by virtue of change or instability. Syer's *team structures* include specifications for the size of a team, team membership, a place and time to meet, team roles and team goals and objectives. Team processes will be norms that include methods of problem solving, decision-making and planning. The team system acts to transform inputs to outputs through its processes. Once established processes are subject to deviation from target values and because of internal and external factors produce 'error states'. *Team processes* can be defined in terms of sets of temporarily ordered activities that consume resources available to the team so as to transform ideas, skills and qualities of team members, materials, problems, etc into discoveries, solutions, proposals, actions, design ideas and products. These input-output transformations are subject to speaking skills, agendas, time management, descriptive feedback, warming up and task methodology.

Smith (1987) refers to *team structures* as *frozen processes.* This helps explain the notion that teams have an inherent capability to evolve their behaviours by freezing and unfreezing processes so as to change team structures. This differentiates capabilities of relatively simple groups of technical resources from teams of people although advances in software agent technologies have begun to blur this differentiation. Hardingham (1994) points to common features of all teams even when they function differently. She identified four stages of the 'Team Life Cycle', namely: *Forming Stage:* the team is under-developed and its members are concerned with who fits is under-developed and its members are concerned with who fits where; *Storming Stage:* the team is experimenting and concerned with how members can work together; *Performing Stage:* the team is mature and concerned with goal achievement; *Mourning Stage:* the team is ending and concerned with breaking up and moving on to new tasks.

Also referring to the lifetime of teams, Foster et al (1996) explained that some teams perform a function then disband after original goals are achieved. But this may not be the case where reoccurring tasks need to be performed, or where new tasks arise of similar type. Hence the same team may be reused. Implicated within the models of teams of Syer and Connolly (1996), Smith (1987), Hardingham (1994), Foster (1996) and Larson and LaFasto (1989) is that structure can be determined in two (possibly complementary) ways, viz.: (i) identify candidate team compositions and structures needed for the achievement of specific objectives (i.e. to adopt a top-down approach to team design and change); (ii) identify candidate team compositions and structures that best co-ordinate the competences of team members (i.e. to engineer and develop teams in a bottom-up manner).

In many practical cases it is assumed that the 'design' and 'development' of teams needs to be ongoing.

3 LIFE-CYCLE ENGINEERING OF TEAMS – MATCHING TEAMS TO BP'S

At the MSI Research Institute in the UK a research programme is underway which seeks to deploy enterprise modelling (EM) and enterprise integration (EI) technologies in support of the life cycle of enterprise teams. The programme comprises a number of projects with joint support from the UK's EPSRC and a consortium of manufacturing businesses. Individual projects are focused on different life-cycle aspects and seek to evaluate developed EM & EI methods for use in different industrial domains. The aim is to use both static and dynamic team models (in visual, tabular and computer executable forms) to structure key aspects of ongoing team-based engineering. This should not 'handcuff teams' but rather improve their initial and integrated design and as appropriate inform their ongoing development so that (a) best practice can be transferred, (b) capital value can be placed on teams and (c) teams design and operation can remain effective within various complex and changing business environments.

Collectively the projects cover the following phases of team engineering: 'specification of team requirements'; 'team design'; 'team implementation'; 'team development'; 'task development' and 'team maintenance'.

A common feature of the projects is their focus on matching (1) *static and dynamic competences of teams* to (2) *goals and task requirements, identified in terms of business process models*. It has been assumed that such a focus can: lead both to intra and inter organisational design of teams; facilitate ongoing team engineering; enable business benefit and analysis with respect to team-based strategy and policy making. This assumption is being tested by the industrial evaluation work. The remainder of this paper is focused on results from one of the projects that has developed a static reference model of teams to inform decision-making during 'team design'.

4 DEVELOPING A NEW TEAM CLASSIFICATION

4.1 Initial Literature Review

In 2001, the authors conducted a literature review of 'team types' listed in Table 1. The study found limits on the reuse of human factors knowledge which stem from (1) significant overlap in the scope and purpose of alternative team descriptions, (2) inconsistent use of terminology and (3) varying levels of abstraction at which properties of teams are analysed.

Table 1: Names of teams attributed by the literature

– Steering Teams	– Tiger Teams	– Virtual Teams
– Planning Teams	– Launch Teams	– Top-Management Ts
– Proc. Improvement Ts	– Work Teams	– Mid-Management Ts
– Self Managed Teams	– Project Teams	– Ad Hoc Teams
– Multi-disciplinary Ts	– Focus Teams	– Work Groups
– Interdisciplinary Ts	– Task Force	– Quality Circles
– Impact Teams	– Natural Groups	– Co-ordination Teams
– Cross Functional Ts	– Functional Teams	– Think Tank Teams

In Loughborough (http://) the authors draw tabulated comparisons between the team types described in the literature with respect to their 'main purpose', 'make-up', information flows', 'focus', 'behaviour', 'leadership', 'degree of autonomy', 'membership time-span' and 'team limitations'.

4.2 Primary Team Selection Actors and Activities

Any enterprise team selection process will involve 'choosing a team with competences to perform a defined set of activities within an acceptable time frame, in order to deliver specified outputs'. But it was observed that no formal method of team selection is reported in the literature. Current best industry practice observed by the authors at collaborator sites is as follows: (i) The team already exists in the organisation and after fulfilling a role with respect to an initial task set it is assigned alternative responsibility for a new task set; (ii) A group of individuals are formed into a team without a well-defined structure or process and it is assumed that a suitable structure and process will emerge over time.

Therefore current enterprise team selection is ad hoc and incompletely planned. Normally though complementary team roles, processes and structures develop over a period of time after tenure trial and error. During forming and storming stages process lead-times and efficiencies will be degraded. It follows that the initial selection of a team can have a major influence on the team's performance, its outputs and the time it takes to perform acceptably. The more suited the initial team design is to performing assigned tasks the sooner it can carry them out satisfactorily (on a continuing basis if required).

It was observed that the *performance of any new team selection method* must suit the needs of would-be 'customers'. In this respect two main customer groups were considered, namely 'team initiators' and 'team designers'. It was assumed that the team initiator customer group would likely operate mainly with a strategic or tactical purpose, with respect to the design or development of one or more business processes. Team designers were expected to use information (about process definitions and outline resource requirements) generated by team initiators and to couple this to knowledge about suitable teams in order to select a viable team type. It was also understood that teams specified and selected by team initiators and team designers might be required to operate for strategic, tactical or operational purposes.

4.3 Knowledge Needed –Enterprise Modelling Views

The study considered general requirements of 'team initiator' and 'team designer' customers. This consideration was cognisant of reported deficiencies of current ad hoc team selection practice used by collaborating companies.

Fig. 1 illustrates the general context of the *team selection process*. The team initiator's role is to identify a need for teams and, associated with this, develop outline requirements definitions and human resourcing policies. This might typically include some form of general task definition. The role of team designers is to determine the composition, structure and other properties of teams. Typically they might detail team objectives, develop descriptions of one or more viable teams and conduct an analysis to differentiate between candidate teams, e.g. in terms of their performance, cost and quality of outputs, 'flexibility', etc. Based on this line of thinking it was observed that common classes of information needed to facilitate team selection include: 'team function'; 'team leadership styles'; 'team size'; 'team composition'; 'team organisation'; 'team resource'; and 'team activities'.

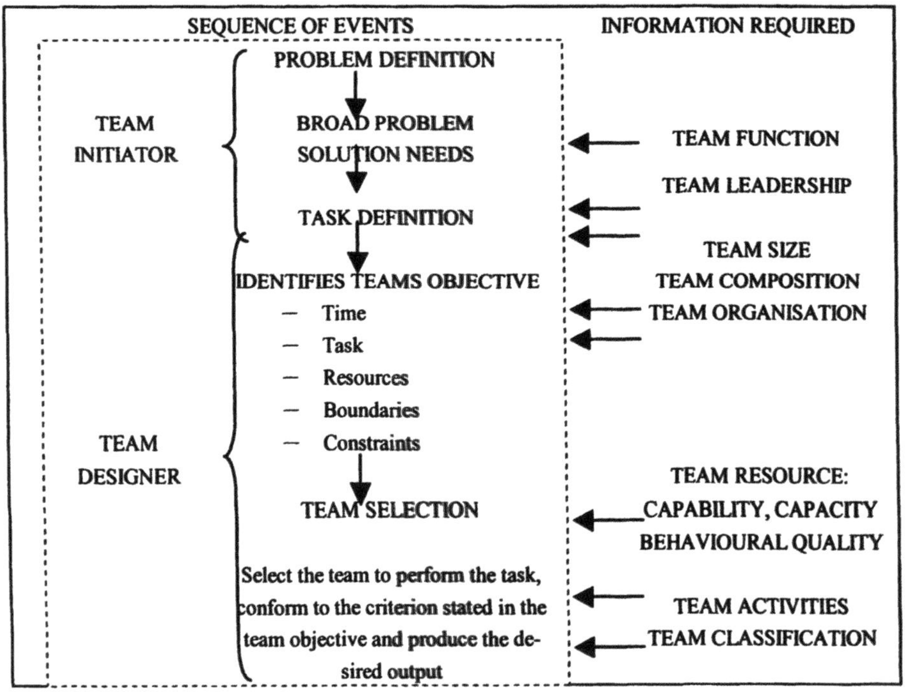

Figure 1: Typical events that impact on team selection

5 TEAM CLASSIFICATION

5.1 Analysis of Team Classifications by Purpose

Previous authors had developed team classifications with reference to their purpose. One such classification (Moris, et al, 2000) distinguishes between Integrated Product Development (IPD) teams currently deployed in aerospace industries. Three main classes of team were observed, so called 'management', 'core' and 'task' teams. Differentiation between team classes was made based on the following views: 'team focus'; 'team composition'; 'team purpose or responsibility'; 'team skill base'; and 'team controls'. However, an observed critical limitation of this and other previous classificacations of teams by purpose is that information about 'team objective' and 'leadership style' is not adequately represented.

5.2 A New and Enhanced Team Classification

A new and enhanced classification of teams was developed with respect to their 'purpose'. Cognisance was taken of the three main classes of enterprise activities (BS ISO 14258, 1998), namely: (i) *Strategic Teams:* that 'run things' and are responsible for WHAT enterprise activities; (ii) *Tactical Teams:* that 'recommend things' being responsible for HOW enterprise activities; (iii) *Operational Teams:* that 'make or DO things'.

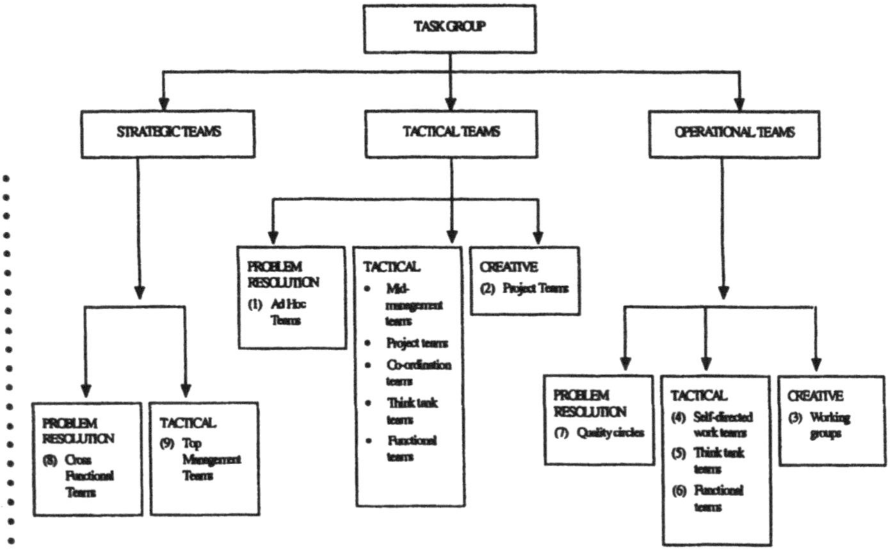

Figure 2: Team Classification Chart

Table 2 lists five broad groupings of questions that require answers during the *team selection process*. Answers to these questions were analysed for teams classified under (i) through (iii), in terms of (a) achieving team goals and (b) determining obstacles and opportunities associated with teamwork. This analysis helped clarify distinctions drawn between teams and led to the development of Tables 3, 4 and 5 which constitutes a new reference model of enterprise teams with respect to their purpose. Results from the analysis were also used to construct the team classification chart shown in Fig. 2.

Table 2: Criterion for Team Selection

Team Function – Why should this team type be formed?
What does it do? What is the overall purpose? What are the critical success factors? What are the priorities? What external information does the team need? What type (size and generality) of problem does the team solve?
Team Performance – How is the team performance measured?
How well does the team achieve its objectives? What time constraints exist on achieving objectives? What makes the team productive? What kind of relationship is there between team members? What kind of relationship is necessary for achieving goals? Are the goals and objectives clearly defined? Must the team be flexible?
Competition – Whom, if anyone, is the team meant to beat?
Will the team be competing against other teams? Are competing teams inside or outside that organisation? What are the benefits of competition? What are the disadvantages?
Team Roles – Are the roles clearly defined in the team?
Can team roles be swapped between team members? Could team members fulfil any of the roles, or are there many 'specialist' roles? To what extent do the interactions; information exchange and joint working of team members need to be orderly, predictable and polished? Is the team structure formal or informal? Are there defined paths for information and data flow?
Team Relationships – What are the relationships between team members?
How important is loyalty to the team? What are the disadvantages of loyalty? Is the team size and membership fixed or predetermined? Do team members belong to more than one team, and is that necessary? Will the team have part-time members? What is the length of the team life cycle? What are the advantages of people staying in a team for a long time? What are the disadvantages?

Table 3: Recognisable strategic team needs and requirements

Items	Needs
Team Function	Teams that run things, management teams. – Sets company direction, determines organisation's mission, goals, objectives and priorities – Serves as the major link between the organisation and the outside world – Creates organisational plans and strategies based on information from the internal and external environment: competitors, market demand, internal capabilities – Problem types are generally wide ranging and generally deal with situations that arise external to the organisation

Table 3: (continued) Recognisable strategic team needs and requirements

Team Performance	– Team generally sets short, medium and long term goals or objectives – Objectives are measurable and generally include increase in ROI, increased market share, training of staff, penetration of new markets – The teams performance is determined by the achievement of goals set, these goals are clearly measurable.
Competition	This team-type sets plans and strategies that will enable their company to compete in any world market. They are competing against similar teams in many organisations around the world Advantage of Competition: to increase the organisational standards and allow the company to compete in the international arena. Disadvantage: In an attempt to compete organisations may take on more than they can handle.
Team Members Roles	Generally cross-functional teams that span all aspects of the organisation. Team member's roles are well defined and the team consists of a large number of specialists. The team structure is formal and somewhat hierarchical.
Relationships	Important relationship factors: – Team size and membership are generally fixed – Team loyalty is important – Team members usually belong to more than one team – The team may also use consultants or specialists – Long team lifecycle
Team Leadership	Delegating style of leadership where leaders respond to proposals and suggestions from the team. The team consists of competent experienced members. This team type is also characterised by an inspirational / charismatic style of leadership. This type of leadership exists in high-risk situations where the teams are highly competent.
Team Types	– Top management teams – Cross-functional teams
Typical Team Structures	– Problem resolution – Tactical

Table 4: Recognisable tactical team needs and requirements

Items	Needs
Team Function	Teams that recommend things, core teams – Develops clear objectives and controls and co-ordinates work throughout the organisation – New product development manufacturing processes, new marketing strategies, cost reduction ideas, new business ventures – Product decomposition and product integration – Develop the day-to-day plans of action that support the strategies and objectives set by the strategic team – Problems are wide ranging but cover issues that are within the organisation

Table 4: (continued) Recognisable tactical team needs and requirements

Team Perform-ance	– Teams objectives are generally project, process or product oriented – Team generally sets short, medium and long term goals or objectives, based n the constraints which are determined by the strategic teams – Team productivity is measured by the rate at which the desired goal is achieved. Objectives include: new product design, project management.
Competi-tion	These teams would not necessarily be competing against other teams but rather against constraints imposed on them by the strategic teams. These constraints include monetary, time, resources constraints. The constraints set by the strategic team would ensure that the tactical team accomplished the desired task in the given time with the resources available to ensure that the organisation remains internationally competitive.
Team Members Roles	These teams can be both cross-functional and functional. In functional teams the members can be rotated. Depending on the nature of the task to be performed the team structure can be formal or informal. For example, if the team's objective is project manager, then the structure is formal and somewhat hierarchical. However, if the team's objective is new product development then team creativity is essential and the team structure is flat and informal.
Relation-ships	Important relationship factors: – Team size and membership is determined by the team's task – Teams can be functional or cross-functional in nature. – Team's life cycle is determined by the nature of the task. It should be noted that these teams are generally transferred from one task to another and generally have long life cycles.
Team Leader-ship	There are two types of team leadership styles that are characteristic of these types of teams: – Directive - teams at initiation, members unsure of tasks, leaders give information and direction – Delegating - teams competent and experienced, leaders respond to proposals and suggestions from members.
Team Types	– Ad hoc teams – Mid-management teams – Project teams – Co-ordination teams – Think tank teams – Functional and cross-functional teams
Typical Team Structures	– Problem resolution – Tactical – Creative

Table 5: Recognisable operational team needs and requirements

Items	Needs
Team Function	Teams that make or do things, task teams – Usually involves a supervisor and those who report to him – Provide ideas for the process – Discuss and propose ways to improve the workplace arrangements, production process and lines of communication – Responsible for a clearly defined area of work, responsible for the whole product or process; planning, performing, implementing and co-ordinating improvements. – Problem types are general and are specific for the team's defined area of work.
Team Performance	– Teams objectives are measurable and can be monitored.
Competition	These teams generally compete against other teams within the organisation or compete against constraints imposed on them by the tactical teams. These constraints include daily throughput, number of defects, lost time, injuries, and daily/weekly/monthly machine operating hours.
Team Members Roles	Since these teams exist within a department the team composition can be both functional and cross-functional depending on the task. The team structure is generally formal with their being a team leader who is responsible for scheduling, planning, co-ordinating and monitoring teams tasks.
Relationships	Important relationship factors: – These team memberships are generally within the division or department – Teams are functional and consist of specialists – Team lifecycle is long term – Team size is dependent on the number of employees in the department or division or business unit – Teams may also use consultants or specialists
Team Leadership	There are two types of team leadership styles that are characteristic of these types of teams: – Directive - teams at initiation, members unsure of tasks, leaders give information and direction – Delegating - teams competent and experienced, leaders respond to proposals and suggestions from members
Team Types	– Quality circles – Self-directed work teams – Think tank teams – Functional teams – Working groups – Cross-functional
Typical Team Structures	– Problem resolution – Tactical – Creative

6 CONCLUDING REMARKS

Available literature about the properties of teams has been analysed and represented in tabular and visual forms which can guide *enterprise team selection processes* and thereby realise an improved initial match between the competences of a team and the enterprise activities and tasks they are assigned. It is understood that subsequent to their implementation, teams will evolve their behaviours, processes and structures, i.e. as they better understand their tasks, interrelationships and the outputs they should generate (Syer, Connoly, 1996). In the context of process-oriented enterprise engineering the ability to reuse human factors knowledge afforded by the new tabulated reference model of teams offers means of starting team task development from a much-improved initial state.

7 REFERENCES

Anantharaman, V. Chong, L. Richardson, S. Tan, C. (1984), *Human Resource Management: Concepts and Perspectives*, Singapore, Singapore University Press.

Barker, J. R. (1993), *Tightening the Iron Cage: Concertive Control in Self-Managing Teams*, Administrative Science Quarterly, 38, pp. 388-407.

BS ISO 14258, (1998), *Industrial automation systems – concepts and rules for enterprise models*, British Standards Institute, Chiswick, London, UK.

Foster, S. F. Heling, G. W. J. Tideman, B. (1996), *Teams in Intelligent Process Based Organisations*. Lansa Publishing BV, Leiderdorp, The Netherlands.

Hardingham, A. (1994), *Working in Teams*. London, Institute of Personnel and Development, (P658.402/HAR).

Larson, C. E. LaFosto, F. M. J. (1989), *Teamwork: What must go right/What can go wrong*. Sage Publications, Newbury Park, California, USA.

Loughborough, http://msiri.lboro.ac.uk

Morris, A.J. Syamsudin, H. Fielding, J. P. Guenov, M. Payne, K. H. Deasley, P. J. Evans, S. Thorne, J. (2000), *Macro- A tool to Support Distributed Multi-Disciplinary Design and Organisation*, Cranfield University, Cranfield, UK.

Smith, K. and Berg, D. (1987), *Paradoxes of Group Life*. Jossey-Bass, London, UK.

Syer, J. Connolly, C. (1996), *How Teamworking Works: The Dynamics of Effective Team Development*. McGraw-Hill.

Tranfield, D. Parry, I. Wilson, S. Smith, S. and Foster, M. (1998), *Teamworked Organisational Engineering: Getting the Most Out of Teamworking*, Management Decision, 36(6), ISSN: 0025-1747.

Analysis of Perceptions of Personnel at Organisational Levels on the Integration of Product, Functional and Process Orientations

A Case Study

Ruth Sara Aguilar-Savén

Linköping Institute of Technology, Sweden, ruth.saven@ipe.liu.se

Abstract This paper presents a qualitative analysis undertaken in a large telecommunications company. The stimulus to this paper is to understand the beliefs and perceptions of company personnel of *Enterprise Integration* (EI). The research presented constitutes part of a wider project into EI within the company at one of its production facilities. The EI project studies a number of different aspects for the integration of product, functional and process orientations. This paper highlights some considerations for Enterprise Integration as a result of the analysis presented. The paper aims to enhance understanding of individual beliefs on EI in manufacturing companies.

1 INTRODUCTION

For enterprises in the manufacturing sector, their value adding processes are changing from a semi-stable state to a highly dynamic one. The result is a forthcoming era of continual change in their economic and technological environments. Coping with this continual change is the major future challenge. Enterprise Integration (EI) is seen by some people as one solution by which this could be accomplished.

Companies are traditionally divided into specialists within a hierarchical organisation, normally referred to as functions. As they become more "process orientated" they must overcome organisational boundaries to integrate the new process orientation into their existing functional organisation. Besides, more and more importance is given to the specific characteristics of products, leading to product-driven manufacturing e.g. focussing on the product life cycles. Thus, one may find companies dealing with three different ways of working at the same time: product orientation, functional orientation and process orientation. These companies need an effective way of integrating these three perspectives with a common understanding of the problem and a common language to communicate on this matter. What do employees and managers understand by these terms? What are the actual or perceived obstacles to achieve integration? What are suitable performance measures to ensure adequate integration?

This paper presents the answers to these questions. Beliefs and perceptions regarding "integration" among employees and managers at different hierarchical levels are analysed. Section 2 presents a description of the whole EI project carried out in the factory as a backdrop for the present analysis in order to understand the context of the interviews and their interpretation. Section 3 describes the methodology followed to accomplish the qualitative analysis. Section 4 shows the analysis as well as the main highlights of the present study. Finally, some conclusions, remarks and further research end the paper.

2 BACKGROUND: CASE STUDY DESCRIPTION

The case study is an EI project done in a large company into the telecommunications sector. The factory where the study was held has around 2.000 employees and is one of a number of production facilities within the company. The factory manufactures products for the Consumer market, which represents the most rapidly changing business segment for the whole company. New products are launched to the market continually with typical product life cycles of between 6 and 18 months. Flexibility, quick response, high efficiency and productivity are seen as some of the key success factors to enable the company to win order in the marketplace. These success factors lead the company, and specifically the factory under consideration, to focus on its *processes* and their relationship to its *functional* departments. Due to the short *product life cycles* the factory realises their importance in achieving the key success factors. New product introduction or product obsolescence affects many aspects of enterprise manufacturing decisions such as capacity, production processes, information system (IS), dedicated resources and operational responsibilities. Thus the company has three key considerations, defined as *orientations,* at this factory in the belief that they could achieve high operational and marketing benefits of working with them at the same time if they could ensure adequate alignment, i.e. *integration.* They think that a "tri-dimensional" approach might offer even higher benefits than simply the sum of each of the individual "dimensions". Nevertheless, to achieve their goal, an integrated manner of working among the three orientations is required. *Integration* means putting together the heterogeneous orientations to form a synergistic whole (similarly defined by Vernadat, 1996). It is not obvious how to achieve this last and specifically how the company's IS could support it. Furthermore the three perspectives have an effect on both the inter- and intra-organisational integration. Many detected conflicts when working together remain unsolved. Hence, the need for an EI project appeared. The goal of the project is to develop a generic model to explain and describe the integration between these orientations. The question to be answered: are the theories in the EI field adequate for this factory.

The literature on EI has been reviewed. There already exists a considerable amount of literature on the matter, explaining different emerging architectures and their related methodologies, concepts and tools; for example, Vernadat (1996) and Bernus, *et al,* (1996). Even comparisons between them (e.g. Kosanke, *et al,* 1997 or Bernus, *et al,* 1996) and great efforts to establish standards have been undertaken (e.g. ENV 40 003, 1990, ISO 15704, 1999 and ENV 12 204, 1995). A first drawback for the project when reviewing the literature on EI was that this concept has emerged from Computer Integrated Manufacturing (CIM) and thus the terminology utilised is often related to automation, automatic control and Computer Science resulting in difficulties when applying them in manufacturing. E.g. the term *orientation* was not found in the literature as managed in the company. Hence the need of deeper analysis of these and other terms and concepts, the integration from these orientations viewpoint and how they relate to the dimensions and views of GERAM, which is based on CIMOSA's cube. All this is further discussed in Savén (2002).

The *orientations* are defined as follows. *Functional orientation (FO)* is a way of working organised in departmental areas each of which group homogeneous functions highly specialised. This is the traditional approach and reflects the current organisational structure. *Process orientation (PsO)* is a way of working focused on the identification of important work's flow. A process is a collection of activities that take one or more kinds of inputs and create an output of value to the customer. *Product orientation (PtO)* is the way of working in which all operational issues regarding the product are followed during its life cycle on a global basis to ensure that they are carried out in a consistent, effective manner, promoting integration between all parties involved with the product.

Despite a number of meetings held in order to let managers and other employees know of the EI project during its first months, the author of this paper realised that still different people at the factory have different perceptions of EI. Specifically the term EI seems to be quite vague for the majority of employees. Words such as *dimensions* or *views* are found to have varied interpretations or even sometimes treated as synonymous. Hence the need emerges to solve these communication problems.

3 METHODOLOGY

Bearing in mind that the qualitative study was one part in a larger case study its purpose was to get an understanding of the perceptions, difficulties and contradictions in employees' mental models that prevent EI development and from them to be able to improve EI approaches including associated theories, methods, techniques and tools. At the same time an unexpected but welcome output of the study was to aid employees to achieve a shared vision and understanding of EI in order to contribute in the EI project.

According to Trost, (1997) if the aim of a research study is to understand the way people reason and/or to establish common patterns in the way they perceive matters, as in this paper, then a *qualitative approach* is recommended. There are three general steps in the process of either a qualitative or quantitative study: data gathering, data analysis and interpretation of the results. The combination of these three steps and the consideration of the study type form what is called the "property space" (Trost, 2001). According to Trost there are 8 study variants. The study described in this paper is classified as variant A, which is a *complete qualitative analysis*. This means that data collection; analysis and the interpretation of the results are qualitative in their nature.

The data collection is based on *interview questionnaires*. The sample is chosen non-randomly using the technique known as *strategic sampling* (Trost, 2001). The idea is to choose the sample in such a way that one may find variations of the answers from those who are interviewed according to relevant variables or categories. Of course this sample is statistically non-representative but perfectly valid for qualitative analysis. The first step in the method employed is to define the variable that has the most significant theoretical meaning. This was established as the organisation's *hierarchical level.* Four hierarchical levels are defined: strategic management, tactical management, operational management and operational. Next another variable is found relevant: *the orientation* in which the interviewees are working in i.e. process, product or functional orientation. Finally, if interviewees are *technical or non-technical* educated is the last category under consideration although later on this is found non relevant. Hence the sample is chosen to select individuals covering all possible combinations of these variables.

The individuals to be interviewed are selected based on the *snowball* method. That is, the first person in interviewed during which that person is asked who is the person(s) he or she considers the most appropriate to be interviewed next to cover the categories considered. Then the interviewer does the next interview according to this suggestion and thus the interviewer proceeds so forth until the whole sample is covered.

The first interviewee was the site's top manager who suggested that himself and those above him in the organisation should be considered as *strategic management*. In this category most of those who are related to the factory under analysis were interviewed. The top manager also considered that all managers in the next level under him formed what is defined as *tactical management* and thus most of them were interviewed. Some tactical managers suggested some of the operational managers to be interviewed and these suggested a number of operators. When selecting among those suggested persons it was considered whether they were technicians or not as well as their involvement in the orientations always bearing in mind to cover as many variants as possible from these variables

It was important too to take into account the idea of creating a work group that would support the EI project. The objective is to establish a work group for the author to work with in which consists of people from different departments, levels, orientations, and type of education with the purpose of getting information, competence and experience to accomplish the EI project. A total of 22 interviews were carried out from February to April 2001. On average each of them lasted around 1.5 hrs. For each interview firstly the questionnaire was sent by internal post together with a covering letter explaining the reasons and goals of the survey. In this way the interviewee could think about the questions. The questionnaire is standardised in the sense that its focus is one subject: EI and it is based on open questions. Secondly, phone calls were made to get an appointment for the interview and make sure that the interviewees wanted to help. This last is found very delicate for the validity of the interview since the company was passing across a critical situation and the confidence on the interviewer is of the major importance. Finally, the interviews were carried out bearing in mind the conditions needed for its reliability (Trost, 1997).

The data analysis is presented in next section as well as the interpretation of the results. For the validity of the findings some meetings with some of the managers at different levels were performed.

4 DATA ANALYSIS AND FINDINGS OF THE STUDY

Before starting the description of the analysis it is worth mentioning that because of limitations in the length of this paper only some of the main findings are presented. As a starting point the percentage distribution of interviewees per category is shown in Table 1. Although there is a strong desire to integrate the three orientations defined the study showed that the perception at the factory is that they feel that they are working 100% only as FO and partially in the other two orientations. People think these last orientations either are working inefficiently or do not exist at all: "...we try but we cannot reach them yet", "...we have some flows but not a process oriented enterprise" (see Table 2).

Table 1: Distribution of interviewees' category in percentage.

Category	Subcategory	Interviewees (%)
Hierarchical level	Strategic management	23
	Tactical management	27
	Operational management	23
	Operational	27
Orientation	Involved in Functional	45
	Involved in Process	18
	Involved in Product	36
Education type	Technicians	82
	Non-technicians	18

Table 2: Interviewees' affirmative perception in percentage of the orientation's existences at the factory.

Subcategory	Orientation		
	Functional	Process	Product
Strategic management	100	50	50
Tactical management	100	60	60
Operational management	100	70	60
Operational	100	75	62
Involved in Functional	100	50	55
Involved in Process	100	71	43
Involved in Product	100	100	100
Technicians	100	69	61
Non-technicians	100	75	61
All	100	70	62

Concerning the need of working with the three orientations at the same time 21 of the 22 interviewees (95%) recognised such a need: "...yes, we need them be cause they complement each other". Only one person doubted that FO helps. When trying to understand in which way they complement each other is four categories of answers resulted: 19% find them necessary because they each have a different *strategic focus* " ...they support in different way the missions and visions of the company". 40% find the need due to *organisational focus* with arguments as: "...it is a natural way to distribute responsibilities" or "...decisions must be taken with different criteria". 32% think that orientations have different *control focus*: "... they control in medium and long term: products, markets, resources and procedures". Finally,

9% stated *other reasons* such as: "...well, technically they are different and thus must exist".

However, Table 3 shows that FO is not seen as an important orientation at any level. PtO is the most important for higher hierarchical levels while for lower levels the PsO is perceived as the most important. This may be explained by the fact that PtO is closely related to the market place and thus has a strong strategic component while PsO is related to operational matters. Only one fifth of the interviewees find all orientations equally important.

Table 3: The most important orientation in percentage of interviewees

	Orientation			
Subcategory	Functional	Process	Product	All three
Strategic management			80	20
Tactical management		33	50	17
Operational management		60	20	20
Operational		83	17	
All hierarchical levels	0	46	36	18

When looking for *means to integrate* the orientations answers are categorised as those related to the: Information System (IS), Organisation, Technology, Mind-set/attitudes and Strategies as Fig. 1 shows. When splitting the answers according to hierarchical levels it is found that *Mind-set* was pointed out at all levels, *Organisation* is pointed out mainly at higher levels while *IS* at lower levels. It is important to mention as well that *Technology* is perceived as a means to integration only in 5% of the cases.

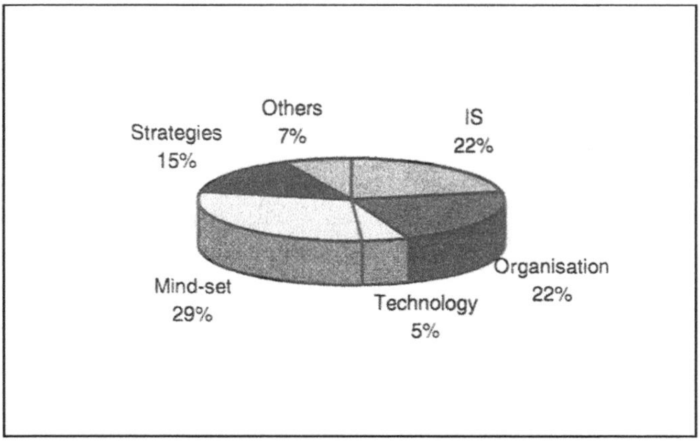

Figure 1: Means to integrate the answers on orientations

Using the same categorisation for the *obstacles*, which prevent the integration of the orientations, it is found that those related to *Organisation* are the ones most pointed out with 29% (see Fig. 2), followed by those related to Strategies, IS and Mind-set with 21.5%, 19% and 17% respectively. *Technology* represents now a higher percentage with 13.5%. When splitting them according to levels: higher levels see obstacles mainly in the *Organisation* and *Strategies* while lower levels in *all of them.*

Concerning what the performance measurements are to assure the integration among the three orientations the following was found:

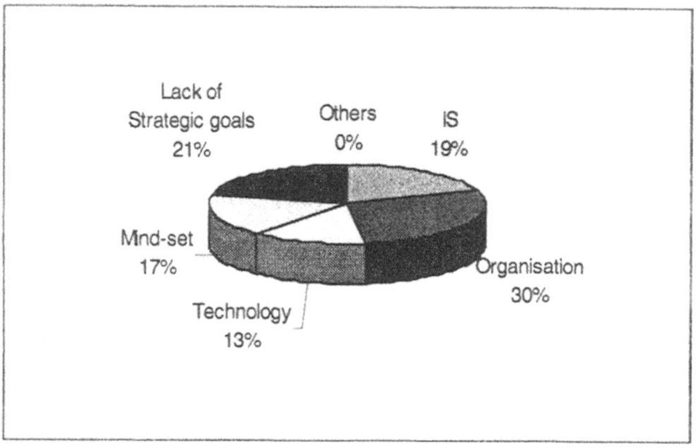

Figure 2: Obstacles to integration

- At all hierarchical levels the company's *key performance indicators* (KPI) are basically considered global indicators of the performance and thus show degree of integration.
- Profits, production yield (percentage of products that are okay) and shipment accuracy (number of orders delivery on time) are mentioned by 90% of the interviewees.
- At low hierarchical levels more importance were given to technical performance measurements such as quality or it surrogate scrap (volume of defective products), blocked invoices (number of invoices blocked due to quality problems when products were delivered to customers), production or utilisation (occupied capacity or load) while tactical and strategic levels focused on financial measurements as profits/losses and costs.
- Other specific indicators must be settled in order to measure both strategic goals for the whole company as well as those per function, process and/or product. A comparison between the planned indicator's values versus the real performance's values is necessary to be done.
- A proposal to control the performance of each orientation is through the cost: cost per function, cost per product and cost per process (planned cost versus real cost).

5 CONCLUSIONS

There is a clear contrast between what the managers of the enterprise define as ways of working and the perceptions of employees who are actually working within the enterprise. While the enterprise managers may suggest certain working models for the employees the employees themselves do not find such models the easiest to implement, operate and manage. Many of the declared models of EI have concentrated on the technological aspects of implementation. This study has highlighted the possibility that organisation aspects, such as people's mind- have a significant role to play. This study indicates that new EI models need to be developed and tested that explicitly addresses these non-technical issues.

People's mind-set/attitudes is seen as the main means to integrate an enterprise although an adequate organisational structure, effective IS and appropriate business strategies are also necessary. Technology is seen more as an obstacle to achieving integration rather than a means to attaining it. This establishes a future research agenda for developing, verifying and validating new EI models.

FO is found important only in relation to the other orientations but not in itself. It is seen as the one that supports the other two taking some decisions or certain responsibilities over the others when necessary. However, PsO is seen as being essential to accomplish operations efficiently while PtO is having its importance based on its strategic nature. PtO uses the other two orientations and might modify them. Therefore all three orientations complement each other and hence it is relevant to research their integration. The integration of the three orientations may use traditional global performance indicators as a means to assess its accomplishment. However, other more specific indicators need to be established.

6 REFERENCES

Bernus P, Nemes L. and Williams T. (1996), *Architectures for Enterprise Integration*, Chapman & Hall.

ENV 40 003 (1990), *Computer integrated manufacturing – systems architecture – framework for enterprise modelling*, CEN/CENELEC TC310/WG1.

ENV 12 204 (1995), *Advanced manufacturing technology – systems architecture – constructs for enterprise modelling*, CEN/CENELEC TC310/WG1.

ISO/DIS 15704 (1999), *Requirements for enterprise reference architectures and methodologies*, ISO TC 184 SC5/WG1.

Kosanke K. and Nell J.G. (Eds.) (1997), *Enterprise Engineering and Integration: Building International Consensus*, Springer-Verlag.

Savén R. (2002), *Integration of Product, Process and Functional orientations: principles and a case study*, Paper submitted to APMS 2002, The International Conference on Advanced Production Management System, 8-13 September, Eindhoven, The Netherlands.

Trost J. (1997), *Qualitative interviews*, Lund: Studentlitteratur, Sweden.

Trost J. (2001), *The questionnaire book*, Lund: Studentlitteratur, 2nd ed., Sweden.

Vernadat, F. (1996), *Enterprise Modelling and Integration – Principles and Applications*, Chapman & Hall.

Challenges to Multi-Enterprise Integration
the EECOMS Experience

William J. Tolone[1], Bei-tseng Chu[1], Gail-Joon Ahn[1], Robert G. Wilhelm[1], and John E. Sims[2]
[1]University of North Carolina at Charlotte, USA, [2]IBM Corporation, USA,
wjtolone@uncc.edu

Abstract: The EECOMS Project is a multi-company and university joint effort to research and develop intelligent, dynamic technologies for integrating supply-chain planning, scheduling, and execution and for enabling the evolution of such multi-enterprise integration solutions. In this paper, we describe several critical challenges to enterprise integration in the form of "lessons learned" by the Project in its effort to develop leading edge multi-enterprise integration solutions. These lessons reflect the human side of enterprise integration, the integral role of security and privacy, and the re-examination/definition of traditional business processes that enterprise integration requires.

1 INTRODUCTION

Advances in Information Technology (IT) have transformed the conduct of business. As IT has matured, more and more business processes have been automated. Recent attention has focused on the integration of individual processes across the business enterprise (e.g. WebSphere, MQSI, Neon, RosettaNet, CommerceNet, OAGIS). Enterprise Integration (EI) refers to the methodologies and technologies that support these efforts. The purpose of this paper is to describe several critical challenges to EI as "lessons learned" by one large-scale effort, the EECOMS Project (NIST ATP 97-05-0020, 1998), to develop leading edge multi-enterprise integration solutions. We begin with a brief overview of the project. Next, we highlight three important lessons learned in regards to Enterprise Integration.

- **Lesson One:** People are Essential Participants in Enterprise Integration
- **Lesson Two:** Security and Privacy are Integral to Enterprise Integration
- **Lesson Three:** Effective Enterprise Integration Often Requires a Re-examination/definition of Traditional Business Processes

We conclude with some reflections on the EECOMS experience, discussing some strengths and weaknesses of a consortium-based approach (involving both industry and academia) to research and develop leading edge enterprise integration solutions.

2 THE EECOMS PROJECT

In 1998, the EECOMS Project was established as the second project managed under the CIIMPLEX joint venture agreement (CIIMPLEX, http://). EECOMS stands for the *Extended-Enterprise Consortium for Integrated Collaborative Manufacturing Systems*. Support for the three-year project originated through a government/private-sector partnership program. Federal support totaling $14.5M came from the Department of Commerce's National Institute of Standards and Technology, Advanced Technology Program (NIST/ATP). The mission of the NIST/ATP program is to strengthen the U.S. economy through high-risk, leapfrog technologies that broaden both participant and national competencies with the potential for broad base diffusion. Private-sector support totaling $15M came from the project's industry partners: BAAN SCS, Boeing, Envisionit, IBM, INDX, Scandura, TRW, and Vitria Technologies.

Key to this government/private-sector partnership was the inclusion of three universities: the University of North Carolina at Charlotte, the University of Maryland at Baltimore County, and the University of Florida. Together, members of the project propose to develop and demonstrate intelligent, dynamic technologies for integrating supply-chain planning, scheduling, and execution and enabling the multi-enterprise integration to evolve in step with changing circumstances. One practical goal was to create the building blocks of a distributed computing environment that accommodate diversity in the processes, practices, and software of supply-chain members. Another was to develop methods, embedded in executing software, for evaluating supply-chain designs and for facilitating collaboratively made changes in those designs. Four research foci were highlighted in this effort. They include: multi-enterprise integrated collaboration support; support for secure multi-enterprise transactions; rule and constraint-based support to knowledge management and integration; and customer scenario identification and design.

The EECOMS project desired multi-enterprise integration solutions that provided not only greater efficiency across supply chains, but also a degree of synergy among supply chain participants and their business processes. In the following, we highlight three key lessons learned from the EECOMS experience.

3 LESSON ONE: PEOPLE ARE ESSENTIAL PARTICIPANTS IN ENTERPRISE INTEGRATION

One of the pillars to the EECOMS research effort was in a technology we described as Virtual Situation Rooms (VSR) (Tolone, 2000). The original objective of the VSR technologies was to create shared information spaces supported by asynchronous and real-time collaboration technologies to provide command and control-like support (following the military situation room analogy) to facilitate the resolution of integration problems by supply chain participants. Early on in the research process it became clear that providing collaboration support merely for exception resolution was insufficient. In fact, it uncovered a fundamental flaw with the current industry held view that enterprise integration is primarily a problem of automation. To underscore the significance of this problem, we offer several illustrations of common occurrences that become problematic when using even the most current automation-centric EI solutions.

3.1 Narrow view of business processes

In general, business processes, while often described as repeatable, are rarely completely prescriptive. Yet, current EI solutions are designed specifically to support activities that are more prescriptive in nature.

Consequence: "Exceptional" activities, while often handled best as part of "normal" business processes, end up removed from these processes. This leads to business processes that are fragmented between prescriptive and exceptional activities though it is more appropriate and effective to handle these activities together (Hammer, 1996). Moreover, as business processes are further deconstructed so that they are more amenable to automation, they are simultaneously becoming more distributed. Timely and accurate awareness to the state, progress, participants, responsibilities and data relative to these processes, is increasingly essential but more difficult to maintain. As processes become more distributed, the lack of planning for human participation usually means that human needs for communication and collaboration media are not considered.

3.2 Disconnect between people and business processes

Current EI solutions tend to remove people from or inadequately incorporate people into business processes. (Billings, 1997) Automation increases the speed at which business data can be processed. However, increased data processing speed alone does not reduce the time needed or the effectiveness of the decision-making. The key to achieving increased quality, effectiveness and speed is the timely and appropriate participation of people.

Consequence: Human participation in business processes becomes increasingly difficult because current EI solutions can cause people to be relegated to ineffective functions and increase the dependency of an enterprise on automated decisions. If effective human roles are not properly maintained, a greater frequency of misjudgments is likely. (Tolone, 1998) But maintaining proper roles is extremely difficult, error prone, and time consuming, requiring answers to questions such as "What data are relevant?" "How should they be represented?" and "Who should be involved?" Unfortunately, decision-makers are often provided too much data, as well as data that are insufficient, untimely, improperly formatted, or simply incorrect. This results in people having inaccurate mental models upon which decisions are made. For example, in the aviation industry as automation was added to the cockpit, there were times when "pilots have simply not understood what automation was doing, or why, or what it was going to do next." (Billing, 1997) Similar challenges face enterprise integration, as people cannot be eliminated from decision-making processes.

3.3 Scope Expansion

Current EI solutions increase the scope of business processes while ignoring the inherent complexities introduced by this change in scope. As a result, the role of decision-makers is extended, for example, up and down the manufacturing supply chain or across a wider range of caregivers in healthcare.

Consequences: First, this expansion fosters a lack of understanding of the impact of decisions on both upstream and downstream activities. Prior to the introduction of EI solutions, the effects of activities were far more localized. Second, this confusion increases the difficulty of assigning management responsibility within and across business processes. That is, it is difficult to answer the question "Who is responsible and in control?" Is it the EI solution or a person? How do we answer this question now that the business processes extend across enterprises? True integration of humans and automation actually requires solutions that eliminate disconnects and seamlessly support the balance of control between automation and human interaction. (Gelernter, 1991) Third, the extension of decision-making scope usually ignores the importance of communication and collaboration to support these newly extended business processes. (Hammer, 1996, Tolone, et al, 1998).

3.4 EECOMS Solutions

Thus, the view that enterprise integration is equivalent to the automation of repeatable business processes is insufficient. Rather, enterprise integration solutions must promote an effective mix of human decision-making and automation. In fact, synergistic solutions require human participation because ultimately it is people that bring synergy to enterprise integration. Automation technologies will never produce benefits greater than the sum of their parts because automation is fundamentally about efficiency and not synergy.

How, then, did this view impact our research? This growing understanding of the human side of enterprise integration affected Virtual Situation Room research and development in four important ways.

First, VSR became an equal participant within the EECOMS integration architecture. Traditionally, collaborative systems, particularly real-time collaboration support, were islands of technology. The VSR research team, however, incorporated the VSR technologies into the integration architecture in such a way that enabled VSR to be an active participant in multi-enterprise transactions.

Second, we began to see collaboration support not as a fixed set of services or facilitates but an evolving, pluggable set based on business process requirements. Consequently, VSR technologies emerged

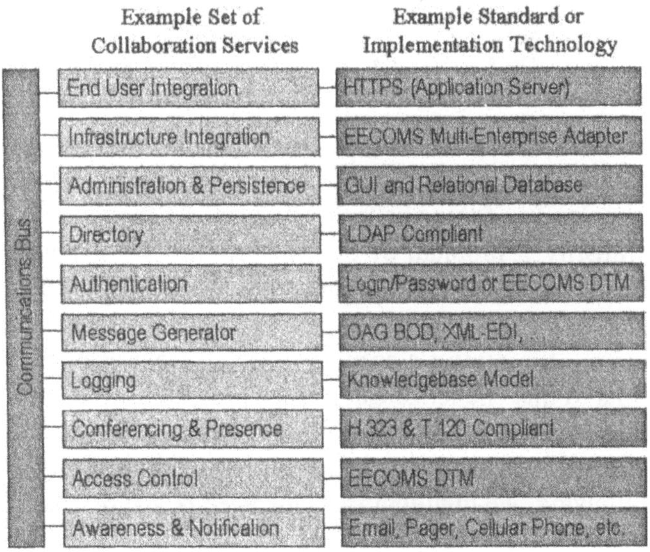

Figure 1: VSR Conceptual Architecture

not as a tightly coupled, monolithic application but rather as a loosely coupled component-based application. Fig. 1 depicts a high-level version of the VSR conceptual architecture and set of collaboration services, as it existed near the conclusion of the project.

Third, collaboration support, and thus VSR, share the same security concerns and problems that face multi-enterprise application integration. Thus, VSR was designed to leverage directly the security results of the EECOMS Project (see the following lesson). Finally, a multi-enterprise collaboration architecture must pervade the integration architecture rather than be a participant within it, i.e. collaboration and integration must be design cooperatively from the ground up so that they may be seamlessly integrated, leveraging common services. Just as security and privacy cannot be add-ons, the same is true for collaboration. While VSR research addressed each of these concerns to some degree, each constitutes a research problem whose magnitude far exceeds the capabilities of a single three-year project; and thus, are a part of a continuing research effort at UNC Charlotte.

To summarize, then, through the EECOMS Project, we learned more deeply that enterprise integration is not solely a problem of automated inter-enterprise transactions (i.e. automatic data synchronization), but truly an enterprise synchronization problem, where enterprises support business processes as the integration of people, applications, practices, and data transactions. This vision is fundamentally different than the automated/data-centric approaches of the past and it provides an appropriate framework for the next generation of EI.

4 LESSON TWO: SECURITY AND PRIVACY ARE INTEGRAL TO ENTERPRISE INTEGRATION

One of EECOMS' principle objectives was to send information across organization boundaries. Over the life of the project, our appreciation for the complexity of this challenge evolved. At the outset, secure integration was primarily a problem of enabling application adapters to communicate securely across enterprise boundaries. While clearly essential to secure integration, this problem is just the first in a series of challenges that must be faced. In fact, the project's security research team was able to attain effective results to this challenge early in the project. Through that effort, though, additional security and privacy issues that are essential to secure integration emerged, resulting in a reformulating of the security research agenda. In the following, we highlight this new agenda and discuss in more detail an open security and privacy issue that emerged near the end of the project.

As the EECOMS Project refined its security agenda, proper authentication and authorization of multi-enterprise transactions emerged as a central issue. Current commercial systems, then and now, do not offer satisfactory solutions to these issues for several important reasons. In today's dynamic work environment with frequent changes in personnel and responsibilities, it is very difficult to manage passwords and access rights within a single organization. It is harder to track users across organization boundaries. Partly due to the challenges of large number of users, most systems do not implement fine levels of access control. An important requirement for integrated multi-enterprise architecture is a model that would allow distributed and scalable management of access rights. Such a model must be easily tied to legal policies where companies decide who and what information should be shared as well as providing an easily traceable audit trail to enforce access policies.

4.1 EECOMS Solutions

During EECOMS we developed a distributed trust management access control model based on digital signatures as well as delegation of access privileges (Chu, Tan, 2000). We believe recent developments in attribute certificates and Privilege Management Infrastructure (PMI) provides the right tools towards establishing such a scalable access control model.

PMI proposes a certificate-based scalable and interoperable authorization solution to enterprise integration (ITU, 2001, Farrel, Housley, 2001). However the roles model in PMI is so primitive that it lacks some advanced components such as role hierarchies and constraints that are core components in role-based access control (RBAC) reference models (Sandhu, et al, 1996). RBAC has been acclaimed and proven to be a simple, flexible, and convenient way of access control management (Sandhu, et al, 1996, Ferraiolo, et al, 1995). Our objective is to investigate how RBAC components can be designed and realized on PMI so that we may enhance authorization services to enterprise integration using a notion of PMI's roles model. In addition, the necessities of security architectures for presiding over the marriage of these two technologies are explored. We also demonstrate the feasibility of the architectures by providing the proof-of-concept prototype implementation.

PMI is a collection of attributes certificates, attribute authorities, repositories, entities involved such as privilege asserters and verifiers, objects, and object methods (ITU, 2001). The attribute certificate binds entities to attributes, which may be the entities' role or group information. PMI introduces its roles model by defining two different types of attribute certificates: *role assignment certificate* and *role specification certificate. Role assignment certificate* has the binding information of an entity and its associated roles, while *role specification certificate* contains the binding information of the role and its associated privilege policies. In RBAC, roles are defined as job functions or job titles within an organization, users are associated with appropriate roles, and permissions are assigned to roles. It is the roles associated with the users that restrict access to objects, not the ACLs on the object. Thus RBAC makes it simpler and more convenient to manage permissions, reducing the complexity of administrative tasks. It also enables centralized and consistent management of access control policy (Ahn, Sandhu, 2000). RBAC and PMI are complementary, positively producing an alternative authorization solution to enterprise integration.

We developed several system architectures for authorization services based on RBAC and PMI. Push and pull modes in handling attribute certificates introduce four different system architectures. Fig. 2 shows one of these architectures. It consists of three components: *privilege asserter*, *privilege verifier*, and *PMI attribute authority*. *Privilege asserter* is a client application working on behalf of an individual. The individual can request and retrieve *role assignment certificate* from *PMI attribute authority*, or request services (such as access requests to protected resources) using this client application.

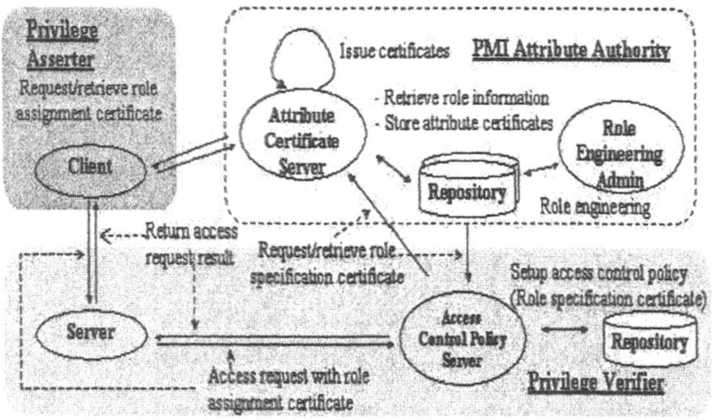

Figure 2: Secure System Architecture (Push mode)

Privilege verifier is composed of server, access control policy server, and repository. Server maintains protected resources or applications. When a client wants to access the server, the server asks the access control policy server whether or not the client has appropriate access privileges. Repository is the data storage for caching the received attribute certificates. As an access control management server, the access control policy server handles access control decisions based on the specified access control policies. Like *privilege verifier, PMI attribute authority* has three components: attribute certificate server, repositories, and role engineering administration. Intuitively, attribute certificate server manages all requests of *role assignment certificate* and *role specification certificate* issuances. After issuing those certificates, it stores them into a publicly accessible repository. Role engineering administration is an entity performing role-engineering tasks such as role management, user-role assignment, policy specification, and so on.

In summary, PMI is an emerging authorization infrastructure, providing an interoperable and scalable privilege management solution through the use of attribute certificates. RBAC can add simplicity, flexibility, and convenience to PMI through its advanced components such as role-hierarchies and constraints. Privilege management is becoming one of the critical components in designing, developing, and deploying enterprise applications. We believe that our work contributes to the enterprise integration as well as RBAC research communities.

5 LESSON THREE: EFFECTIVE ENTERPRISE INTEGRATION OFTEN REQUIRES A RE-EXAMINATION/DEFINITION OF TRADITIONAL BUSINESS PROCESSES

The EECOMS project planned to demonstrate its multi-enterprise technology by implementing various integrated customer scenarios. One can view these scenarios as expanded use cases integrating various enterprise applications to support business processes. During the project it became obvious that (a) multi-enterprise integration as the "gluing" of existing enterprise-level processes in most cases leads to a less than optimal integration solution, and (b) the combination of the human collaborative technology, automated business rules, and the underlying security technology could provide a novel, synergistic solution to multi-enterprise integration.

5.1 EECOMS Solutions

As a result of this observation, a team led by our customer technology partners identified and designed several customer scenarios. Most notable among this group due to its unique integration of project technologies and design for multi-enterprise integration came to be known as Scenario X. This scenario extended the well-known "available to promise" business process in the following ways. First, it allowed a human team to determine at design time the cost and supply in a series of "what-if" fulfillment to promises. Thus, this multi-enterprise integration scenario runs counter to the trend of gaining efficiency through automation and the elimination of human participation. Rather this scenario was designed with people as central to its effective and efficient execution. Second, this scenario leveraged input from multi-directional enterprise rules thus enabling the human team the ability to role-play during "what-if" analysis (e.g. participate as: the buyer in a multi-tier supply chain; the collector of critical business rules from third and forth party enterprises; etc.). Third, though not necessarily an extension to the "available to promise" process, demonstration of Scenario X (and others) was completed within an environment that leveraged the requisite security and privacy advances identified and developed through the project. By combining these results iteratively, the multidiscipline product design team functioned as an integral participant within a secure multi-enterprise integration process and understood more efficiently and effectively the availability consequences of their designs, actions and plans.

6 EECOMS PROJECT REFLECTIONS

As we look back on the EECOMS experience we reflect on a very successful and rewarding experience. Through the research and development efforts, a better understand of the research problems, and requisite solutions, were gained. The challenges we faced were not unlike those faced by many large research and development efforts. Yet, our greatest challenges were also our most unique and at the same time our greatest resources. The partnership among industry (including competitors), government, and academia constantly challenged the project while simultaneously providing a rich and diverse background of expertise and experience upon which the project constantly drew.

We confronted early a problem that faces many large research and development projects like EECOMS, specifically those with many commercial partners and universities. This problem is a heightened tendency to create new or abandon all together sound project and business processes. This lapse of project management is done under the guise of reducing overhead, fast tracking, breakthrough thinking, and freedom for research. Yet, this decision can cause extreme trauma to the very people that it is suppose to help. Individual project members often find it very difficult to perceive and react timely to new or changed technical requirements, inter-disciplinary dependences, risk mitigations, or priority scope modifications. When changes occur in project scope and direction, e.g. EECOMS security research effort, good project management practices and good business processes allow a clear direction and a firm team commitment to the change. Abandoning sound processes make the measurement of what is accomplished and what is needed difficult, if not impossible.

In addition to the diversity of our partnership and our adherence to sound project management practices, we also found our decision to drive research and development from a business scenario prospective to be immensely beneficial. Our approach was based on the use case approach advocated by Jacobson (Jacobsen, et al, 1992) and the Unified Software Process (Jacobsen, et al, 1999), although adapted somewhat to focus on multi-enterprise use cases, or what we called customer scenarios. Lesson Three summarizes some of the lessons learned from this approach.

Finally, one of the most beneficial aspects of the EECOMS experience was the taking of research results to an independent technical advisory board and to conference room deployments for review. These regular deployments at partner sites and advisory board reviews provided a valuable source of feedback from both customers and research experts, respectively. These efforts play an invaluable role in enabling partners to commercials research solutions more quickly.

7 SUMMARY

In this paper, we presented three "lessons learned" by the EECOMS Project about the challenges to multi-enterprise integration. These lessons reflect the human side of enterprise integration, the integral role of security and privacy, and the re-examination/definition of traditional business processes that enterprise integration requires.

The EECOMS Project completed its operation in 2001. While the Project Partners' commercialization plans are proprietary and confidential, it can be generally stated that the migration of research results, which began within a year of the Project's inception, is continuing and in specific instances having significant impact on the quality and effectiveness of Partner solutions.

8 REFERENCES

Ahn, G. Sandhu, R. (2000), *Role-based Authorization Constraints Specification*, ACM Transactions on Information and System Security, 3(4).

NIST ATP Project, 97-05-0020, (1998), *EECOMS: Extended Enterprise Consortium for Integrated Collaborative Manufacturing Systems*, CIIMPLEX Consortium. See www.ciimplex.org.

Billings, C. (1997), *Aviation Automation: The Search for a Human-Centered Approach*, Lawrence Erlbaum Associates, Publishers, Mahwah, New Jersey.

CIIMPLEX http://www.ciimplex.org.

Chu, B. Tan, K. (2000), *Distributed Trust Management for Business-to-Business e-Commerce Security*, In *Proceedings of the ACME 2000 International Conference.*

Ferraiolo, D. Cugini, J. and Kuhn, D.R (1995), *Role Based Access Control: Features and Motivations*, In *Annual Computer Security Applications Conference*, IEEE Computer Society Press.

Farrell, S. Housley, R. (2001), *An Internet Attribute Certificate Profile for Authorization,* PKIX Working Group.

Hammer, M. (1996), *Beyond Reengineering*. Harper Business, New York.

ITU-T Recommendation X.509, (2001*), Information Technology: Open Systems Interconnection - The Directory: Public-Key And Attribute Certificate Frameworks*, ISO/IEC 9594-8.

Gelernter, D. H. (1991), *Mirror Worlds or: THE DAY SOFTWARE PUTS THE UNIVERSE IN A SHOEBOX...HOW IT WILL HAPPEN AND WHAT IT WILL MEAN.* Oxford Univ. Press.

Jacobson, I. Booch, G. Rumbaugh, J. (1999), *The Unified Software Development Process.* Addison Wesley.

Jacobson, I. Christerson, M. Jonsson, P. Övergaard, G. (1992), *Object-Oriented Software Engineering: a Use Case Driven Approach.* Addison-Wesley.

Sandhu, R. Coyne, E.J. Feinstein, H.L. Youman, C.E. (1996), *Role Based Access Control Models*, IEEE Computer 29 (2).

Tolone, W. J. Chu, B. Long, J. Wilhelm, R. G. Finin, T. Peng, Y. Boughannam, A. (1998), *Supporting Human Interactions within Integrated Manufacturing Systems,* International Journal of Agile Manufacturing, Vol. 1(2).

Tolone, W. J. (2000), *Virtual Situation Rooms: Connecting People Across Enterprises for Supply Chain Agility*. Journal of Computer-Aided Design, Elsevier Science Ltd. Vol. 32(2).

Practices in Knowledge Management in Small and Medium Firms

Raul Poler Escoto[1], Angel Ortiz Bas[2], Guillermina Tormo Carbó[1], and David Gutierrez Vañó[1]
[1]Escuela Politécnica Superior de Alcoy, Spain, [2]Universidad Politécnica de Valencia, Spain, gtormo@omp.upv.es

Abstract: Changing market environments and increased competition require significant changes in the way of working in all industrial enterprises. These changes become even more pronounced in the small and medium enterprises and are even less likely to met by theses organisations, due to their specific nature. This paper describes a project in the textile industry that has been carried out during one year and a half in an area situated in the Comunidad Valenciana (Spain).

1 INTRODUCTION

Two hundred years after the industrial revolution that changed the establish world order, we are again under a big transformation that needs the adaptation of the enterprise to the new environment. This requires for all the firms to update their enterprise concepts as well as their business and operational models. Even more important they have to enhance their management knowledge that allow them to compete effectively in an environment which is more and more uncertain and very prone to changes.

One of the possible actuation frameworks, for giving answers to some of the aspects which enterprise must face, is that of Enterprise Engineering and Integration projects. Enterprise engineering and integration frameworks are valuable to structure, plan and guide improvements in enterprises (Vernadat, 1996; Bernus et al., 1996; Kosanke, Nell, 1997).

Enterprise Engineering and Integration projects will improve the flows of materials, information, decisions and control throughout the organisation and thereby provide flexibility and adaptability to chance. Those projects will identify how to connect the functions with the information systems, physical resources and human resources, with the aim of improving communication, co-operation and collaboration. Thereby, the enterprise will function as a whole and every part of the enterprise will contribute to the strategy (Ortiz, 1998).

Over the last year's different enterprise engineering and integration architectures and methodologies have been developed (CIMOSA, PERA, GIM, IE-GIP...). Their analysis suggests that one of the critical elements for the success in EEI projects are related with Human Resources and Organizational aspects. In this sense, the EEI architectures and methodologies do not go deeply into how to manage the transformation adequately. They do not face the required change management needed to help the people and the organisation to assume such difficult projects. Likewise, they do not identify the knowledge that is needed for the people involved in the transformation, to carry out the different business processes in the enterprise.

The present research has been developed under the framework of IE-GIP architecture and methodology, trying to go deeply into management change and Knowledge Management for improving the result of IEE project, focusing on team roles.

2 IE-GIP METHODOLOGY

IE-GIP (*Enterprise Integration – Integrated Management of Processes*, Spanish acronyms) (Fig. 1) is an EEI proposal, which combines methodological aspects of PERA (*Purdue Enterprise Reference Architecture*) and architectural aspects of CIMOSA (*Computer Integrated Manufacturing – Open System Architecture*), introducing additional elements wherever necessary (Ortiz, 1998).

The IE-GIP methodology has the objective to co-ordinate the work and information flows throughout the organisation for which it is necessary to reach either horizontally and vertical integration and to ensure physical integration.

Horizontal integration will be achieved through the analysis of the sequence of the activities that are carried out as part of the business process leading to the desired result. For each activity the principal elements must be identified; the functionality to be provided, the information needed, the resources (humans and machines) involved and the relations to the organizational framework.

Vertical integration will be obtained by identifying the relationships between the different enterprise processes. Relations that can take place among processes at the same level or at a different level of an enterprise operation (Strategic, Tactical and Operative).

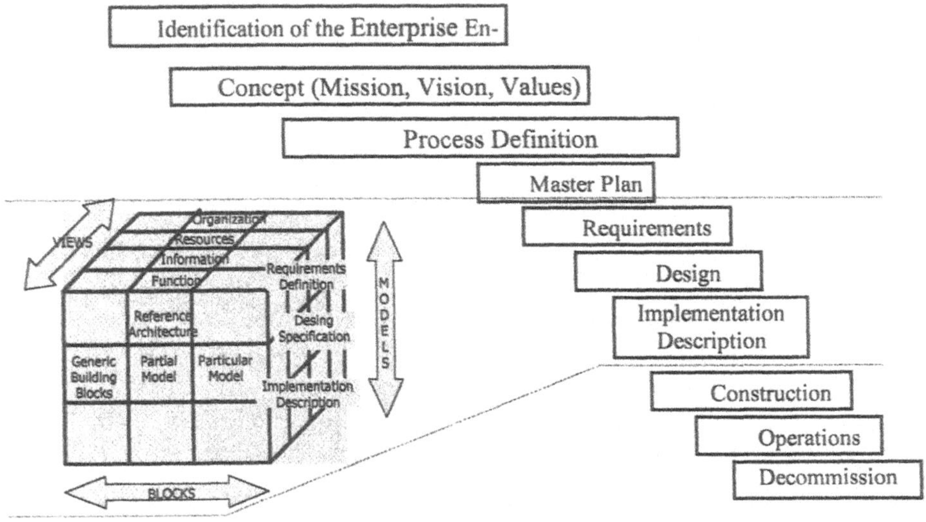

Figure 1: IE-GIP Methodology

The analysis of the different IE-GIP views (Function, Information, Resource and Organisation) will identify the kind of knowledge embedded in them. But modelling in this context means creating an *explicit representation*, usually computable, for the purposes of understanding the basic mechanics involved. One often uses that understanding to measure, manage and improve the process.

This kind of modelling is based on a relatively predictable view of the business environment and all its elements can be expressed in a deterministic way. The model is based on the assumption that all relevant knowledge in an enterprise, including tacit knowledge, can be stored in computerised databases, software programs, rules and procedures (Malhotra, 2002).

However nowadays, the business environment is unpredictable and unstable and it requires a dynamic model rather then a static one. One that is based upon ongoing reinterpretation and creation of knowledge to adjust better to the unpredictable future. Yet, process models do not capture the soft elements involved in the business processes as culture, tacit knowledge, human behaviour…

As has been pointed out earlier, one of the most critical and important aspects in EEI projects are the Human resources. These are treated by conventional EEI architectures as resources. Required personal competencies can be defined for running a certain activity, but no considerations about experience, motivation, know-how, culture, training, … are proposed.

The dynamic representation of knowledge will provide a more realistic model integrated within human and social interactions.

Processing of knowledge through the machinery of information technologies may still be represented by simplified, highly routine, and, structured forms that allow pre-definition, pre-programming and predetermination of data inputs for achieving pre-specified performance outcome. In contrast, human sense making is an active, affective, and dynamic process influenced by culture, motivation, leadership, commitment, creativity, and innovation of individuals and groups.

3 THE RESEARCH PROJECT CONTEXT

The Intermediate regions in Spain are composed mainly of SME type organisations and the most representative industry in our area is the textile industry (more tan 50% compared with the other industries). These organisations have an average of about 30 employees.

Research carried out by the OCDE (Organization for Economic Cooperation and Development), (2001) show a number of development obstacles due to the size of the firms:

- Difficult to increase the professional competence,
- Functions such as R & D cannot be carried out efficiently inside the firm,
- Outsourcing is common practice in the industry and the resulting attitude between firms usually is local competitiveness versus co-operation,
- Local services even of advanced suppliers are poor in terms of quality and quantity,
- Poor communications infrastructures,
- Shortage of skilled labour,
- Inadequate links between universities and firms, both on the employer and employee level.
- However these organisations have several success factors as well:
- Great entrepreneurs,
- Experienced and specialised workforce,
- High flexibility.

In addition, the Textile association, the Technological Textile Institute (AITEX) and the University that act like a specialised centres, providing training and co-ordination for the firms. However, it is very necessary for the individual firms action focus on:

- Address process modernisation by modifying manufacturing practices and focusing on product differentiation through design and quality,

- Update business and operational models, concepts and traditional management knowledge to allow for effective compete in an environment which is more and more uncertain and changing,
- Change management practices. E.g. introduce innovation procedures,
- Increase intra and inter co-operation networks within the industry.

4 PROJECT RESULTS

In this project five important SMEs have participated. In each of them the same pilot domain and research methodology has been developed. The research project focussed on the identification of the soft issues. The objectives were to promote improve communication, co-operation, and co-ordination within this enterprise so that the enterprise behaves as an integrated whole, therefore enhancing its overall productivity, flexibility and capacity for management of change.

The focus of the project was on increasing competitiveness of local regional firms, in four different directions:

- Increase productivity,
- Manage and promote innovation,
- Share knowledge through collaboration and co-operation,
- Stimulate the growth of new firms.

Several methods in the frame of the IE-GIP methodology have been applied in the project. Using an action-research approach (interviews, questionnaires, own observation, ...)

Five different human related aspects have been identified and evaluated in terms of barriers and of potential solutions. These aspects are:

People and Mangers aptitude, Company Organisation, Culture and Technology. The main results obtained so far are described below.

4.1 Intra Organisational Barriers

People: Lack of trust in others and their capabilities, lack of sharing knowledge. Reasons: Self-satisfaction, Personal fear and anxiety (vulnerability and inadequacy).

Managerial aptitudes: Lack of leadership and lack of ability to change,

- High-level management frequently separated from the day-to-day business of the enterprise. They look at the organisation mainly from the perspective of numbers and financial statements,
- Companies are operate as static machines rather than a living (evolving) systems,
- Managers don't want to lose control,

- Top managers present innovation initiatives with the result that they are only marginal effective,
- Middle managers have a critical role because they are in the centre of vertical and horizontal information flows coming from top management and front-line employees as well as from their fellow managers,

Organization: Current processes and structures do not provide for knowledge flow. Organizations are viewed as rigid hierarchies rather than communities of practice. E.g. Initiatives, which are driven from the top, have only marginally results.

Culture. Lack of commitment. Employees do not feel to be involved in the firm and do not care much about their work, except to make a living.

Technology. Lack of trust in technology due to past failure and incomplete systems.

4.2 Potential drivers and structural facilitators

People - Labour climate. Co-operation and collaboration vs. competition.

- Teamwork and team roles will help to run these long-term projects. Based on Belbin Methodology (Belbin, 1993).
- Leadership should be shared among the team.

Culture. Deep change comes only through real personal growth and through *learning, unlearning and re-learning*. An explicit reward system will improve people commitment. Internal and external equity should be applied in this reward system (Adams Theory, (Adams, 1976)).

Managerial System. As Senge, (1996) argues, "We need to think less like Managers and more like biologist".

- Focus is on Middle managers because they act as a bridge between the vision and ideals of the top management and the realities of business of the front line workers. They synthesise the tacit knowledge of both top management and front-line employees, make it explicit, and incorporate it into new technologies, product and programs.
- Adopt middle-up-down management.

Organization and relation system. The process of changing a relationship is very complicated. It required a sense of openness, a sense of reciprocity, even a kind of vulnerability. One must be willing to be influenced by another team person in the discussion and debate for fixing objectives and methods.

Technology. Some technology applied works in high levels of the organisation but it doesn't work in front line workers. A new way should be looked for capturing the front line worker knowledge (tacit knowledge mainly).

5 CONCLUSION

In each of the five important SMEs the same pilot domain and research methodology has been developed. In the following points the main points found, for the success of the project, are briefly described:

- Start with defining a "PILOT" Domain in each firm involved in the project. This domain should be composed by top manager, middle manager, supervisor and front line workers. Experimentation is critical to testing knowledge capture, codification, and transfer methods, encompassing both quantitative and qualitative measurement process.
- Be aware in order to make the group work (Compliance or Commitment). Personal enthusiasm is the initial energise of any change process and enthusiasm feeds on itself.
- Organizational structure (or lack of it) can have a direct affect on the collaborative. Being clear about what are the roles and responsibilities of each member.
- Analysis what are the self-reinforcing processes and the limiting processes that take place and can keep the growth of the pilot group? It should be paid special attention to the limiting processes (an important invest of time is required). As Senge, (1996) says: "All growth in nature arises out of interplay between reinforcing growth processes and limiting processes".
- Members of the pilot group need enough support, coaching, and resources to be able to learn. It is worthy to given them enough control over their schedules to give their work the time that it needs.
- Training people in new skills, but those skills that they can need in the day-to-day real work.
- Look for a way of measure the correlation between results and actual behaviour. In the short term it is better to start with easy and actable goals. No traditional metrics can be utilised.

In this project, the way of building a new view (*Knowledge View*) is still being analysed, the necessary phases and the role that humans should play in each of these steps with the purpose of giving guidelines that allow us to complete successfully this kind of projects.

All of these are the result of an action-research approach and have been validated and are still being refined in enterprise projects.

6 ACKNOWLEDGMENT

This research has been developed in the framework of a project funded by the Government of the Comunidad Valenciana, as part of the program for the development of new projects of Research and Development, Emergent teams of R&D and it is titled "Identificación y Modelización de Procesos, Determinación de Parámetros Indicadores, Gestión del Cambio y Gestión del Conocimiento en Integración Empresarial", Ref. GV00-134-11.

7 REFERENCES

Adams, J., Freedman, S, (1976), *Equity Theory Revisited: Comments and Annotated Bibliography*. Academic Press, N.Y.

Belbin, M. (1993), *Roles de Equipo en el trabajo*. William Heineman. London.

Bernus P. Nemes L. William T.J. (1996), Architectures for Enterprise Integration. Chapman & Hall.

Kosanke K. Nell J.G. (Eds.), (1997), *Enterprise Engineering and Integration: Building International Consensus*, Proceedings of ICEIMT'97, International Conference on Enterprise Integration and Modeling Technology, Torino, Italy, Oct. 28-30, Springer-Verlag.

Malhotra, Y. (2000), *Why Knowledge Management System fail? Enablers and Constrains of Knowledge Management in Human Enterprise*, http://www.yogeshmalhotra.com.

OCDE, (2001), *Examen de la OCDE sobre las comarcas centrales de la Comunidad Valencia*, Universidad Politécnica de Valencia. ISBN: 84 – 9705-031-2.

Ortiz Bas, A. (1998), *Propuesta para el desarrollo de Programas de Integración Empresarial en Empresas Industriales. Aplicación a una Empresa del Sector Cerámic*, Tesis Universidad Politécnica de Valencia.

Senge P. (1996), *The ecology of leadership*, Leader to leader, No 2.

Vernadat F.B. (1996), *Enterprise Modeling and Integration: principles and applications*. Chapman & Hall.

Component-Based Automotive Production Systems

Richard H. Weston, Andrew A. West, and Robert Harrison
Loughborough University – MSI Res. Inst., r.h.weston@lboro.ac.uk

Abstract: The paper explains how 'process aware machine components' have been developed so that they can be reused as building blocks of 'in production' assembly and transfer machine elements used by a global consortium of companies making automotive products. It explains how computer executable models of components, alternative composition of machine components and responsive production lines provide common representations of production requirements and candidate solutions for reuse by members of distributed engineering teams. The findings are generalised in a philosophical discussion about 'change capable' systems and the current and future role of EM and EI in producing 'pro-active systems' capable of rapid 're-composition' and thence rapid and effective 'behavioural change'.

1 PRODUCT AND PRODUCTION SYSTEM DECOMPOSITION IN SUPPORT OF GLOBAL MANUFACTURE

The paper considers engineering processes used by a global consortium of businesses. A range of car engine types (known as I4/I5) was rationalised to facilitate mass customisation. Fig. 1 illustrates the product decomposition agreed by a number of car manufacturers who normally compete for business. For the new I4/I5 engine product range these manufacturers will collaborate and use these product modules as standard components of different makes and models of car.

The I4/I5 product decomposition provides new opportunities to standardise and geographically distribute production and logistical activities involved in engine manufacture. The collaborating partners will enact these simplified production and logistical processes globally and thereby will supply I4/I5 engine variants in Europe, the Americas and Asia.

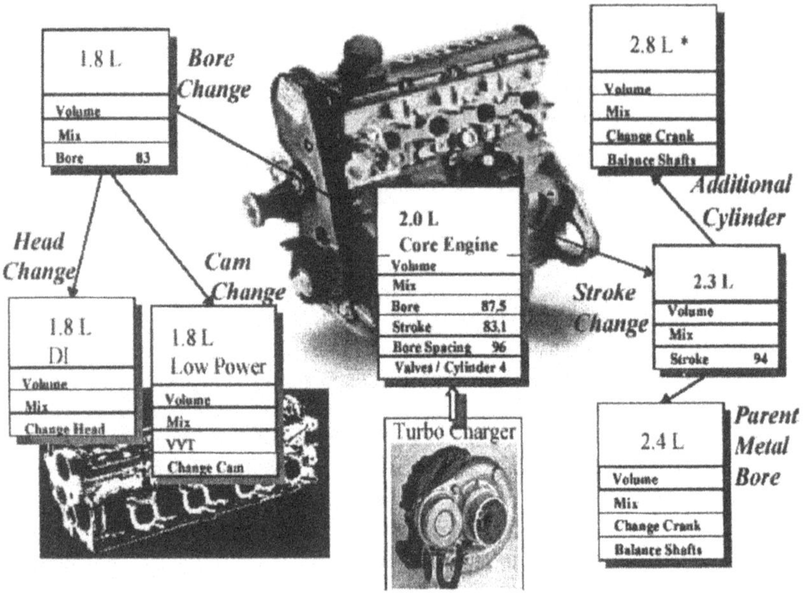

Figure 1: I4/I5 Product Decomposition and Rationalisation Achieved by the Consor-

1.1 Stage 1. Manufacturing System Decomposition

To achieve standardisation of engine product design and to rationalise associated production processes, the collaborating car manufacturers have organised and managed the set of large-scale engineering projects (Fig. 2) (Harbers, 1996). It

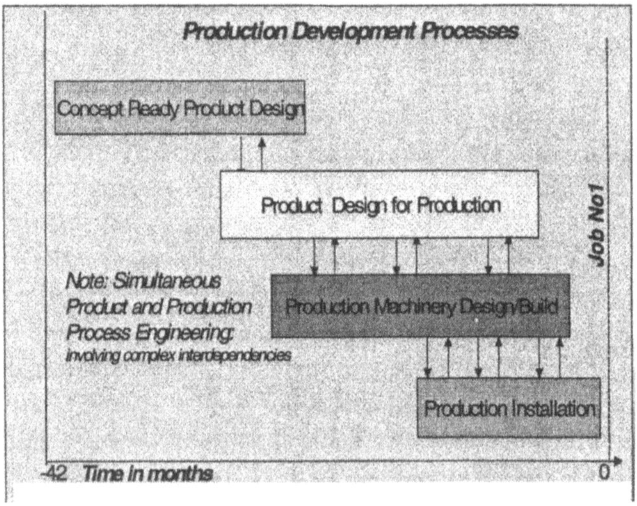

Figure 2: I4/I5 Product and Production Machinery Engineering Processes

has been necessary for many activities involved in these projects to proceed in a concurrent manner, in order to compress lead-times. The I4/I5 engineering programme needed completing over a 42-month timeframe. At completion products will be sold to customers.

Naturally the engineering activities and processes involved are complex in their own right. Also activities and processes cannot wholly be decoupled from each other. Dependencies take a variety of forms typically involving cause and effect relationships that may change in space and time and have circular knock-on effects, one on another.

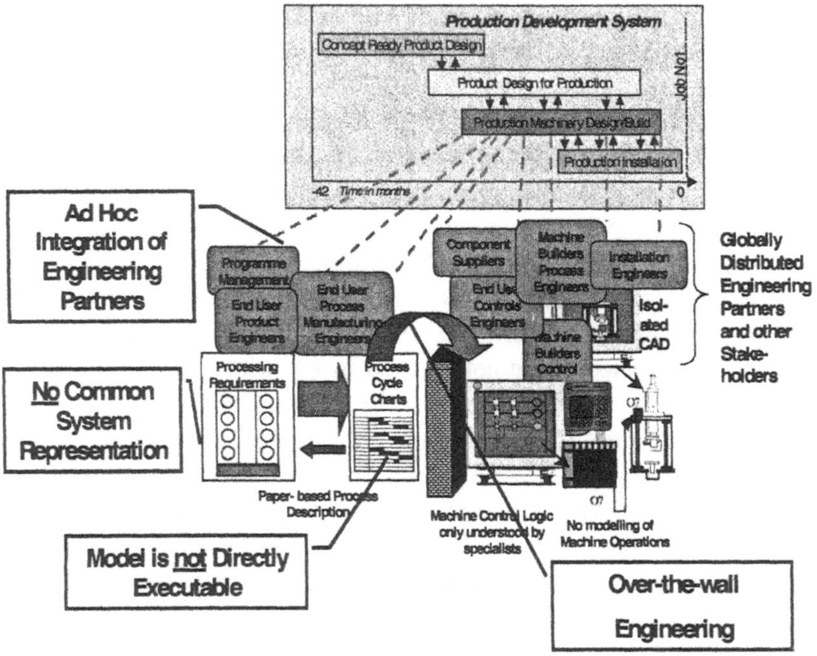

Figure 3: Current Production Machinery Design and Build Practice and Actors

The picture of dependency networks that link concurrent engineering processes to their products (which in this case are production and logistical processes needed to make automotive products) is further complicated because: (1) Necessarily engineering processes will be resourced by various teams of people, often with responsibilities aligned to different organisational units; (2) Engineering processes have a finite lifetime (up to 42 months in this case). Because of interdependencies making late but necessary change to the design and implementation of products, production processes (and associated production machine systems) can invalidate earlier engineering decisions and actions. This will result in re-engineering work and can impact significantly on costs and lead-times. I4/I5 production processes alone can involve over 5000 production line unit operations. Production facility development activities and flows are even more complex. Consequently change can propagate through and impact on a very large number of production engineering activities.

Essentially ad hoc integration methods and mechanisms are currently employed to support the people who enact I4/I5 engineering activities. As illustrated in Fig. 3, machine and control system design and implementation is carried out by people who have various engineering roles and perform these roles using a fragmented set of specialist tools and methods that include computer aided design (CAD) tools, ladder logic programming tools and structured design and diagnostic and coding methods. The integration methods deployed to co-ordinate the use of these heterogeneous tools will at best be informally and uniquely specified (as best practice) in those companies responsible for production machine design and build. They do not utilise a *common* representation or visualisation of production machines throughout the lifecycle of machines, i.e. their design, analysis, implementation, testing, maintenance and reuse lifecycle. Nor is there any overall computer-executable model capable of supporting 'what if' analysis of alternative machine designs and behaviours (i.e. machine state changes over time). Consequently current production machinery design and build processes used in the automotive sector are not realised in a flexible way, in the sense that little support is provided for the reconfiguration and reuse of production machinery as products. The present engineering process used to design and build production machinery is also largely paper-based.

The automotive manufacturers and their partner machine builders are already aware that use of a suitable production system decomposition can help reduce the levels of complexity involved in: (a) engineering new products like I4/I5 and their production systems and (b) achieving and managing product and production facility change during engineering and production process lifetimes. However there are many possible 'entities' that could be 'modularised' and there will be multiple views (held for example by designers, builders, and managers within product manufacturer, machine builder and technology vendor partners of any global consortium) of what constitutes a 'good' production machine decomposition. Different decompositions can have conflicting business, technical and social implications. In practice any adopted decomposition will be a compromise, particularly because of uncertainty about future instances of change that might occur during the lifetime of products, production systems and production machines. Fundamental reasons to compromise will also arise because some entities cannot be broken down effectively, possibly because of natural couplings to other entities. The machine builder will have concerns about what will be acceptable to other customers during the lifetime of machine system modules (viewed as a product) and other uncertainties caused by a changing base of enabling methods and techniques. While the product manufacture will be concerned primarily about the implications of product change and/or change in production methods requiring new product mixes to be realised.

Consequent on these factors machine builder partners of the I4/I5 consortium have continued cautiously but progressively to modularise their relevant machinery by producing mechanical and control system elements that can readily be configured and programmed so that they meet specified user (manufacturer) production requirements adequately well. This has helped end user car manufacturers to standardise production processes and machine builders to design, make and reuse 'modules' of production machines.

End user manufacturers have continued to rationalise the engineering and production processes they deploy. Typically their processes are now described in terms of well defined units of 'operation', 'activity' or 'task'. The automotive industry at large has also understood a need to select and deploy suitable types of team (comprising people and technology) to resource production and engineering activities. However like their machine builders generally end user manufacturers have rationalised and standardised their processes and systems asynchronously from individual new product developments. Normally rationalisation, simplification, modularization and standardisation have come from understanding best custom and practice over a number of product iterations and represent a compromise solution that is acceptable to key parties over the lifetime of a number of products. This state of affairs can be seen as a natural one. However, the current outcome is far from optimum in the face of requirements for reduced product lifetimes, increased product variety and the need for multiparty consortia to come together in partnership to satisfy a specific global product and service requirement. A concrete but simple illustration of the deficiency of the status quo is illustrated on considering the projected relative lifetimes of I4/I5 engines and I4/I5 production lines, where the latter is likely to be around twice that of the former. The implication is that either over its lifetime 'production line utilisation' will be low or that production lines must be reconfigured around halfway through their lifetime to enable their reuse when producing the next product generation. It is observed that the current engine assembly production machine decomposition has ignored at least one important view, i.e. the view that automotive production machines need to be reused for more than one product and that embedding of a suitable change capability into a new generation of production machines might save tens of million US dollars during the lifetime of a single production line.

2 RESEARCH STEPS TAKEN TOWARDS REALISING A NEW GENERATION OF COMPONENT-BASED PRODUCTION MACHINES

Early in the I4/I5 product and process engineering programme the partner companies recognised that significant potential business benefit could accrue from realising an improvement in the 'change capability' of production machines. It was understood that advances in distributed computing component technologies and enterprise modelling might be harnessed to facilitate: (a) Faster and significantly lower cost first-off machine design and build; (b) Significantly faster and lower cost mid-life machine 'recomposition' and 'reconfiguration'; (c) More effective reuse by machine builders and their technology suppliers of machine modules at various end-user sites.

A programme of research was funded by the partner companies and the UK research councils. The collective aim was to prototype and industrially test a new generation of 'change capable' engine assembly production systems. Much of the research work has been carried out at Loughborough University, in parallel with the conventional I4/I5 product and production system engineering work of the industrial partners. Research assumptions being tested are that:

a) A suitable 'to be' decomposition of automotive production systems into mechatronic components can be identified, practically realised and supported on an industry-wide scale. Machine components need to be reused in 'composable production systems' that provide plant specific capabilities needed to assemble customer specified quantities of car engine types.
b) Key properties of production operations, mechatronic components and various production machine systems composed from the components, can be adequately modelled. The purpose being to match models of composed machines to models of production needs (i.e. production activities and processes) within a virtual environment for machine engineering.
c) Engineering activities needed to design and build component-based automotive production systems can be represented and simulated with sufficient realism to facilitate improved engineering process design and resourcing by engineering teams. By modelling engineering activities the aim was to predict benefits arising from using standard production machine components and to facilitate the design of a virtual environment (and supporting infrastructure services) with capa-

bility to support and co-ordinate the distributed interworking of teams of engineers.

d) A suitable and effective set of engineering tools specified under (3) can be prototyped and their use demonstrated and appraised by consortium partners.

3 SAMPLE OF RESEARCH RESULTS

CIMOSA modelling concepts were deployed to structure the capture of multi-perspective views of 'as is' practice used by I4/I5 engineering partners as they define production and logical processes and design and make car engine assembly plants at various sites around the globe. The approach allowed multiple views of current practice to be captured in a standardised and coherent fashion. The data captured has been used in numerous ways by MSI researchers. For example, *iThink* dynamic system models have been generated using this data that simulate the distributed operation of machine engineering processes amongst partner companies. Also workflow models of engineering activities in machine builder companies have been generated using the *iFlow* tool with a view to achieving improved engineering management and control. The modelling studies have contributed towards the development and documentation of new multi-perspective understandings about engineering processes, structures, resources and services needed to realise car assembly plants. Ongoing research is developing the use of this multi-perspective knowledge with a view to specifying a semi-generic (domain) architecture for component-based engine assembly machines.

The research has also specified and developed an exemplar set of engine assembly 'machine components' that conform to the multi-perspective domain architecture. Component design has been achieved in two complementary ways, namely:

a) By achieving a next step elemental decomposition, of existing mechanical modules produced by machine builder partners, and embedding a suitable control system decomposition into mechanical elements based on using Fieldbus and distributed object technologies.

b) By embedding common runtime models into mechatronic and virtual components so that (i) individual component behaviours and interlock and exception conditions can be programmed, controlled and monitored and (ii) the collective behaviours and conditions of composed groupings of components can be matched to production process needs and controlled and monitored in conformance with the multi-perspective domain architecture.

Another main focus of research has been on specifying, developing and appraising the industrial use of an exemplar distributed engineering environment (Harrison, 2001) Fig. 4 illustrates some of the tools and concepts embedded into this environment. An initial set of engineering tools and infrastructure services has been prototyped. A base level of support is provided for distributed teams of interdisciplinary engineers and managers (from machine customers, builders and technology providers with responsibilities for engineering, operating, maintaining and changing production machines.

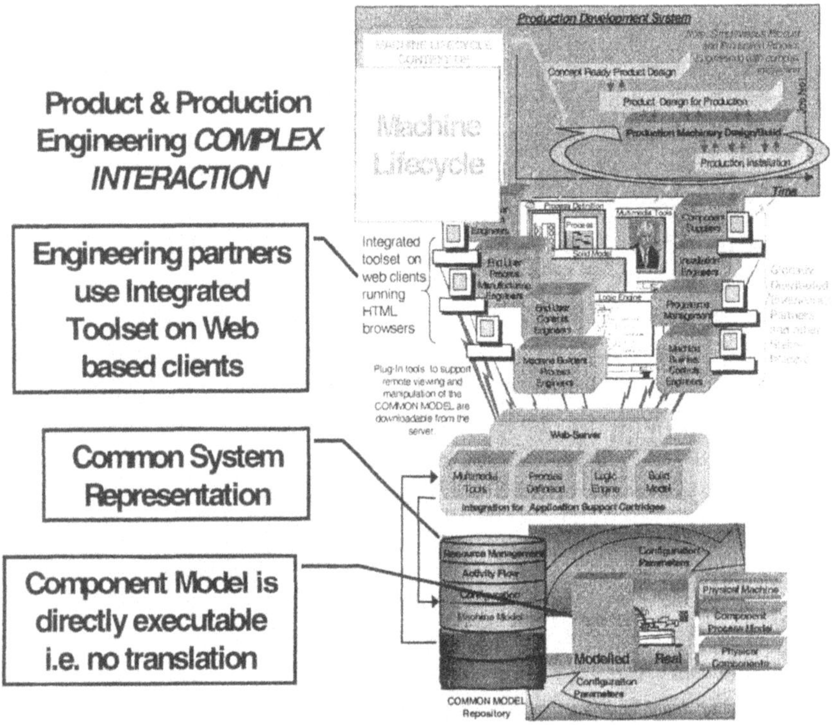

Figure 4: Characteristics of the Prototype Distributed Engineering Environment

Fig. 5 shows in concept how it has proven practical and effective to maintain consistency and coherence between runtime behaviours and exception conditions encoded by models of production processes and individual machine component behaviours by using engineering tools that collectively form a distributed (virtual) engineering environment. By such means many views of components and composed systems can be created and manipulated by suitable software tools located at their point of use. For example, models of machine system kinematics (in the virtual environment) can be 'connected' to actual machine behaviours and feedback data. Thereby it has proven possible to create multiple visualisations of virtual and real machine behaviours. The multi-perspective domain model provides a contextual framework that connects the generation and use of these modelling views to engineering actors as they carry out engineering activities pertaining to the lifecycle of the machine.

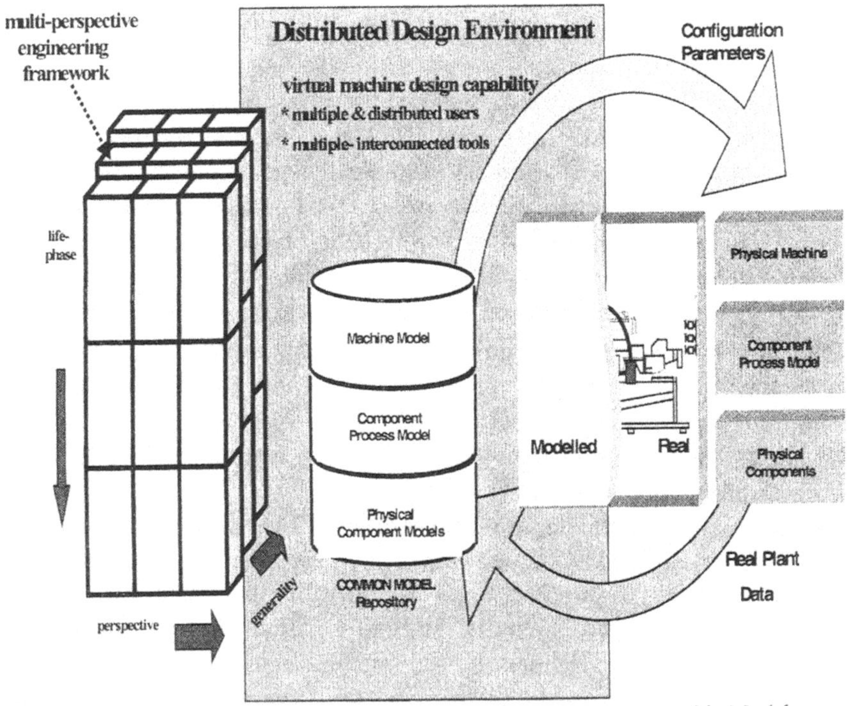

Figure 5: Example Mechatronic Component and its Integral Executable Models

Prototype testing and ongoing development of the new component-based approach to engineering engine assembly machines is ongoing in MSI laboratories. In the first quarter of 2002 prototype-testing work had begun at the promises of two machine builder partners. The concepts and their first generation implementation show significant promise from both practical and theoretical perspectives.

4 SOME GENERAL OBSERVATIONS AND LESSONS LEARNT

As a consequence of moving to the 'to be' component-based engineering approach it is anticipated that significant cost savings would result from new I4/I5 plant installations around the globe and in other domain systems. For example, a 50% savings in commissioning time is expected with typical cost benefits equating to circa US$20M. Improved machine operation, maintenance and minor change will potentially save US$50M during the machine lifetime through halving downtimes. Also 'mid life' product change will save between US$6M and US$15M. However, significant research and development resource has been expended in developing the new approach (order of US$2M). Much greater product development resource will be needed to develop the approach into an industry-wide technology that can be supported in the field but although some of the concepts used in developing the multi-perspective domain architecture, the machine components and the virtual environment can be reused in other engineering domains, generally any new domain application of the concepts will be distinct and require the capture of new multiparty domain knowledge. Use of project methods, many of which are based on public domain concepts like CMIOSA concepts can facilitate this process. But generally current generation enterprise modelling tools have proven deficient in their lifecycle engineering scope, coverage of modelling views and ease of integration with other tools, and this will limit reuse of the project methods.

The research pointed up new areas where enterprise modelling, engineering and integration research is needed. One key area of unsatisfied need is that of managing dependencies between different aspects of engineering and production processes. Although commercial enterprise modelling tools, data management and versioning tools and advance of transaction processing tools can be deployed generally they only address part of this problem. A coherent rather than fragmented technology is needed so that various types of change can be managed effectively. New research is needed to address inadequate interprocess co-ordination. Two co-ordination concerns could not be effectively satisfied, namely: (1) where a process with finite lifetime is executing and essentially dynamic (or runtime) change is needed. Typically solution to such problems requires rich process descriptions and improved commercial modelling tools with capability to capture and reuse state data (present and past) within change processes; (2) Where process descriptions are distributed into autonomously executing components but should operate coherently (but in a change capable way) with respect to a component group and/or backbone process execution.

5 REFERENCES

Harbers, W. O., (1996), *Ford Automation Strategies and Needs*, Automation Research Corporation: Automation Strategies Forum, Boston, Massachusetts, June.

Harrison, R, West, A.A., Weston, R.H. Monfared, R.P. (2001), *Distributed Engineering of Manufacturing Machines*, Journal of Eng. Manuf., Part B2, Vol. 215, pp 217-231, March.

The MISSION Project

Demonstration of Distributed Supply Chain Simulation

Markus Rabe, and Frank-Walter Jaekel

FhG-IPK, Germany, Frank-Walter.Jäkel@ipk.fhg.de

Abstract: Supply chains bring specific tasks to simulation. As long as the simulation is performed at a very high level, the simulation can be done in the traditional way. But, for detailed simulation the competence from the single chain elements has to be incorporated into the models. This can be done best by the local engineers. Up to now, integrating local models into one complete model was time consuming and error-prone. Even more critical, local maintenance of partial models was inhibited. A new approach solves this problem, and furthermore provides encapsulation, if supply chain partners do not wish to publish details of their node to other partners. The interfacing description generates XML files, which provide a specification of each supply chain node and its interfaces, too.

1 INTRODUCTION

Global Enterprises have to face new ways of distributed work. Within this huge field, MISSION focuses on the Manufacturing Engineering process and, furthermore, on simulation. The global approach is enhanced by the integration of three regions from Japan, Europe and USA. The E.U. and U.S. partners have defined a common architecture, called the MISSION General Architecture. The European demonstrator architecture is one instance of the MISSION general architecture (MISSION http://).

The demonstrator illustrates the connection between different simulation models as well as between software tools providing information for the simulation process. The software components can be executed on different computers at different locations.

The demonstrator illustrates how the MMP can be used to bridge the gap between different simulation model islands. In addition, the bridge between the simulation and the necessary information available within different software tools is realised. The demonstrator is based on existing tools and methodologies.

The MISSION methods and software have been applied within a supply chain scenario, the "MISSION Enterprise". This enterprise is world-wide distributed. It has a main assembly facility in which electric motors are assembled. The necessary components –housing, rotor, stator, control cards and bearings are supplied by specialised Manufacturing Shops, placed in Spain, Germany, Japan and USA (Rabe, Jaekel, 2000).

2 THE DISTRIBUTED SUPPLY CHAIN ENVIRONMENT

When setting up a supply chain, elements of this chain have to be selected and arranged to achieve both an effective material flow and a smooth organisation. Furthermore, the interfaces regarding material and order flow need to be specified. If enterprises or parts of them are integrated in multiple supply chains, the specification work can be significantly reduced by describing the manufacturing or logistics system by a re-usable template, and store it within a library for later use (Mertins, et al, 2000).

Additionally, manufacturing systems are often similar in various aspects with respect to the supply chains. Therefore, structuring the templates in an object oriented class structure saves modelling effort and at the same time supports additional transparency as well as some standardisation. MISSION includes a template library approach, which allows the design of reusable templates for distributed scenarios. On that base supply chains can be tested applying simulation, which requires simulation models. In total, six major groups of information are necessary for the MISSION templates (Rabe, et al, 2001):

- **Description of the template behaviour.** The template behaviour has to be described in order to get a clear understanding about the template. Obviously, each simulation template connected to this template has to fulfil this description. An informal description is possible but a more formal description in XML is recommended.
- **Referred simulation models.** One or more simulation models can be referred which satisfy the template specification. The simulation model has to be created within a simulator, which has an interface to the MMP.
- **Parameter descriptions of the template.** Each application template has a set of attributes. A subset of these attributes can be directly linked to simulation parameters. Each simulation parameter requires a

link to an attribute but not each attribute does necessarily link to a simulation parameter.

- **Description of exchanged objects.** An important issue is the data exchange between the different components within a distributed scenario. This is described by exchanged objects. Within the supply chain scenario the exchange objects describe the interface between the different companies.
- **Visualisation of the template.** Each application template needs one or more possible representations within graphical views or animations. These representations are used for static or dynamic visualisation of the supply chain.
- **Input and output segments.** The suitable input and output segments need to be connected before an evaluation can be done, E.g., if a container of finished goods leaves a factory and enters a railway system, it is of utmost importance to know at which station this railway system is entered. Furthermore, the container might have to leave the factory at different locations, depending if he will be fetched by railway or by a truck.

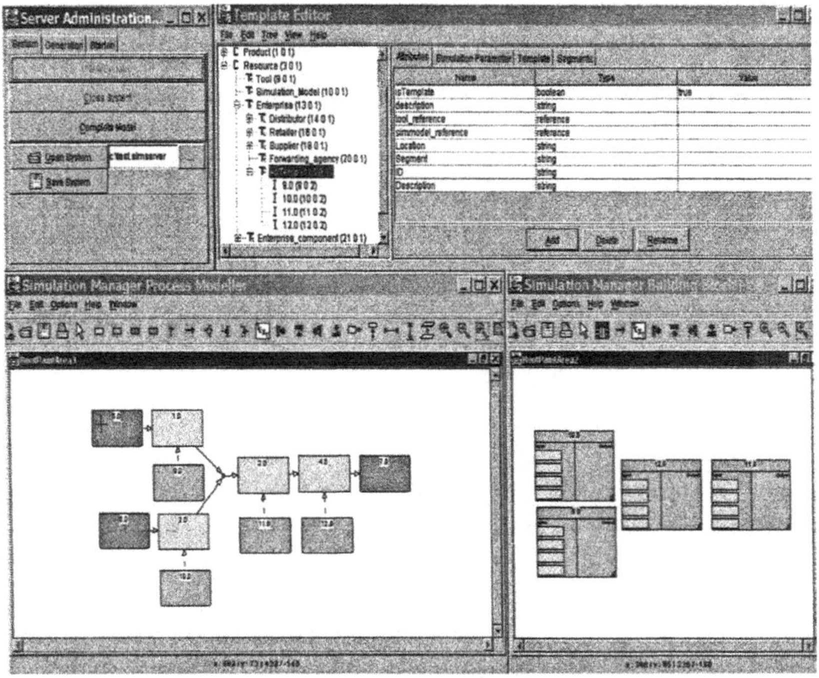

Figure 1: Components of the Simulation Manager

The template library content is not fixed. The user can add classes and attributes at any point of time to fulfil the project requirements for a specific supply chain.

The preparation and application of distributed simulation scenarios are supported by a Simulation Manager (Jaekel, Arroyo Pinedo, 2000) (Fig. 1)

This tool enables the construction of template libraries (predefined simulation models) as well as the graphical modelling of the scenario.

The template library and the graphical representation of the simulation scenario (within the demonstration: Supply Chain scenario) within the simulation manager is the base for the generation of a XML description of each interface between the companies within the supply chain.

During the generation of the interface files the exchange objects play an important role. The exchange objects describe the data which one template exchanges with the other ones. MISSION provides a first reference structure for the se objects. The reference structure includes object definitions as well as the necessary attribute descriptions. E.g. all exchange objects have the attributes about their current position in the scenario. As the demonstration scenario is a Supply Chain, in this case the interface files describe the interfaces between the individual enterprises.

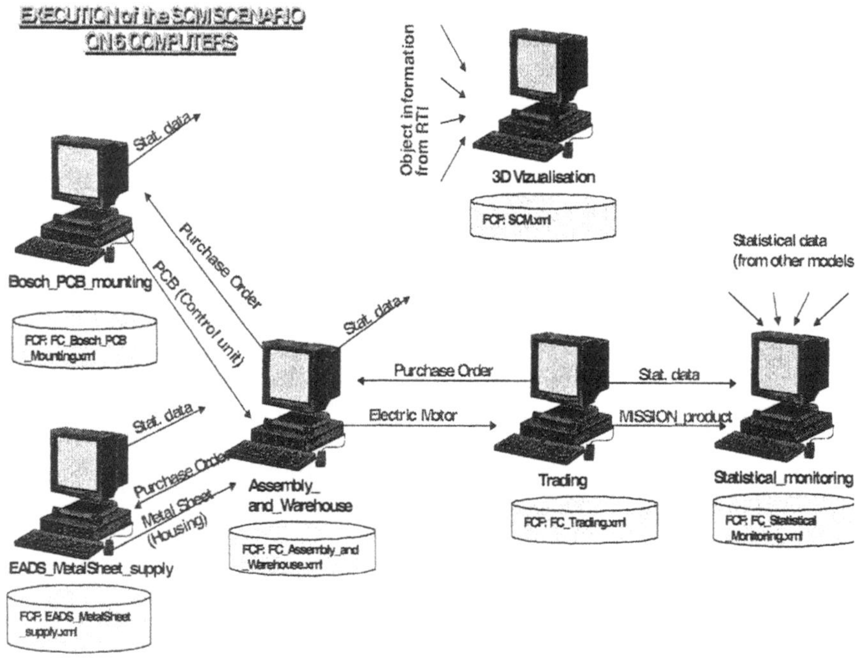

Figure 2: MISSION Supply Chain Scenario on six Computers

The MISSION Supply Chain scenario includes up to 6 personal computers (PCs) with Microsoft Windows NT or 2000 as operation system (Fig. 2). Each computer serves one simulation model. The connections between the simulation models are configured by the interface descriptions generated by the simulation manager.

To achieve the benefit of evaluating the whole supply chain scenario, information has to be gathered from all components, and then evaluated. This is done by the monitoring component (Fig. 3). It collects information about the exchanged objects and processes this information statistically. Results are, e.g., the whole production time, the order time, the manufacturing time and the procurement time.

Similar to the Monitoring component a VR component can be connected within the MISSION scenario, also. It represents online the movements of objects within the supply chain.

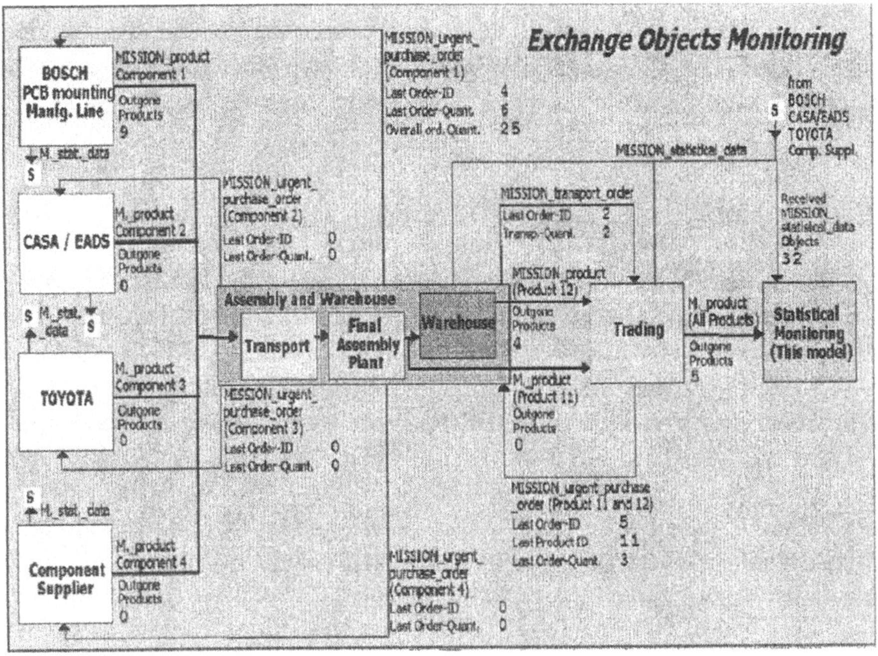

Figure 3: Monitoring Component (Overview Screen)

3 BENEFITS OF A DISTRIBUTED SIMULATION ENVIRONMENT

Among the most important issues, which need to be tackled when modelling supply chains, the followings are addressed by the MISSION project:

- Knowledge is distributed, as local engineers know the local rules and environment best. Therefore, best modelling results are expected with decentralised models, which are built and maintained locally.

- Simulation models, which are needed for dynamic evaluation, contain detailed rules and strategies of the company. Engineers might not wish to make these known to others, especially not outside of the enterprise. Therefore, mechanisms for information hiding are required.
- Complex enterprise software systems like control systems are a critical factor within the enterprise processes. It is necessary to get a test environment to check the software before it will be really installed. Moreover the number of parameters of such tools is very high and has to be calibrated within an experimental environment before they are used within the daily work.

Separate distributed models are adequate to fulfil these requirements. They are built and maintained locally, and joined for evaluation purposes, only. The interface description, which is necessary to evaluate the complete supply chain, can be used for purposes of process engineering, evaluation by simulation and specification of the supply chain control system.

4 TECHNICAL BASE

The project MISSION develops an environment for integrated applications of simulation and non-simulation tools from different vendors. Three years ago the High Level Architecture (HLA) (IEEE, 2000, Kuhl, et al, 1999) was selected as the base for the MISSION architecture. The HLA satisfied many requirements for distributed simulations. Within military applications of the HLA, for each new model a new simulator will be programmed, typically. Therefore, a flexible interface for simulation models is not required for military applications. This is completely different within civil domains, where the total effort spent to one simulation study is extremely low, compared to defence applications. The dependency of the interface description to the HLA-RTI from the specific simulation model is a critical disadvantage for regular civil applications of HLA.

Within MISSION an approach has been established that allows flexible configuration of interfaces of tools within a simulation scenario. This approach is based on a template library approach and the generation of a federate configuration file for each simulation interface as well as the generation of the federation execution file for the HLA-RTI in parallel. The major advantage of this approach is that changing simulation models or adding completely new ones requires only changes of the configuration, but no reprogramming. Of course, those interfaces need to be related to a specific application field then, which is manufacturing systems (including virtual enterprises and logistics) within the MISSION project.

5 CONCLUSION

An environment has been developed to combine different simulation models and real systems. This combination can be done on „engineering level“, i.e. by building blocks, business-process models and graphical editors.

There is a predefined Template Library, which includes simulation elements as well as exchanged objects. This library is object oriented and may be enriched by the end user.

There is a ready-to-use supply chain model as a building-block system. Any building block may be replaced by another (newly created or existing) model to achieve very detailed results.

The MISSION project was completed in December 2001. All components of the project were successful. New technologies were discovered. Some components, in particular the Supply-Chain-models, are applicable in the present status immediately.

Corresponding discussions with interested industrial companies have been initialised. During the project additional research activities were discovered for further research activities. Corresponding projects are expected to start in Winter 2002/03.

6 ACKNOWLEDGEMENT

The European Module of the MISSION project (ESPRIT PROJECT N° 29 656: Modelling and Simulation Environments for Design, Planning and Operation of Globally Distributed Enterprises) was carried out with financial contribution of the European Commission under the specific RTD Programme, Esprit Project 29 656. Partners within this European Module are Bosch (D), EADS/CASA (E), Sisteplant (E), vr-architects (A), ProSTEP (D), Fraunhofer IPK (D), Loughborough University (UK) and Coventry University (UK).

7 REFERENCES

IEEE P1516.1/D5, (2000), *Draft Standard for Modeling and Simulation (M&S) High Level Architecture (HLA) - Ferderate Interface Specification.*

Jaekel, F.-W. Arroyo Pinedo, J.S. (2000), *Development of a Demonstrator for Modelling and Simulation of Global Distributed Enterprises.* In: Mertins, K. Rabe, M. (Eds.): The New Simulation in Production and Logistics. 9th ASIM Dedicated Conference on Simulation in Production and Logistics, Berlin.

Kuhl, F. Weatherly, R.; Dahmann, J. (1999), *Creating Computer Simulation Systems.* Prentice-Hall. An Introduction to the High Level Architecture, ISBN 0-13-022511-8.

Lee, Y.T. Umeda, S. (2000), *Information Model of the Supply Chain Simulation.* In: Proceedings of the EUROSIM (2001), Conference, Delft.

Mertins, K. Rabe, M. Jaekel, F.-W. (2000), *Neutral Template Libraries for Efficient Distributed Simulation within a Manufacturing System Engineering Platform.* Winter Simulation Conference (WSC), Orlando (USA).

MISSION http://www.ims-mission.de

Rabe, M. Jaekel, F.-W. (2000), *Simulation for Globally Distributed Enterprises.* 12th European Simulation Symposium 2000 (ESS), Hamburg.

Rabe, M. Garcia de Gurtubai, G. Jaekel, F.-W. (2001), *Modelling and Simulation for Globally Distributed Enterprises.* In: Proceedings of the EUROSIM 2001 conference, Delft.

PART 4. INTEROPERABILITY OF BUSINESS PROCESSES AND ENTERPRISE MODELS

Integration is the timely and meaningful exchange of information among software applications. It requires the error-free transfer of information, a total agreement on its syntax, and the correct understanding of its semantics. The Internet and its associated standards have addressed successfully the first of these requirements. Syntax and semantics, on the other hand, remain as elusive today as they were ten years ago. These are resolved typically by proprietary, de facto, or standard-interface specifications, which, in theory, should have solved the problem, but have not because the costs of development and custom implementation remain prohibitively high, ranging from 2 to 5 times the software costs. Estimates range from $200B to $500B for the manufacturing industry alone.

This section of the proceedings addresses the issue of interoperability from several points of view. Two workgroup reports address the issues of systems requirements (Nell) and the role of ontologies (Goranson) from an integration point of view. Discussions were on life-cycle-based system engineering and how to interoperate across the different engineering life-cycle phases and between their different processes in the enterprise. Emphasis was on product development and production processes development. The second group addressed the barriers of enterprise integration and examined the new leverage that ontologies might provide. The group agreed that such an approach could overcome the most severe of these barriers. A number of actions and proposals have been outlined, which may be taken up especially in NIST activities.

These group reports are followed by a vendor view on the integration (Payne) that considers not only the technical issues, but addresses the human and management aspects of integration as well. The author favours the business process approach and discusses briefly a platform for collaboration.

Three papers address the role of standards in interoperability. Chen and Vernadat analyse standards that significantly contribute to achieve enterprise interoperability. A brief overview of standards in enterprise modelling and engineering states their role to standards related to enterprise interoperability. The focus of delaHostria is on the use of an application integration framework that facilitates the construction of a set of manufacturing application system models and the compilation of the standard interfaces, to support interoperability. The MultiView program (Engwall) aims to achieve a high degree of interoperability of the IT systems for complex engineer-to-order systems, products and processes over their life cycle. Developing data standards for the integrated digital environment and providing a single schema for seamless integration of the data sets and a framework for data access and communication.

The aspect of infrastructure support is addressed in the two papers of Cardoso (Workflow Quality of Services - QoS) and Li (Product Data Management - PDM). The first paper describes a workflow oriented QoS model with four dimensions (time, cost, fidelity, and reliability) intended to support quality management. The second paper is concerned with agent-based integration of heterogeneous PDM systems found in the virtual enterprise environment. It presents an infrastructure with agent-based services to support distributed management of documents.

In his paper Obrst addresses the use of ontologies to support semantically interoperable B2B electronic commerce. Describing the nature of B2B and presenting arguments towards why B2B needs ontologies, the paper concludes with the interaction of ontologists and domain experts in the building of ontologies for business. In addition, some of the tools available for developing ontologies are identified.

The Editors

Kurt Kosanke
CIMOSA Association Böblingen, Germany

James G. Nell
National Institute of Standards and Technology (NIST), Gaithersburg, Maryland, USA

System Requirements: Products, Processes and Models

Report Workshop 3/Workgroup 1

James G. Nell[1], (Ed.), Em delaHostria[2], Richard L. Engwall[3], Myong Kang[4], Kurt Kosanke[5], Juan Carlos Mendez Barreiro[6], and Weiming Shen[7]
[1]National Institute of Standards and Technology, USA, [2]Rockwell Automation, USA, [3]R.L. Engwall and Associates, USA, [4]Mitretek Systems, USA; [5]CIMOSA Association, Germany, [6]AdN Internacional, S.A., Mexico, [7]National Research Council, Canada, nell@nist.gov

Abstract: see Quad Chart on page 2

1 INTRODUCTION

The work of the group is summarized in the following Quad-Chart (Table 1). It identifies the approach taken to resolve the issues in domain of interoperability of both processes and models and proposes a concept for planning such collaborations. In addition it states some ideas for future work for testing and enhancing the proposed solutions.

1.1 Problems and Assumptions

The workgroup discussed how to satisfy customer expectations for high quality, low price, fast delivery, agilely produced, and environmentally clean products. The core concept toward creating the approach and recommending future work is to be based on best practices of systems engineering.

Some benefits for enterprises to have this capability are:

- Reduced need for physical prototyping because engineers can evaluate product and production system operation using electronic-simulation techniques, including interference, and safety aspects.

Table 1: Working Group Quad-Chart

EI3-IC Workshop 3 ***Interoperability of Business Processes and Enterprise Models***	**Workgroup 1:** ***System Requirements and Models and Processes***	**2002-February-6/8 Gaithersburg MD, USA**
Abstract: Enterprises use stovepipe tools that limit interoperability, traceability, consistency, and complicate data sharing and impact satisfying customer expectations for high quality, low price, fast delivery and environmentally clean products. Focus of the workgroup in creating the approach and recommending future work is based on best practices of systems engineering and the holistic integration of people, processes and systems & technology.	**Major problems and issues:** – How to define interoperability of processes and models? – How to model interactions between all activities of the enterprise -, production-, and product life cycles? – How to achieve concurrent use of product design and production system eng. data; focus on design optimization through electronic prototyping and production simulation for decision support during operation? – How to provide synthesis of data dictionaries?	
Approach: – Use GERAM life-cycle concept to identify relations between different processes and models – Use systems engineering approach in designing, manufacturing and supporting a product/system including the enterprise as a component of the system – Identify interoperation needs among and between product, process, and enterprise life cycle structures – Identify interoperation problems and propose solutions – Propose interoperability measurements (quality, cost, time) – Analyze existing standards like Systems Engineering, EPISTLE, Application Integration Frameworks – Find a way to go from architectures and frameworks to a software strategy for interoperability	**Results:** – Interoperability: on-time transfer of understood information between processes. – Distinguish between enterprise processes, product processes and production processes. – Metrics for interoperation quality: number of conversations needed--not needed--to get understanding with minimal or no loss, delays, synchronization success **Future work:** – Define product design, system engineering and operational processes such that their process data can be used for both design optimization and production decision support. – Analyze and model exchange of information during product and production process design with emphasis on human-oriented information exchange – Define set of required standards – Identify elements of an enterprise-wide data framework and provide methodology for its creation.	

– More timely access to information from production, use, and support functions for product-definition processes.

- Optimized parallel processes to enable solutions for the overall production system or for general problem (such as a battlefield) solutions.
- Improved production processes due to use of more efficiently coordinated knowledge and business rules.

But to scope the effort, the workgroup decided to accept certain existing assumptions:

- An enterprise-reference architecture embodying life-cycle concepts would provide a basis for starting the analysis (ISO 15704, 1999, pre EN ISO 19439, 2002).
- An enterprise-wide or greater viewpoint will be used.
- Analyses will be done using the principles of systems engineering--the system-design type, not the computer-systems type.
- Interoperability will refer to information transfers among enterprise processes. Specifically, the parts of the processes that send, receive, interpret, and respond to information. This could include interoperation enablers such as humans, resources, machines, and material.
- Integration at the process-model and resource-model levels is the correct approach, as opposed to integrating applications, services or product data flows. This mirrors the implicit common denominator of the enterprise-modeling community. Moreover, the group asserted that for practical advance in virtual enterprises, existing model and model integration paradigms must apply (Vernadat, F.B. 2001, ISO 18629, 2002, ISO 15704, 1999 (see Annex for GERAM)).

2 THE APPROACH

Following general convention, the group assumed an architecture consisting of virtual enterprise components that may be companies or relatively independent operations of a company. Each component is composed of processes, resources, and products. Each of these things, since they have structure and have order; that is, lower entropy; contain information. All have information ports into and from which information flows hopefully effecting communication or a form of understanding. Some of these process components are automated objects, some are exclusively human, and humans operate some of the components. Some of the information exchanges are explicit, well-formed messages, perhaps from machine to machine.

Human-to-human communication contains much tacit information. Models capture the mechanics of the process within each component. And because models also capture the "process" of component-to-component interaction, those models define the structure, the nature, and timing of the information exchanged.

More specifically the group agreed on the following facets for the approach presented in Table 1 to better enterprise interoperability:

- Exploit the many concepts of GERAM that will guide the analysis and modeling of enterprise components.
- Use a wide-scope, systems-engineering approach to avoid islands of improvement.
- The group was reminded of the history of ICEIMT, in which certain solutions, once in place, have tended to block facile transition to further improvement. Also, the mentality required to solve some of these problems is greater than the mentality used to create them. Aware of this difficulty and aware that the problems probably were created unintentionally, the group agreed to use extra care to avoid such de-enablers.
- Using an enterprise viewpoint, identify interoperation needs among product, process, and other enterprise structure, and infrastructure.
- This is more rigorous than simply saying that everything has to talk to everything else. There probably are far fewer things that need to interoperate than there are things in the enterprise.
- There also probably is more information issued by enterprise things than information used, thereby taking up valuable bandwidth and compounding the interoperation problem.
- Propose a set of metrics to measure relative interoperability.
- These metrics will serve to add credibility to user-executives that must decide whether or not investment in this domain is worthwhile. The solution to these problems needs to be advocated aggressively at the executive level. Improvement will only accrue if the problem is addressed as the problem of becoming a "lean" enterprise is addressed. That is, continuous improvement over a long term.
- In formulating a standards approach, which becomes versed in the key enterprise and system analysis, numerous standards that have been issued recently will be analyzed (ANSA/EIA 632, 1999, IEC 62264, 2002, Epistel, ISO 15704, 1999, ISO 15745, 2002, ISO 15926, ISO 1600 (2002), pre EN ISO 19439, 2002).

3 THE NATURE OF INTEROPERABILITY

The group tried to reach consensus on various concepts relating to improving enterprise interoperability. We agreed that we are talking about interoperability among enterprise processes. Processes contain activities that use enablers such as buildings, humans, machines, information, and material. Each process has information ports to transmit or receive information. These ports are sensors of some sort, and the information that passes through them can range from one bit to huge STEP files, to spoken words. There needs to be medium transducers to convert from a form of information used in one process to the form used in the other one.

We felt that we needed to understand the distinction between process data and product data, whether it should be treated differently, and if new representation methods or standards are required in interoperability analyses. Product data is an easier thing to represent because it is mostly nouns describing attributes of a product. Process data is mostly verbs that are functional, behavioral and time related in nature.

The product, process, and enterprise complex of information must be managed by a systems-engineering discipline that helps to manage the disparate enterprise and process system data and human interfaces available anywhere that could impact end product or system operation. Humans are required in this information system to help manage and coordinate knowledge and business rules. This could provide the capability to realize some significant benefits to the enterprise.

This data, when integrated, provides capability to trade off parallel processes to optimize product creation, factory throughput, or problem solution such as a battlefield scenario. Information generated during one part of a product-life cycle has little chance of interoperating with information used or generated by a process in another part of the product-life cycle. This is because users select tools in the different processes that optimize the output of that particular process and does not pay conscious attention to the information needs in the remainder of the enterprise. So there is a need to consider, in addition, information transfers among the product and process mix regarding design, producibility, and supportability tasks to maximize process interoperability and, hence, enterprise flexibility and agility.

In simulations, both of these data sets need to be available in a format usable by simulation applications. With necessary data at the correct time, we can simulate with good accuracy. Electronic prototypes for simulations are useful to demonstrate such phenomena as: producibility, interference, assembly, tolerance, test, and operation safety. Vendors can use such integrated information to visualize how their part is used in the end system. To accomplish effective simulation, users would need data from computer-aided design and computer-aided engineering systems. These information sources may be disparate and quite distributed, but they must interoperate.

We could also create an integrated, configuration-management model in the end system with parts not organized by system but in space by a Cartesian, (X, Y, Z) coordinate system, or a polar (r, θ) system.

Interoperability needs to be analyzed to the transaction level to ascertain a transaction's nature and to define requirements for needed infrastructure capability. This includes analyzing timing, nature, and context of messages.

4 INTEROPERABILITY METRICS

To evaluate how well interoperability is going the group felt we need some metrics to report the quality of an instance of interoperability. The key aspects of interoperability are:

- Number of conversations needed (or not needed) to get understanding or desired behavior
- Synchronization of message; that is, did it occur just-in-time, with the correct information

From here, the metrics discussion migrated to exchange characteristics to improve interoperability. Basically, the source of the information must be trusted. While this is somewhat of a soft issue, trust can be statistically predicted using the success of prior transmissions from the same source. We also can use concepts such as a product's trustworthiness--communicated from prior use, use by other users, product reviews, and the producer's reputation for trustworthiness.

5 SYSTEMS DESIGN

Given the need for interoperability what can be done to improve the interoperability process in enterprises? The group felt that good system design techniques should be applied to the problem.

- Define the entire system
- Design (consciously) the entire system so that it is more interoperable
- Use ANSI/EIA 632 (1999) Process for engineering a system

Other good engineering design axioms apply to this problem. Basically the group recommends that rather than design processes in human-logical chunks, partition the system for interoperability. For example the following axioms may help (Suh, 1990). Partition the system such that:

- Amount of information transferred is smallest at the partition interfaces
- All the functional requirements for a process design must be independent of each other
- Try for a design that requires a minimum of information to execute each function

6 JOINED MEETING OF WORKGROUPS 1 AND 2

Of most importance to me was the joint discussion between our two workgroups outputs that highlighted the key issues between small stand-alone point solutions and the bigger interoperability problems

The group considered a holistic integration of lifecycles for the enterprise, products, and processes. However, the viewpoints and the nature of the information that each group needs to use are different. For Group 1, looking at wide-angle-strategic things and product representation, the type of information is mostly noun oriented, while for group 2, considering mostly process models, thinks in terms of concepts that are largely verb-oriented. Group 1 views the enterprise from the top downward and Group 2 views it from the bottom upward.

That, in itself, is not a problem. The problem arises in the orientation—verb-oriented concepts are difficult to merge or integrate with noun-oriented concepts. Group 2 had investigated taking the complexity out of the models by removing elements and information relative to the ontology. Then there would be a light-weight mechanism backed up with the heavy-weight ontology stuff. This approach would alleviate the problem that enterprise models have—a high cost and limited reusability. An approach such as this would appeal to users, especially small-to-medium sized enterprises.

The problem of matching model composition poses problems for the simulation benefit seen by improving interoperability. Simulation requires us to amass large quantities of information at the correct time from various sources—namely from the modeling domain of groups 1 and 2. We need all of the information to capture enough meaning, the semantics, plus at the same time resolve different methods and rules for representing that meaning, the syntax. Group 2 is designing an example by creating an information artifact and forcing the artifact to combine noun and verb techniques into a singularity of meaning.

An approach to the solution is to develop a generalized method for integrating or federating the business-process-application level to the lower-tier information systems of the enterprise, the application software. At present, software vendors do not have the same view as process builders and entities are not the same granularity, nor, as stated before, do they focus on the same parts of speech. The integrating issue in matching this up is probably the ontology. Modeling in the past captured data and relationships, not the meanings. The ontology can provide the meaning. Ontologies were discussed in detail in the Workshop 2 Workgroup 1 report.

7 PROPOSED RESEARCH PROJECTS

The following research issues have been identified, which should read to be proposals rather than questions.

- How to combine the noun and verb oriented models into a singularity of meaning quickly enough to allow sophisticated simulations of enterprise integration?
- What do we need to do with the models or the processes to allow interoperability?
- Can we define better the model matching challenges: smaller to bigger, bigger to smaller, and models employing different languages?
- Can we define an enterprise system design theory for interoperability?

8 CONCLUSIONS

WG1 and WG2 discussed in a joint session the information objects mismatch issue between verbs and nouns. Agreement reached that WG2 must assess the broad needs of WG1 holistic integration framework before demonstrating a small pilot production application to test the reusability of what they develop. It was suggested that we need a generalized method for mapping business application levels to software information systems levels.

9 REFERENCES

ANSI/EIA 632, (1999), *Processes for Engineering a System*, http://global.ihs.com

EPISTLE (European Process Industries STEP Technical Liaison Executive) published by POSC/CAESAR Association.

IEC 62264/FDIS, (2002), *Enterprise-control-system integration*, ISO TC 184 SC 4 JWG 15.

ISO 15704, (1999), *Guidelines for enterprise-reference architectures and methodologies*, TC 184 SC 5 WG 1.

ISO 15745/FDIS, (2002), *Industrial automation systems and integration: Open systems application integration frameworks*.

ISO 15926, *Integration of life-cycle data for oil and gas production facilities*, TC 184 SC 4.

ISO 16100/FDIS, (2002), *Software-capability profiling*, TC 184 SC 5 WG 4.

ISO 18629/DIS, (2002), *Process Specification Language*, TC184 SC 4 JWG 9.

pre EN ISO 19439, (2002), *Enterprise Integration--Framework for Enterprise Modeling*, CEN TC 310 WG1.

Suh, Nam P. (1990), *Principles of Design*, Oxford University Press, NY, NY.

Vernadat, F.B. (2001), *UEML: Towards a unified enterprise modelling language*, Proceedings 3rd Conference Francophone de modelization et simulation (MOSIM'01), Troyes, France, April-25/27, pp. 3-13

Ontologies as a New Cost Factor in Enterprise Integration

Report Workshop 3/Workgroup 2

H. Ted Goranson[1], (Ed.), Bei-tseng Chu[2], Michael Gruninger[3], Nenad Ivezic[3], Sern Kulvatunyou[3], Yannis Labrou[4], Ryusuke Masuoka[5], Yun Peng[6], Amit Sheth[7], and David Shorter[8]

[1]Old Dominion University, USA; [2]University of North Carolina, USA; [3]National Institute of Standards and Technology, USA; [4]FIPA and PowerMarket, USA; [5]Fujitsu Laboratories, USA; [6]University of Maryland Baltimore Campus, USA; [7]University of Georgia, USA; [8]IT Focus, UK, tedg@sirius-beta.com

Abstract: see Quad Chart on page 2

1 INTRODUCTION

The following Quad-Chart (Table 1) summarizes the work of the group. It identifies the approach taken to address the issues of infrastructures for virtual enterprises exploiting agent technology and proposes future work on ontologies and thereby addresses the issue of model complexity and costs.

2 BACKGROUND

The benefits of modeling are widely recognized. They are significant enough to have driven widespread implementation and the support of substantial theoretic, implementation and user communities. Initially, the benefits of modeling were in identifying and clarifying a process so that it could be formally improved. These models formed the basis of a reusable knowledge store on processes and the science behind process engineering. Since then, two major extensions to this start greatly extended the benefits.

Table 1: Working Group Quad-Chart

<table>
<tr><th>EI3-IC Workshop 3
Interoperability of business processes and enterprise models</th><th>Workgroup 2:
Ontologies as a new cost factor in enterprise integration</th><th>2002-February-6/8
Gaithersburg MD USA</th></tr>
<tr><td>Abstract:
The workgroup focused on key barriers to enterprise modeling for process and system optimization. Overcoming these long-lived barriers requires some new approaches and the workgroup settled on the introduction of ontologies. Several problems and new approaches were explored. Some reasoned speculations resulted, together with proposals for testing their validity. The problems concerned lowering the cost of building and changing models; building and using component libraries; linking to non-process (like data and product) models; and furthering of the agent notions of prior workshops.</td><td colspan="2">Major problems and issues:
– The cost of enterprise modeling is too high - to model, store, maintain, validate, change, and to inter-operate with other models.
– The modeling function has unpredictable costs.
– To accomplish the type of integration needed the models are too complex to use.
– The latest trend in modeling and interoperability is the need to create ontologies. We need to determine what they are, are not, and how we make them share information.</td></tr>
<tr><td>Approach:
– Study the model and ontology relationship with respect to cost, effectiveness, and ease of use
– Follow the promising thread of the previous two workshops regarding knowledge bases and autonomous agents
– Consider the difficulty of representing product and process information and moving that data among a product's life-cycle phases.
– Use a paradigm in future work that assumes that systems and their models will be able to self-integrate and share information among each other once a context is defined.</td><td colspan="2">Results
– Model complexity is due to the semantic content that overloads models. Move semantics into ontology.
Future work:
– Study the role of ontology to reduce the complexity of models
– Assess the cost versus benefits as the amount of formal rigor is added to ontologies
– Explore the process of separating the semantic aspects of models from the non-semantic aspects
– Introduce ontologies into an agent's use of knowledge needed to accomplish its functions.
– Explore a security scheme that doesn't prevent sharing information but requires an ontology to assign a context code to each parcel of information to be shared.</td></tr>
</table>

One extension is the combination of the process and the model into a control model. The model contains the extra functionality to apply its intrinsic algorithmic representation of the process to actually do the work or control the work of the process. The other development is so-called "enterprise integration," where the processes are interlinked in a global framework. Then many of the engineering techniques can be applied at the system level for optimizations and benefits otherwise impossible. The benefits of process modeling in these two contexts are significant.

This sketch of the history indicates a problem. The early process modeling methodologies were relatively lightweight. Usually they were targeted for use by process owners as a tool to represent their own activities and to engineer them. Over time, as the utility of these models have grown, and the benefits have ballooned, the modeling tools and methodologies have become increasingly complex, formal and costly.

The problem is now not a matter of benefits. These are very large. But so are the costs and risks of modeling. As a result, modeling is not widely employed and enterprise modeling by integration even less so.

- It costs too much to do the modeling in the first place. The tools and methods are so complex and constrained now that two experts must be involved: the process expert and the modeler.
- It costs too much to develop, maintain and practically use libraries or repositories of processes, including components associated with control and best practices.
- It costs too much to validate the model in terms of the interactions among others.
- It costs too much to adapt a model once conditions change. This includes the re-engineering, reintegration and revalidation.
- It costs too much to integrate with other methods. Some methods impinge on the control function, such as the selection of components and their validation in a software engineering sense. Some other methods are "outside" the process scope, such as strategic, financial/legal and marketing concerns.

With each of these costs are attendant risks that increase the worst-case cost. So not only are costs high, but they are unpredictable as well. The group determined that this cost issue is the greatest single barrier to gleaning the benefits of enterprise integration. Additionally, the group concluded that the root of the problem was in "overloading" the modeling function. On the one hand, modeling methods and tools should be intuitive, flexible, easy to use and experiment with. They should be as diverse as the applications, which is to say very diverse and domain-specific. But on the other hand, they should be extremely rigorous and standardized. But accommodating the latter, the former is compromised.

The group proposed taking a look at separating these two needs — allowing the actual modeling tools to be lighter weight, closer to the problem, easier to tinker with, more diverse and specialized. The technique is to propose the use of ontologies for the formal rigor and multiple benefits of standardization. Those benefits are primarily found in ease of integration and enabled component libraries for reuse.

3 ONTOLOGIES

Ontologies are a relatively recent evolution in the modeling and representation family tree. As with many animals in this family, definitions differ. Most definitions share two features: ontologies are formal, explicit conceptualizations and their role is as abstract specification of shared concepts. The latter characteristic places it in the position of empowering the collaboration of models and modeling activities — ontologies are to models in a rough equivalence to what models are to the real world. Except ontologies generally attempt to define a complete domain, and models often focus on elements within a domain.

To support this role, ontologies are generally more abstract, formal and axiomatic than models. Some ontologies are more quick and dirty, and the boundary between ontological engineering and modeling (or knowledge engineering) is fuzzy. But in general, ontologies support the work of representation (in models and the like) and models support work on the real world. As noted, the group explored the possibilities of overcoming the cost barriers of enterprise process models by apt use of ontologies. The general idea is to remove a lot of the restrictions and overhead from the modeling layer into the ontology layer.

Expected benefits will occur on the modeling layer as noted. But greater efficiencies and capabilities are expected at the ontology layer as well. There is universal concurrence in the group that overhead associated with the mechanics of ontologies for the expected needs will be less than similar functions currently supported at the model level. These needs are centered on knowledge capture, but include reuse needs.

4 SUGGESTED ACTIONS AND PROJECTS

The discussion of the workgroup was centered on a number of high value problems. These were considered in turn, and specific actions identified for tests, research or further discussion. Each of these is presented below, in no particular order.

4.1 Project 1: Component Reuse and Tweaking

Problem: This project began with a discussion of the relative advantages of ontologies. The group noted that ontologies are proliferating in apparently much the same way as modeling and representation methods. Some in the larger community question whether anything fundamental is solved by the introduction of ontologies. Perhaps the diversity and need for harmonization will be just as great but effectively complicated by an additional layer, so the speculation goes.

The group noted that the diversity of ontologies is fundamentally different than that of models and model methods. The latter has two primary dimensions of diversity, a "vertical dimension" of different levels of rigor, and a "horizontal dimension" which reflects the different needs of various domains. By definition, ontologies don't have this horizontal diversity.

The question then focused on the problem of the proliferation of ontologies in this vertical dimension. At the highest level of formal rigor are approaches like PSL (Process Specification Language); toward the middle of the spectrum is DAML (DARPA Agent Markup Language); toward the bottom are many web-oriented taxonomies. The workgroup believes there is a simple cost-benefit function at work here: the "cheaper" approaches provide accessible benefit. One presumes that the market uses the cheapest approach that solves the problem, and that general perceptions are that the benefits of additional rigor rise less fast than their costs.

The workgroup believes that the cost/benefit curve is reversed, that the benefits of additional rigor grow faster than the additional burden they impose. But this has not been yet demonstrated. A project is proposed that would do just this: audit some benefits of rigor in the context of costs. If the benefits of additional rigor are shown to be much larger than the additional cost, some collapse in the ontological spectrum of diversity are anticipated.

Approach: The project has a narrow scope. An existing set of many integrated models of processes is presumed to exist. The project does not care how they can to be in this stable, functional state. The focus is on two processes, illustrative of many in the enterprise. To be interesting, these are presumed to be in different virtual enterprise components (meaning different companies), they involve some transactions/interactions that involve "soft" elements (trust, uncertainty, tacit knowledge), and the two models were created using different modeling approaches.

The following action will be compared in three modes: without ontological support; with "medium-weight" ontological support; and with PSL ontological support. Suppose that the first model (M_1) must be changed in response to a change in the real world; either a change from "below" (within the process) or "above" (from the larger enterprise context). M_1 uses some means to peruse a component library supported by an "intermediate" ontology, O_0. The mechanism of this perusal is not in the scope of this project, and has been broken out as a separate project below (see "Component Pattern Strategies").

M_1 selects a model component that is a near match for its need. O1 provides a means for mapping that component into M_1, identifying the tweak needed, and supporting the tweaking itself. O_1 and O2 support a similar process in M2, which must adapt in response to the changed M1. Then, validation of model interoperability occurs, again supported by the ontology level.

Expected Benefits: The experiment specifically evaluates:

- The cost of making the model changes
- The cost of revalidating the model within itself for consistency
- The cost of revalidating the interoperability of the two models

Specifically excluded from the exercise is validating the correctness of the model(s) in representing the real world. That was judged to be a relatively straightforward process, which will not be affected by the introduction of ontology. Instead, the focus is on the validated interoperability of the models. There is some expectation that some substantial portion of the integration now supported by model frameworks will be better handled by this mechanism.

Second order insights expected from the experiment are:

- Suggestions for a new paradigm for enterprise model integration
- Insights into enterprise component libraries and reuse strategies
- Indications of research needs
- Perspectives on standards tasks
- Links between process model integration and popular parallel trends, many web-based
- Better understanding of dynamic, multidimensional configuration management of large systems
- Possibly guidance on new lightweight, intuitive model freedoms

The experiment is expected to be explored as a "paper" experiment before being fielded on a test bed.

4.2 Project 2: Component Pattern Strategies

Problem: This second project is a companion to the one just described. It focuses on the characterization of the various constructs that are evident in existing and developing enterprise-modeling methods and tools, recognizing that the collection of constructs evident in each method/tool are in some sense coherent. They reflect a particular intellectual approach to the problem of enterprise modeling, and hence the form in which they are characterized and represented to the end-user (the enterprise modeler) reflects this view of how the world is. In this sense the 'constructs' represent potential stepping-stones between the putative common ontological underpinning whose cost-benefit will be assessed in Project 1, and the enterprise model to be constructed.

The project focuses on how such stepping stones might be deployed in the (probably Web-enabled) tools that a modeler might use to access component libraries, to choose components from different libraries derived from various methods and tools, and to advise on tweaking strategies to support model functionality and interoperability. This approach will complement the lingua franca approach being developed in the UEML project for an Enterprise Model Interchange Format (EMIF).

Approach: Therefore this project focuses on the boundary between M1 and component libraries (or frameworks) through support of the ontologies. The basic idea is that dividing the "old" modeling function into two layers (one formal, one lightweight) is not a trivial job, and some attention must be given to intuitive access of higher level functions. In a functional sense, an intermediate or bridging layer is proposed. The purpose of the layer is to provide a formal metaphor for indexing components and for presenting and managing that index to modelers.

It should also provide placeholders for the propagation of constraints derived from the ontology (or construct to construct mappings), e.g. valid and necessary relationships, restrictions on involved objects and/or their attributes.

This notion grew in importance as the group worked through some scenarios. Given the existing investments in and fragmented usage of existing methods and tools, it is clearly necessary to find ways of increasing tool adoption and encouraging a stronger enterprise modeling community to emerge. This is the issue that a stepping-stone-based component pattern strategy will seek to address. The group also reviewed a collection of de facto and de jure standards that will guide this effort.

Expected Benefits: The specific aims of the project are to:

- Define the functionality of such an interface, including the possible reuse of metaphors from software and system engineering (and similar)
- Explore the use of constraint representation and propagation to support semantic decomposition, semantic aggregation, and more ambitiously semantic mappings
- Explore the direct applicability of existing and emerging standards to support this interface. Certainly to be included will be the standards for model constructs (ENV 12204, under revision), EDOC, BPMI.

A test scenario is likely to result in an additional activity of the project noted above. That scenario may be centred on M1 locating component fragments from two independent (stepping-stone) repositories, combining these in some predictive or computer-assisted way, and then tweaking as already described.

4.3 Project 3: Extending the Previous Workshops' Approach

Problem: Workgroup 1 from the first workshop (in Paris) proposed a new perspective on enterprise integration that focused on integration model frameworks to support knowledge management. In this view, "registering" a model in a context was conflated with "situating" an element of information in the enterprise context as knowledge. Some promising notions were developed.

Workgroup 1 from the second workshop (in Singapore) started from this notion and explored a more rigorous approach involving agents. Key notions from that work involved rationalizing transactions among agents in the virtual enterprise context with process state management and process adaptability. The combined insights of the two workshops showed some promise toward a new approach to agent based virtual enterprise system optimization.

This project extends the work of those two groups by introducing ontologies to the existing notion of actor-model pairs (See "A Merged Future for Knowledge Management and Enterprise Modeling Agents" and "Advanced Virtual Enterprises: Needs and an Approach" in this volume for more detail.) A fine level of granularity is defined by the constraint that an actor can support only one message operation, the messages being based on standard speech act transactions. A set of meta-actors can change processes.

A high value problem concerns the means by which these meta-actors can operate. In the early vision of the Singapore workshop, these processes were allowed to be ad hoc. But in the context of the present workshop the suggestion was to employ a component library as outlined in the two projects above. This project extends the work of those two in the specific case of the agent model.

Approach: What is new in this project is the introduction of global process ontologies for agent evolution. An interesting feature is the discrimination between component libraries for processes that do work and the processes that improve those. The group expects that this distinction will ease the problem of identifying the nodes where standardization can support the entire system. In particular, one set of "pattern" indices might be maintained for process components and another set of indices for:

- Means of ordering searches among components in libraries
- Measuring algorithmic fit between needs and those components
- Drawing from a set of tweaking, assembling and validation processes on those components to meet the requirements
- Perform all of this by autonomous agents working at the bottoms-up level

Expected Benefits: The result of the project is expected to be an agent-based self-evolving index strategy that supports the knowledge management paradigm of the Paris workshop in a virtual enterprise context.

4.4 Project 4: Interfacing with Product Life Cycle Information

Problem: This project arose from interaction between the two work-groups. Workgroup 1 focused on the global problem of integrating the integration frameworks of product and process information through the product and enterprise life cycles. This was seen as a significant, important problem that has been vexing enterprises for decades. Piecemeal solutions appear to be making the problem worse.

The combined workgroups suggested that the new element of ontologies might provide a new tool. Several approaches were suggested. One that seems promising was to consider product data in the context of process data. This is a departure from the current approach, and especially in the case of the U. S. Department of Defense. That default approach posits a substantial distance between the two, in part because the military sees product data as a manufactured item itself that is bought and delivered. Product data carries certain legal assumptions, like ownership and portability that are not as explicit in the case of process information.

Process modeling has other fundamental differences from product data modeling. Modeling of processes is much more difficult, owing to the need for control, and the explicit management of state. Moreover, processes tend to be more self-referential, for instance processes are expected to evolve more opportunistically than products and the means by which a process adapts is a process itself.

In the combined group session, some suggested that the theoretic foundations related to process representation have developed much more rapidly than similar foundations on the data side. This was not a universally held position, but arguments in favor cited the advances in situation logics, the widespread implementations of agents and the new techniques of software engineering that have no counterparts on the data side.

Approach: The approach on the product modeling side has been centered on creating standards that are at the same "semantic level" as the data. The PDES/STEP standard is of this type, but it seems an unwieldy solution that is not widely employed, and does not gracefully extend to process information. The suggested approach adopts two new elements:

- To leverage ontologies where possible — in fact, to move as much of the focus of standardization efforts from the domain-specific to the generic. There is growing appreciation for this approach already.

- To "situate" product data in the context of process data. This is the novel notion of the workshop. Generally, every element of product data, especially data that is maintained in parallel with the product, exists to support a process. But the group offers a more specific notion of bonding, resulting from the overarching notion of enterprise integration.

A primary motivation behind enterprise integration is to provide a philosophy for creating enterprise-wide frameworks. These frameworks tend to use normal transaction boundaries as leverageable division within the framework. The information passed at these transaction boundaries corresponds well to the information, which concerns the product data community. Therefore, one would expect process-oriented integration to show promise for integrating all the information in the enterprise.

This project will:

- Take a fresh look at the problems of life cycle product data integration by defining the actual needs. This would go well beyond the simple call for interoperability and define the particulars of why data needs to interoperate, in what context, and for what ends.
- Examine the possibilities of emerging process integration frameworks to address this requirement. Particular attention will be given to the emerging notion of ontology-supported, scalable component-based frameworks as outlined in the workshop. Some detailed attention will be paid to context-based use strategies, and new notions of integration-supported knowledge management. (The benefits of knowledge management will be added to the benefits of "old-fashioned" integration in the problem statement.)
- Develop an approach of process-context based life cycle integration that is a strong balance of addressing needs and leveraging existing efforts in the standards community
- Run some "paper experiments" to evaluate the relative benefits with a focus on hidden infrastructure costs, lowered costs of adoption and increased benefits (particularly scalability)

Expected Benefits: The project is not expected to result in an action plan for life cycle product and process information integration and knowledge management.

4.5 Project 5: Supporting a New Paradigm for Trusted Systems

Problem: The workgroup recognized that the new approach to enterprise integration would not only address traditional concerns, but also provide new types of benefits not previously recognized. Problems of security came up, as they always do. The basic problem is that integration is all about appropriate sharing. Current paradigms of security are based on engineered denial of sharing. Usually, the integration framework operates behind a wall of multilevel access gates, which are designed and imposed relatively independently.

The topic arose as a desire to engineer security needs into the integration strategy. In the course of discussing this approach, a radical new approach was incubated. This project was defined to take that relatively speculative notion and to explore it in some detail to discover associated practicalities, boundaries, benefits and constraints.

Approach: The underlying integration paradigm of the workgroup is one based on dynamic, context-sensitive integration. Perhaps this will be agent-supported. Certainly, it will be ontology-driven. In this view, local context and (transaction-oriented) reference frameworks apply to incoming information. It is registered in the local context for local use, and in the system context for global analyses and optimization.

In other words, models are "decoded" for their value in the global context of the enterprise. The advanced case has individual models that may be structured in different ways, so a different model's perspective very literally needs the global framework to "decode" it. In the normal case, the models are obfuscated by the accidents of domain preference and expediency. But the models could be obfuscated deliberately. The group proposed just this.

This new paradigm is not based on denial. Everything is shareable. Some information is modeled using situation or state-dependent ontologies. "Sense"" can only be made of each component by situating it (via ontology-based integration). The net effect is that no piece of information needs to be protected, but no one can use a piece without being in the proper context. That context may include familiar certification mechanisms but can allow other types of possibly more robust protections. The obfuscating methods may include familiar numeric processing, but certainly will support additional or alternative methods.

This project is more speculative than the others noted. The method of the project is to convene a workshop of experts in logic (especially situation theory), non-deterministic abstraction (category theorists), ontologies, process engineering and advanced thinkers in enterprise integration.

Expected Benefits: The goal of the project is to explore the idea, map possibilities and define a research roadmap. A secondary goal is to increase the benefits of system-level, ontology-based integration by adding in support for trusted systems.

From Integration to Collaborative Business

Management Through Business Processes

Mike Payne
VITRIA-EMEA, UK, mpayne@vitria.com

Abstract: This brief paper outlines some of the areas needing consideration when creating and delivering an integration backbone to a business. It contrasts the data based approach against the more effective process based approach and highlights some of the non technical considerations in making a strategy that works. The thoughts are drawn from seeing numerous customers and partners put together frameworks with a greater or lesser degree of success across Europe over the last three years.

1 EVOLUTION OR REVOLUTION

The purpose of this brief paper is to make some observations grounded in experience of the use of business-process management as the keystone to successful business integration. This section deals mainly with the approach to the business, whilst the later sections deal with other aspects of successful integration projects.

The last five years have seen an explosion in the complexity of business systems and the speed with which changes occur to businesses and markets. One of the key drivers to successful organisations is that they are able to see at anytime key parts of their business. Some of the examples of where this failed badly were the abuse of trading systems, the over stocking of supply chains and the failure to absorb and capitalise on mergers and acquisitions. Other pressures were hugely changing competitive landscapes, requirements to deploy new services quickly and the drive to create business-to-business linkages.

Factors that influence significantly the success or failure of integration strategies hinge on whether the people within the business recognise the need to change the way of creating solutions and deploying frameworks and that it is supported at board level. Applying a data or system-oriented approach constrains the potential of the business to genuinely free itself from the infrastructure and to focus on the real business goals.

This occurs because much of the system delivery is done through information technology groups, consultancies or application vendors, most of who have little knowledge of the way in which business processes work within the business. The stakeholders in the business can end up feeling the solution has been foisted on them and in some cases extensive in fighting has broken out within the organisation.

The approach that works in introduction of any new strategy is that it must be an inclusive process to minimise the resistance to and the orphaning of the new solution. There is an opportunity is to use the initiatives as change agents.

So key requirements for any successful strategy have to be:

- Clear business goals including the ROI
- Business level sponsors
- Executive level sponsors
- Objective evaluation of technology
- A pragmatic approach to building the business-process infrastructure

The following sections review the following areas:

- Whether to buy applications from one vendor or many
- What are the characteristics of business-process integration
- Cost rationale
- Collaborative process platforms

2 ONE STOP SHOP OR BEST OF BREED

The IT industry has a habit of going through fads. One that occurred recently was sourcing all software from one application provider. On one hand this works well since there is one point of contact for all issues on the other hand there are a number of drawbacks. These are:

- Vendor lock-in
- Dependency on the vendor's domain expertise i.e. Retailing
- Failure to achieve true integration between the modules that make the solution resulting in data re-entry for example between customer account systems and accounts receivables
- Application fit to the business i.e. do you have to modify your business to meet the application capability
- Ready availability of skills in the correct products

- Dependency on the vendors product cycle for new functions and features

There are others, but this serves to illustrate one of the first hurdles. Even when the business is entirely using one application for the enterprise such as SAP there are tremendous 'integration' issues. A concrete example is where a multi-site implementation usually means not all parts of the business move to the same software revision level at the same time. This causes issues around software version dependencies and incompatibility between environments. This is exacerbated when the integration is tightly coupled to the underlying systems and the business processes held inside the applications.

The alternative approach is to use a best of breed approach. This however necessitates a more considered integration strategy and approach, which is application independent. The key to success is to drive the selection, and delivery from what the business is trying to do. To select a package on functionality is a little unwise, generally speaking most packages can fulfil 80% of all requirements anyway.

Drawbacks to best of breed approaches can be:

- Higher support costs through the need for more contacts.
- Requirement for more integration.
- More complex implementation.

However by articulating the requirements as business processes allows a much cleaner communication with the end users and the opportunity to create an extensible process framework.

Many implementations typically start with a number of processes that may span one or more systems. If a process approach is taken then it is possible to 'wrap' existing applications and to deliver new services without the high cost and high risk of re-engineering legacy systems as well as avoiding a big bang approach.

Gradually over time the integration backbone gets extended throughout the enterprise. The phases that typically are encountered are:

- Internal integration of a few small systems
- Integration of more systems through extensions to existing business processes or introduction of new processes.
- Integration on-line with partners and suppliers.
- Self- regulating business processes.
- On-line collaboration.
- Real-time view of the business processes and the real state of the business.

Clearly depending on the business requirements and the history brought to bear, many IT teams would feel safer with a one-stop shop approach. Vitria have customers who have deployed in many diverse and complex configurations, and built on that or created new values add services around existing infrastructure.

3 CHARACTERISTICS OF A BUSINESS PROCESS DRIVEN INTEGRATION

If the view of integration through business process is considered, some of the things that become immediately obvious are how to manage the diversity of business processes. In the real world business processes can be:

- Dynamic
- Complex
- Long lived
- Self-regulating
- Involve humans
- Span highly distributed systems
- Act both synchronously and asynchronously

The integration backbone must be able to handle all of these options in a manner that allows a rapid deployment, using common skills and metaphors and adhere to known standards. It should also support the concept of a stateful process. This becomes particularly crucial when you have transactions that as part of a process must be delivered once and only once or are running over unstable infrastructure such as a wide area network.

The advantage of this approach versus a data approach is reduced costs to deliver and maintain as well as infrastructure transparency since the process is independent of the systems and is loosely coupled to the data.

One other key advantage in the approach is to externalise the business logic that has historically been secreted inside the applications. This allows a much cleaner and more effective discussion about whether that really is the business logic. It also removes the dominant application syndrome, which is where one application is the centre-piece of the implementation. Examples are where say the ERP system is held to be the core application, when in fact that is not the case. This tends to cause business logic and workflow to be placed inside this application, which negates the benefit of using a process, since much of the logic is invisible at a higher level.

The other consideration when building out the business processes is volumetrics. Once a third party is allowed access to the shared business process whether via the web for some kind of customer self-service or via a bonding to another business control over transaction profiles has gone. This requires several things to be in the backbone:

- Horizontal scalability
- Multi-threading capability

- Vertical scalability
- Statefull processes

The backbone must support these to allow the business to cope with unknown events, unpredictable volumes of activity and long-lived processes that may run over a number of weeks.

4 COST RATIONALE

"GartnerGroup advises its clients that use of an integration broker will *reduce application implementation costs by one-third*.

Greater savings occur after initial development.

Maintenance: a single change to an interface ripples across other interfaces, use of an integration server can *reduce costs by two-thirds."*

One of the challenges faced by businesses deciding on integration strategies is to create a compelling business case. In most instances the reasons fall in quantitative areas such as cost savings through lower maintenance, reduced time to delivery, increased component re-use and lower staff costs due to using commonly available skill-sets. The other area is far more qualitative and more strategic, this typically involves a leap of faith and covers items such as ability to integrate information from other businesses quickly, a real-time view of the business and feeding through to improved customer satisfaction.

One risk is that for the initial project where staffs have just come off training there is higher start-up cost than adopting a point-to-point solution. The challenge is to create a return on investment model that has tactical deliverables to senior management and the business, whilst not being the hardest part of the integration. This helps address the emotions of accepting new technology by the business and builds confidence in the technology by the delivery team.

Implicitly all stakeholders know that creating an infra-structure backbone is a good thing, but are very wary of ivory tower approaches to key projects. This leads back to the earlier statement that driving integration through business processes allows gradual implementation and quick delivery, with minimised risk and maximum flexibility.

5 PLATFORM FOR COLLABORATION

It has been discussed earlier that most businesses using integration are taking the first step on a road, which ultimately leads to interactions with other parties, both customers and partners. This is becoming more prevalent across multiple industries where tight collaboration is required due to initiatives such as cost reduction drives, vendor managed inventory or straight-through processing.

Clearly if the integration has been set-up through a process approach then it is transparent to the businesses what the systems topology actually is. There is a need for the integration backbone to have certain key capabilities such as authentication, repudiation, non-repudiation and protocol management.

This structure allows one business to offer information through processes to third parties with a wide variety of systems capability. Examples include communication at a basic level via secure ftp, through EDI to public standards such as RosettaNet, but all captured from a process viewpoint.

Across many industries there are drives to standardise inter-company and intra-company communication using predefined message formats and process flows. The best understood at the moment is EDI, with it's defined transaction sets and process flows.

The process flows are split into public and private. The private processes access the internal systems to provide information to the public processes. It is these processes that the third parties interact with. This approach allows complete transparency of businesses, network topology and systems maps.

More and more customers are beginning to go down this route. One advantage is that new technologies such as web services can be incorporated very easily as enablers working under the control of the processes. This means the fads and vagaries of the technology marketers will not place at risk a business.

The cornerstone of successful integration projects is to drive the strategy and delivery from a business process. Anything else will result in increased risk and cost and will fail to deliver. Practical experience suggests the emotional response to short-deadlines of engaging in point-to-point solutions is the wrong one. Some customers in the UK spend over 1.5M Euros on maintaining their point-to-point legacy interfaces. There is no value add to the business, just raw cost.

6 CONCLUSION

The general discussion in this paper has been around integration through business processes. Practical experience in Europe as part of Vitria over the last three years has re-enforced the belief that process management is the key platform to building virtual businesses and secure collaborations. This process-based approach is a fundamental characteristic of the globalisation that has occurred over the last ten years and to fail to adopt this thinking is to risk the long-term survival of the businesses.

Integration through business processes succeeds because it allows the creation of a framework on which the business can perform operationally and develop strategic plans for adapting to future change. It also makes available critical business data in real-time to enable a fast responsive business.

It is safe to say the era on non-return of IT investments is over. All IT spending must justify its incursion through measurable value added to the core business goals. This in the integration field can only be achieved through a business process approach; failure to do this leaves the company susceptible to nimbler adversaries. So the winner will be the company with access to core information and resources in the most efficient way.

For further information on Vitria see http://www.vitria.com

Enterprise Interoperability: A Standardisation View

David Chen[1], and François B. Vernadat[2]
[1]LAP/GRAI, University Bordeaux I, France [2]MACSI-INRIA & LGPIM, ENIM/Université de Metz, France, chen@lap.u-bordeaux.fr

Abstract: Standards that significantly contribute to achieve enterprise interoperability regarding both intra and inter-organisational environments are identified and reviewed. The concept of interoperability is firstly clarified and defined in comparison with some adjacent notions such as portability, compatibility and integration. A brief overview of standards in enterprise modelling and engineering is then given to better state their prerequisite role to standards more directly related to enterprise interoperability discussed next. Future needs to improve interoperability are discussed as part of the conclusion.

1 INTRODUCTION

With the globalisation of commerce, distribution and manufacturing, co-operation between enterprises of different sectors and cultures is significantly increasing. It is not only limited to sub-contracting and co-operation with suppliers and customers better known as 'supply chain' or 'extended enterprise', but also concerned with 'virtual enterprise' that can form and dissolve very quickly. Nevertheless, globalisation does not mean unification and homogenisation. Instead, there exists a strong demand to preserve cultural identity and particular ways of working within companies. This leads inevitably to some interoperable situation rather than to tight integration. Furthermore, at a company level, the need for flexibility and reactivity, the inspiration of employees to work in a more autonomous way using their own methods and tools, oblige enterprises to change from traditional hierarchically based organisation to smaller, autonomous, distributed unity structure.

Within this context, interoperability within and between enterprises becomes a necessity and a key success factor of competitiveness.

Without standards there will be no interoperability. The objective of the paper is to identify issues and review standards relating to enterprise interoperability in intra and inter-organisational environment. It focuses on standards in the area of manufacturing activities and is presented from the point of view of users who design and implement enterprise systems. In section 2, the paper will present some basic concepts and definitions to clarify the concept of enterprise interoperability. The link and difference between interoperability and integration are tentatively discussed. Section 3 will provide an overview on standards in the area of enterprise modelling and engineering. Although these standards do not directly deal with the interoperability issues, they contribute to improve the ability of interoperation as well. Section 4 reviews standards and projects that significantly support the achievement of enterprise interoperability. A synthesis summary will be given in the last section to conclude the paper.

2 ENTERPRISE INTEROPERABILITY

The term 'interoperability' is increasingly used in enterprise engineering and its related standardisation activities. Generally speaking, interoperability is a measure of the ability of performing interoperation between two different entities (be they software, processes, systems, organisations...). According to the Oxford Dictionary, interoperable means 'able to operate in conjunction'. The word "inter-operate" also implies that one system performs an operation on behalf of another system. Originally, the concept of 'interoperability' comes from software engineering. From this point of view, interoperability means that two co-operating software systems can easily work together without a particular interfacing effort. It also means establishing communication and sharing information and services between software applications regardless of hardware platform(s). In other words, it describes whether or not two pieces of software from different vendors, developed with different tools, can work together. The ISO 16100, (2000) standard defines the manufacturing software interoperability as the 'ability to share and exchange information using common syntax and semantics to meet an application-specific functional relationship through the use of a common interface'.

Interoperability is however not only concerned with software applications. It may happen between any two entities in a heterogeneous or homogenous networked environment. TOGAF (The Open Group Architecture Framework, (Open Group, 2000)) defined interoperability as: '(1) the ability of two or more systems or components to exchange and use shared information, and (2) the ability of systems to provide and receive services from other systems and to use the services so interchanged to enable them to operate effectively together'.

Enterprise Interoperability is therefore concerned with interoperability between organisational units or business processes either within a large distributed enterprise or within a network of enterprises (e.g. supply chain, extended enterprise or virtual enterprise).

The concept of interoperability has to do with the concept of portability. TOGAF (Open Group, 2000) has defined portability as: (1) the ease with which a system, component, data, or user, can be transferred from one hardware or software environment to another; and (2) a quality metric that can be used to measure the relative effort to transport the software for use in another environment or to convert software for use in another operating environment, hardware configuration, or software system environment. In other words, portability is the ability of data or system to be moved, and interoperability is the ability of software or systems to understand and use information coming from other software or systems. The notion of interoperability is also linked to the concept of compatibility, which is a related term, at least for entities shared or in co-operation in the interoperable environment.

The ability for different systems to work together may be characterised at various levels of co-operation (e.g. physical systems, application, business and networked organisation). Clearly, interoperability has the meaning of co-existence and co-operation, while integration relates to the notions of co-ordination and unification. Vernadat, (1996) defines interoperability as the ability to communicate with pier systems and access the functionality of the pier systems, while integration is a broader concept embracing communication, co-operation and co-ordination capabilities. Thus, interoperability must be achieved to achieve real integration. The difference between integration and interoperability has been further clarified in ISO 14258, (1999) - Concepts and rules for enterprise models. This standard considers that there are three ways to relate models (*entities*) to one another: (1) Integration: there is a standard format for all constituent systems. Diverse models are interpreted in the standard format. This format must be as rich as the constituent system models. (2) Unification: there is a common meta-level structure across constituent models, providing a means for establishing semantic equivalence. The meta-model is not in an executable entity as it is in the integrated situation but a model mapping mechanism. (3) Federation: the federated model scenario may exist if no agent successfully or globally can impose requirements for semantic equivalence across all models of an enterprise. According to ISO 14258, (1999) the federated situation is the most probable scenario for full interoperability wherein most models will not be in a standardised or common form because it is not economically feasible to put them in such a form. The advantage of the federated situation is that it not only allows to collectively provide a (complex) service, but also to preserve the independence and autonomy of its components, to be open to dynamic change in composition and distribution. In such a situation, interoperability requires that co-operating entities be dynamically accommodated rather than having a predetermined meta-model. This assumes that the concept mapping (or semantic unification) is done at an ontology level, i.e. a semantic level (ISO 14258, 1999).

3 ENTERPRISE MODELLING AND ENGINEERING STANDARDS

Enterprise modelling and engineering are prerequisites to Enterprise Interoperability. This section briefly presents an overview of standards related to enterprise modelling and engineering. The main standards for this area have been developed by CEN TC310/WG1 (European Standardisation Committee), ISO TC184/SC5/WG1 and to a lesser degree OMG (Object Management Group).

Concerning enterprise modelling, one can mention:

ENV 40003, (1990)	Modelling Framework for Enterprise Integration,
ISO 14258, (1998)	Concepts and Rules for Enterprise Models
ISO 15704, (1998	Requir. for Enterprise Reference Architecture and Methodologies
ISO/IEC 15288, (1999)	Life-Cycle Management System / Life Cycle Processes

These standards define concepts and principles in terms of architecture and methodology for enterprise modelling and engineering. They do not deal with the representation of how an enterprise is structured or operated.

Enterprise modelling languages that can be used within the frame of the standards mentioned above have also been a concern for standardisation. For example:

ENV 12204, (1995)	Constructs for Enterprise Modelling
ISO 18629, (2001)	Process Specification Language (PSL)
ISO 10303/11, (1992)	EXPRESS
ISO/IEC 15414, (2000)	Open Distributed Processing (ODP) - Enterprise Language
ISO/IEC 15909, (1997)	High-level Petri nets

Among them, ENV 12204 and ISO 15414 support multi-view enterprise modelling. PSL, Petri nets and EXPRESS are formal languages that can be directly implemented on computers. Future work in this area is related to the initiative to develop UEML (Unified Enterprise Modelling Language) (Vernadat, 2001; IST-34229, 2001). An enterprise model built using a specific language (for example IDEF0) can be translated into another one (e.g. GRAI nets) via UEML constructs used as a neutral format.

Standardisation also contributes to the development of reusable models (or partial models) that represent parts of enterprise structures in terms of processes, information, resources, ... The use of these models in enterprise engineering can shorten design delays and increase modelling consistency. Main reference documents are:

ISO TR 10314, (1991)	Reference Model for Shop Floor Production Standards
ISA-dS95, (1999)	Enterprise-Control System Integration
ISO 15531, (2000)	Manufacturing Management Data Exchange (MANDATE).

These documents mainly focus on the function and information aspects. MANDATE also deals with the resource data definitions.

Concerning application development, standards are concerned with system programming such as ISO 13281.2, (1996) - Manufacturing Automation Programming Environment (MAPLE), and OMA (Open Management Architecture) by OMG, (1992).

4 ENTERPRISE INTEROPERABILITY STANDARDS

Enterprise Interoperability is concerned with communication and cooperation between software components, processes, organisation units and humans. To make interoperability happen, exchange of concepts is a key issue. Thus, terminology must be agreed and semantic equivalence established.

4.1 The terminology issue

One of the main obstacles to interoperability arises from the fact that 'the systems that support the functions in many enterprises were created independently, and do not share the same semantics for the terminology of their process models' (Schelenoff, 2000). Without explicit definitions for the terms, it is difficult to see how concepts in each application correspond to one another. Simply sharing terminology is not sufficient to support interoperability; the applications must share their semantics, i.e. *meanings* of their respective terminologies. In TOGAF, it has been underlined that 'interoperability can only be achieved when information is passed, not when data is passed' (Open Group, 2001). This implies that information must be correctly interpreted and understood.

Two projects of interest, which deal with the terminology issue to support enterprise interoperability are: (1) ISO PDTR 16668, (1999) - Basic Semantic Register (BSR) and (2) ISO CD 18629-1, (2001) - Process Specification Language (PSL). In the PSL project, it has been proposed to define a formal semantic layer (called PSL ontology). Within the PSL ontology, the semantics of terminology is specified using KIF (Knowledge Interchange Format), which is itself an ISO standard. In the BSR project a different approach is used to approach the terminology issue. It consists in providing a methodology to establish standard terms by creating semantic equivalence rather then to define the terms themselves. The project has defined three main components of BSR as: (1) *Semantic Components*, (2) *Semantic Units* and (3) *Bridges*.

The semantic components are Units of thought used in everyday life. They may be named by the use of single or multi word terms. They will be used to specify Semantic Units. For examples: Delivery, Actual, Latest, Person, PurchaseOrder, BillOfMaterial, Date, Identifier, etc. The Semantic Units are equivalence of semantically complete data element concepts, i.e. the property of an object class with full qualification. They are the basis for the specification of data elements in information systems. For examples: GoodsDelivery.Latest.Date, Sales.Information.Contact.Telephone.Number. Finally, the Bridges allow to establish links between a Semantic Unit and its equivalence in various directories.

4.2 Message-based interoperability

Interoperability requires that data stored in software system on one machine can be sent and "interpreted" by another software system on another machine and for different purposes. To make this happen, standards on message format and transfer are needed. Today, Internet technology seems to provide the best perspective to support intra and inter-enterprise interoperability. Standards relating to Internet technology are mainly *de facto* standards promoted by the World Wide Web Consortium (W3C). Among various approaches, XML (eXtensible Markup Language) (W3C, 1998) is the most promising. XML is considered particularly adapted for data exchange as it has the promise of making data "self describing". In other words, XML offers 'a non-proprietary and inexpensive way to promote **reuse of data** by providing a way to locate it (semantic search), and by providing a standard way to transform and move it between applications' (CIO, 2000). Coupled to XML, some other W3C specifications are also critical to interoperability. For instance (just to name a few): (1) SOAP (Simple Object Access Protocol) allows applications to communicate directly with each other through Internet by defining a simple, extensible message format in standard XML

(W3C, 2000), (2) WSDL (Web Services Description Language) is an XML format for describing services accessible on the web as a set of endpoints operating on messages containing either document-oriented or procedure-oriented information (W3C, 2001), (3) ebXML (Electronic Business XML Initiative) enables to exchange XML based messages through the web, (4) J2EE standard allows to benefit security features for authentication and access control in Java environments, and (5) RosettaNet defines XML based business interfaces to support interoperability among enterprises such as supply chain management.

4.3 Manufacturing software interoperability

According to ISO TC184/SC5/WG1, standards must define only what is necessary so that the software developer knows with what the software must work. Ideally, standards should enable interoperability and still protect innovation, efficiency of approach, and migration capability (Nell, 2001b). One standard dealing with manufacturing software interoperability is ISO DIS 16100, (2000) - Manufacturing software capability profiling. This is an on-going project carried out by ISO TC184/SC5/WG4. This standard considers that one of the most important aspects that makes interoperability operational is to describe precisely and concisely the capability of software. The capability is defined in terms of potential functions. The standard will specify a standard way of representing "what I do and what I need". The first part of the standard aims only at defining a framework for software interoperability. The ultimate goal of the project is to allow to: (1) select appropriate software, (2) substitute one software component with another, (3) migrate software to another platform, and (4) verify software to a capability profile. Manufacturing software packages concerned with this standard are CAED, NC programming systems, PDM (Product Data Managers), MES (Management Executive Systems), CAE (Computer-Aided Engineering) systems, etc.

Concerning application integration, standards are concerned with the IT based services, which are built upon the seven layers proposed by ISO 7498 OSI model (ISO/IEC 7498-1, 1994). Generally speaking, these services contribute to improve software interoperability. The main approaches are:

ENV 13550, (1995)	Enterprise Model Execution and Integration Services (EMEIS)
ISO 15414, (2000)	Inform. Technology - Open Distributed Processing – Ref. Model
ISO 13281.2, (1996)	Manufacturing Automation Programming Environment (MAPLE)
TOGAF (Open Group, 2000)	Technical Reference Model
OMA (OMG, 1992)	Open Management Architecture
OAGIS (OAG, 2001)	Open Applications Group Integration Specification

According to its developers, OAGIS better supports inter-enterprise business application interoperations, i.e. communication and interaction between companies. In other words, OAGIS is a horizontally based approach that is applicable across industry sectors. Today, OAGIS covers financial transactions, ERP to ERP interoperation, Supply Chain management, etc. On the other side, the reference model of OMA better known as CORBA (OMG, 1992), focuses on intra-enterprise interoperations of software objects that share a same infrastructure.

4.4 Manufacturing and business process interoperability

The ISO CD 18629-1, (2001) – Process Specification Language (PSL) originates from a NIST project aiming at developing a formal language to facilitate process interoperability (Schelenoff, 2000). The goal of PSL project is to deal with the translation problem between two inter-operable process applications. The project has considered that when the interoperation is limited to a small number of processes, the translation can be done at a one-to-one basis. However, when interoperation becomes generalised and widely applied, the development of translators between native formats of those applications and PSL can facilitate the interoperability. PSL will act as an intermediate neutral format. The view hold by PSL is that using this intermediate format can reduce the number of translators from $n(n-1)$ to n in the case where there are potentially n processes that can enter into interoperation. This point of view is also shared in some related domains, for example: (1) ISO 10313-1, (1993) known as STEP in product data modelling, and (2) the UEML initiative in enterprise modelling to develop a Unified Enterprise Modelling Language (Vernadat, 2001).

ISO TC184/SC5/WG1 has also considered a potential standardisation effort to deal with process interoperability focusing on the definition of interaction scenario. The project is an ISO New Working Item (NWI) entitled "Rules for manufacturing process interoperability". It intends to specify the mechanism (processes and meta data) that an enterprise can use to represent and present, in a standard way, information required to establish communications for enterprise-process-interoperability (Nell, 2001a).

5 CONCLUDING REMARKS

To meet new industrial challenges, there is a shift from the paradigm of total integration to that of interoperation, which holds the promise for more flexibility. Relevant standardisation activities focusing on interoperability have just started and most work remains to be done. It has been found that

Internet-based technology standards play an important role to move and transfer data/messages more easily. These approaches have been mainly developed by non-institutional organisations such as OMG, W3C and OAG, and remain *de facto* standards. Standards elaborated by ISO and CEN have focused more on modelling aspects dealing with the specifications of resources (e.g. software profile) and processes (PSL) as well as their related semantic and syntax problems. One of the issues for the future is to establish the link between these two communities and to map these standards to a consistent framework.

Few standards exist that directly relate to Enterprise Interoperability per se. However, it is important to note that interoperability is not only a problem of technology. It implies a better and mutual understanding between partners involved in the interoperation. Cultural inertia will limit the effectiveness and use of standards to design interoperable systems. Consequently, an entity (be it a company or a department within a company) must actively engage to a self-adapting process in terms of working procedures and culture so as to facilitate maximum exchanges of information with the outside world.

6 REFERENCES

Chen, D. Vernadat, F. (2001), *Standardisation on Enterprise Modelling and Integration: Achievements, On-going works and future perspectives*, In Proc. of IFAC 10th Symposium on Information Control in Manufacturing (Invited session), Vienna, September 20-22.

CIO Council, (2000), *Enterprise Interoperability and Emerging Information Technology (EIEIT) Committee*, www.cio.gov/Documents/cio_eieit_xml_workgroup_jul_2000.html.

ENV 12204, (1995), *Advanced Manufacturing Technology - Systems Architecture - Constructs for Enterprise Modelling*, CEN TC'310 WG1.

ENV 13550, (1995), *Enterprise Model Execution and Integration Services (EMEIS)*, CEN, TC 310 WG1.

ENV 40003, (1990), *Computer Integrated Manufacturing - CIM systems architecture framework for modelling*, CEN, TC 310 WG1.

ISA-DS95, (1999), *Enterprise-Control-System Integration*, ISA DS95.01-1999, Instrument Society of America.

ISO 10303-1, (1993), *Industrial automation systems and integration - Product data representation and exchange - Part 1: Overview and Fundamental Principles*. TC 184 SC5 WG1.

ISO 14258, (1999), *Industrial Automation Systems – Concepts and Rules for Enterprise Models*, TC 184 SC5 WG1, April-14 version.

ISO 15704, (1998), *Requirements for Enterprise Reference Architecture and Methodologies*, TC 184 SC5 WG1, N423.

ISO CD 18629-1, (2001), *Industrial automation systems and integration, Process Specification Language (PSL), Part 1: Overview and Basic Principles*, TC 184 SC4/SC5 JW8.

ISO DIS 10303-11, (1992), *The EXPRESS Language Reference Manual*, TC 184 SC4 WG5, N35.

ISO DIS 13281.2, (1996), *Industrial Automation Systems - Manufacturing Automation Programming Environment (MAPLE) - Functional architecture.* TC 184

ISO DIS 15531-1, (2000), *Industrial automation systems and integration -Manufacturing management data exchange - Part 1: Overview and fundamental principles,* TC184 SC4 WG8, N138 R3.1.

ISO DIS 16100, (2000), *Manufacturing Software Capability Profiling, Part 1 - Framework for interoperability,* TC 184 SC5, ICS 25.040.01.

ISO PDTR 16668, (1999), *Basic Semantic Register (BSR) - Rules, Guidelines and Methodology,* TC 154 WG1, N007.

ISO TR 10314, (1991), *Reference Model for Shop Floor Production Standards, Part 1 - Reference model for standardisation, methodology for identification of requirements.*

ISO/IEC 10746-3, (1994), *Information Technology – Open Distributed Processing – Reference Model – Architecture,* ITU-T Recommendation X.903.

ISO/IEC 15414, (2000), *Information Technology – Open Distributed Processing – Reference Model – Enterprise Language,* ITU-T Recommendation X.911, Version 303, ISO/IEC JTC 1 SC 7 WG 17.

ISO/IEC 7498-1, (1994), *Information Processing Systems, Open System Interconnection (OSI) Reference Model, The Basic Model,* ITU-T Rec. X.200 (1994 E).

ISO/IEC CD 15288, (1999), *Life-cycle management System, Life Cycle Processes,* ISO JTC 1 SC 7, N2184.

ISO/IEC CD 15909, (1997), *High-level Petri Nets - Concepts, definitions and graphical notations,* Committee Draft, October, Version 3.4.

IST - 2001 - 34229, (2002-08-12), *Unified Enterprise Modelling Language (UEML),* Thematic Network, Annex 1, Description of Works, EC IST Project.

Kosanke K. Nell J.G. (Eds.), (1997), *Enterprise Engineering and Integration: Building International Consensus,* Springer-Verlag, Berlin, pp. 613-623.

Nell, J.G. (2001a), *Requirements for establishing manufacturing-enterprise-process* interoperability, TC 184 SC5 WG1 New-work-item proposal, N433 rev.2.

Nell, J.G. (2001b), *Enterprise Representation: A different paradigm for designing process-interoperability standards,* TC 184 SC5 WG1, NIST, Gaithersburg, MD, USA.

OAG, (2001), OAGIS: *Open Applications Group Integration Specification,* Open Application Group, Incorporated, Release 7.2.1, Doc. No. 20011031.

OMG, (1992), *Object Management Architecture,* Version 2, Open Management Group.

Open Group, (2000), *TOGAF: The Open Group Architecture Framework,* Document No. 1910, Version 6.

Schelenoff, G. Gruninger, M. Tissot, F. Valois, J. Lubell, J. Lee, J. (2000), *The Process Specification Language (PSL): Overview and Version 1.0 Specification,* NIST, Gaithersburg, MD, USA.

Vernadat, F. (2001), *UEML: Towards a Unified Enterprise Modelling Language.* Proc. 3ème Conférence Francophone de Modélisation et Simulation (MOSIM'01), Troyes, France, 25-27 April, pp. 3-13. (Invited paper).

Vernadat, F.B. (1996), *Enterprise Modelling and Integration: Principles and Applications,* Chapman & Hall, London.

W3C, (1998), *XML, Extensible Mark-up Language,* W3C XML 1.0, February.

W3C, (2000), *Simple Object Access Protocol (SOAP) 1.1,* W3C Note, http://www.w3.org/TR/2000/NOTE-SOAP-20000508/, May.

W3C, (2001), *Web Services Description Language (WSDL) 1.1,* W3C Note 15 March, http://www.w3.org/TR/2001/NOTE-wsdl-20010315.

Interoperability of Standards to Support Application Integration

Em delaHostria
Rockwell Automation, USA, em.delahostria@ra.rockwell.com

Abstract: One of the key challenges in the design, implementation, operation, and maintenance of an industrial automation system is the selection of appropriate standards to facilitate system integration. A significant aspect of the integration challenge is determining whether the standards-conformant components making up the system will operate with each other. The interoperability of these components depends on which interfaces are used in their implementation.

This paper addresses how such interoperability issues can be addressed, early in the system requirements-capture and design stages and throughout its life-cycle, by an application integration framework. The framework provides an approach to identify (a) the necessary integration model views of a control system application and (b) the applicable standard interfaces within these model views. Given the set of required interfaces and the set of selected options for each interface, automation system suppliers can assess design and implementation choices that meet the application's requirements.

1 SITUATION

In deploying an industrial automation system that meets the requirements of an application, the necessary functions may be enabled either as designed-in functions or as add-on functions. The cost of enabling any function, during a production activity, is clearly dependent on the flexibility of the underlying control architecture used and the resources employed to realize the automation control system. The choices of interfaces, made at design time, for the resources that implement the functions determine how well these functions can be performed in a coordinated manner at run time.

In selecting these resources, it is important to insure that the resource interfaces provided are not only compatible to realize system integration but also interoperable to achieve application interoperability. This situation can be difficult to realize especially if the resources are supplied from multiple vendors. One approach to realizing a high level of integration and a fuller degree of interoperability is to select interfaces that conform to international and industry standards. However, it is also equally important to verify that that the standards specifying the interfaces are compatible and interoperable, given the requirements of a manufacturing application. This aspect is key to the practical use of an application integration framework.

1.1 Model view of a manufacturing enterprise

In this paper, the model view of a manufacturing enterprise consists of an organization whose main activities are partitioned according to the enterprise functions described in the IEC 62264, (2002) draft standard. The key areas of interest cover the activities in the design, development, operation, and maintenance phases of a manufacturing application process lifecycle. These activities are very interdependent with similar activities in the design, development, production, and distribution phases of a product lifecycle. In most cases, the production phase of a product's lifecycle overlaps with the operation and the maintenance phases of a manufacturing application's process lifecycle. However, the resources employed in the process can be considered to have their own lifecycle, since some of these resources are consumed while others break down and then either repaired or replaced.

This particular but partial model view is derived using a methodology specified in ISO 15704, (2001), an international standard defining a generalized reference architecture of a generic enterprise. In this standard, the activities performed during an enterprise's operational phase usually include all those activities associated with the lifecycle of a product. In some cases, some of these product lifecycle activities, such as product distribution and field support are each accomplished by a different enterprise in the supply chain.

In practice, different standards are used to describe the lifecycle-specific activities. For instance, a description of an integrated product and manufacturing process planning activity could be realized if the product data, expressed in conformance with the ISO 10303, (1999) standards, can be used to drive the design, development and operation of a manufacturing process, expressed in a standard specification language like ISO 18629, (2002). The integrated description can be generated if the related definitions used in product data exchanges (ISO 10303, 1999), in production process sequences (ISO 18629, 2002), and in the production information exchanges (ISO

62264, 2002) were harmonized to facilitate the construction of an integrated product and process model view.

1.2 Model views of an enterprise's manufacturing system

During the operating phase of a manufacturing process lifecycle, a required function can be performed if it is already enabled or it may be installed, configured, and executed on demand. A key condition to be satisfied is that interfaces of the resources used to perform the function are configured to work with the corresponding resource interfaces of the other functions involved in a target manufacturing application. These conditions can be illustrated using various model views of a target application. The manufacturing application model view provided by ISO 15745, (2002) relates the process to the products produced and to the manufacturing resources utilized.

The application integration framework defines the elements and rules for composing this model view of a manufacturing application. A typical model view of an integrated manufacturing application consists of one or more processes that are enabled by a group of resources of various types – machinery, devices, personnel, software, materials, utilities, and other forms of equipment. The information exchanged between the resources can also be modeled as one or more software units that include various types – data items and structures, databases, communication protocols, and other software components that handle the data items. In the ISO 15745, (2002) standard, the specific interfaces between the resources are enumerated and their compatible options are summarized in a concise statement of interoperability, expressed in terms of XML statements and associated schema (XML, 2000/2001).

2 FRAMEWORK FOR INTEGRATION AND INTEROPERABILITY

To understand how the processes in an application cooperate with each other, a set of interfaces are assumed to be configured to enable the flow of materials, information, or any other resource needed to accomplish the requirements of an application. An example of such an application is a material handling application where the objective is to move raw materials, in-process goods, and finished goods according to the needs of an enterprise's manufacturing and business processes.

2.1 Framework elements and rules

In ISO 15745, (2002), a manufacturing application is modeled as a set of manufacturing processes, resources and information exchanges for the purpose of expressing the integration requirements. Integration models are expressed in terms of the Unified Modeling Language (ISO/IEC19501, 2001) conventions and are used to identify the required interfaces and to show how these choices meet the functional and performance requirements of the application (see Fig. 1 below).

In terms of UML class diagrams, an integration model of a *Manufacturing Application* class consists of a set of *Manufacturing Process* classes, a set of *Manufacturing Resource* classes, and a set of *Manufacturing Information Exchange* classes. Further, a generic *Manufacturing Resource* class represents a set of sub-classes - *Manufacturing Automation Device, Communications Network, Material or Finished Part, Manufacturing Software, Equipment and Machinery*, and *Manufacturing Personnel*. A *Manufacturing Information Exchange* class is further modeled as a set of sub-classes that facilitate the exchange of information structures between a collection of software objects that create, exchange, process, and store information items. These software objects are the sources and destinations of the information involved in the information flow with the rate and volume being gated by the number of transactions needed to support the material and control flows involved in a process.

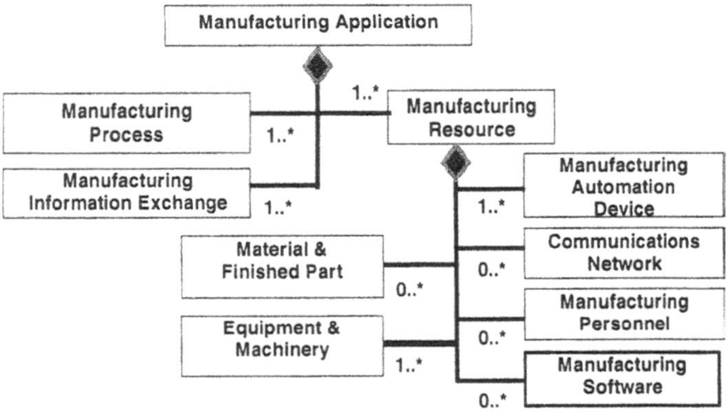

Figure 1: Integration Model View: Manufacturing Application

For example, in a material handling application, the timely delivery of a desired amount of a particular item to a certain destination, requires a set of “flow channels” that are interoperable. These "flow channels" may transport incoming raw material to a work-in-process area or finished goods from work-in-process to inventory or finished goods from inventory to shipping docks.

An integration model view is associated with each major class composing a *Manufacturing Application* class. A UML package consisting of a class diagram, a sequence diagram, and a deployment diagram represents each integration model view, per the ISO 15745, (2002) standard. A sequence diagram shows the series of transactions between the cooperating objects and the interfaces through which the object classes exchange material, information, or energy. These interfaces are further identified in the deployment diagrams and in the class diagrams. A set of integrated activities can only support the desired flow of materials, information, and other manufacturing resources, if the interfaces are interoperable, i.e. whatever is exchanged across the interface can be identified, recognized, understood, and properly handled by both resources collaborating via the interface.

2.2 Integration models and interfaces

The use of standard interfaces to elaborate an integration strategy is an obvious and proven approach. Integration of incompatible resources from different suppliers can be realized using interface adapters, converters, or gateways, but only to a lower degree of integration. A higher degree of integration can be achieved when the common interfaces are configured to interoperate in an optimal fashion, with very minimal use of gateway devices. In addition, use of interfaces based on open system standards – either formal international standards or specifications generated through industry consensus, will favor a broader source of suppliers.

A natural extension of such an approach is the use of an object-oriented integration model view that identifies the interfaces exposed by the component object classes of an application system model and then describes their compatible configurations in a systematic and web-ready form.

2.2.1 Process integration and functional interoperability

A manufacturing process consists of one or more activities, where each activity is associated with the performance of a specific function. Each function has to be performed in the right sequence, at the right time, at the right place, on the right target, by the right resource, with the expected outcome, in order to insure the desired flow of material, information, and energy. When the rate, volume, and quality of the process output meet the production goals, the activities of the manufacturing process are considered to be well integrated and the functions involved being interoperable.

Using the ISO 15745, (2002) application integration model, the various types of process interfaces are identified, chosen and configured in a compatible manner to sustain the flow type and the flow rates per the requirements of an application. The attributes of each process interface type – input, output, setup, monitoring, can be defined using the generic definitions in ISO 18629, (2002) and the activity-specific definitions in IEC 62264, (2002). The inter-process exchanges and their synchronization can be described in terms of the generic and activity-specific attributes. The set of configurations of the selected process interfaces can be expressed as a set of XML files and the corresponding schemas. A set of XML-based configuration statements (XML, 2000/2001) can serve as a reference set of interoperable configurations for such types of application. This reference interoperability statement can then be used to select the resources and the information structures necessary to support the required flows between the processes.

2.2.2 Resource Integration and performance interoperability

Following the ISO 15745, (2002) framework, the resources that implement the processes enable the material flows by coordinating the activities and operations using associated information exchanges. Using the framework's resource integration model view, the various types of exposed resource interfaces are similarly identified, chosen, and configured to support the reference interoperability statement for process integration. Each resource, such as, a device, a machine, or a person, may have one or more types of interfaces – input, output, setup, monitoring, power, and environmental. The use of common standard interfaces, such as those promoted by IEC, ISO, or some industry organization, enables the resources to transfer materials and information with each other.

For example, a device's communication network interface may conform to specific clauses in the IEC 61158, (2000) fieldbus standards, where the parameters in each clause are set to a range of values, as required by the application process. The values are chosen in order to support the bandwidth, response time, and coverage requirements of the application. These communication interface configurations, common to the inter-operating devices may be denoted in an XML file that is organized by its corresponding XML schema (XML, 2000/2001).

A set of XML files and related schema serves as a reference statement of interoperability for the group of resources used in a manufacturing application process. These statements of resource interoperability support the statement of process interoperability. A reference statement can be re-used in similar applications to verify if the proposed resources can coordinate and perform the functions needed to conduct the material and information flows at the desired rate, volume, quality, and cost.

2.2.3 Information exchange integration and data interoperability

In a similar manner, the ISO 15745, (2002) integration framework defines an information exchange model view that exposes the software interfaces and the other information handling units that enable the flow of information required to control the manufacturing process. The information handling objects are distributed among the nodes on a communication network. The nodes are distributed among the devices, which in turn are distributed among the machines, equipment, materials, and operators.

A transaction between two objects occurs at the information exchange interfaces that have been configured to support the rate, volume, and response time of information exchanges. The different types of messages being sent from one object to another object are associated with a set of interfaces that define the services, protocol, and data structures. In general, the interfaces used to setup the messaging paths may be different from those used to convey the messages. The message origin, destination, size, structure, rate, latency, freshness, fidelity, security, and persistence determine the type of interface to be used. These information exchange interfaces are chosen to support the manufacturing process requirements.

For example, the OPC-DA, (2000) Data Access specification defines the generic interface services that may be used to access the data structures and their current values in a particular OPC server used in a material handling system. The meaning of the data structures and the specific formats can be the definitions specified in ISO 13374, (2002) for monitoring the machinery used in a process.

The particular settings for each software interface specification that is used in the application process to meet the information exchange requirements can be summarized in a set of XML files and related schema (XML, 2000/2001), as recommended in ISO 16100, (2002). The set of XML files serves as reference statements of interoperability for the software components that exchange the data structures and handle the corresponding semantics. These statements of information exchange interoperability support the statements of process and resource interoperability.

2.3 System integration and application interoperability

Practical system integration within an application is realized when the processes, the resources, and the information exchanges involved are all served by interoperable interfaces. The combined sets of XML files and related schema (XML, 2000/2001) that represent the various types of reference interoperability statements form an application interoperability statement or profile. The interoperability profile captures the application requirements in terms of the different types of interfaces and their specific configurations.

For example, in designing a material handling application, the selection and configuration of the interfaces for the material transport equipment, escort memory subsystem, and inventory control and management subsystem depend on several time-dependent and site-specific variables. Examples of these variables are the type of unit loads, number of orders issued, number of shipments, finished goods inventory, work-in-process inventory, throughput, average time to fill an order, and other production related parameters. Unless the material handling interfaces are configured to support the flows demanded by the production goals, the various subsystems could not transport the unit, batch, or streaming loads as needed by the application. Furthermore, if the information handling interfaces are not interoperable, then the information flows will not match the material flow and will result in loss of tracking and control.

2.3.1 Uses of interoperability profile and integration models

Interoperability is the ability of two or more systems or applications to exchange information and use the information that has been exchanged (note that the applications may be wholly resident on the same system or distributed across multiple systems).

The availability of a reference application interoperability profile enables an application developer to verify if a proposed resource implementation (i.e. a product) from a supplier provides all the required functional interfaces and whether these interfaces can be configured to operate within a specified performance range.

If each device, equipment, machine, or software being procured has a required interoperability profile, then matching the offered profiles with the reference profiles can facilitate the interoperability verification process.

The set of UML diagrams describing an integration model view - class, sequence, use case, state, and deployment, also provides reference descriptions of the application's detailed process behavior and resource attributes. These diagrams can be helpful in the design, operation, and maintenance of a manufacturing application.

2.3.2 Safety, environmental, and security considerations

Other application considerations can be included in both the interoperability profile and the integration models. For example, the reduction in the amount of scrap and waste, as well as, the efficient use fuel and energy, both contribute towards a more environmental-friendly design of the manufacturing process. The supporting interfaces and the associated criteria can be treated within the integration framework, as well. Product components that do not require special disposal procedures upon product obsolescence can be noted in an extended interoperability profile.

The safe and secure operation of a manufacturing process depends largely on the types of interfaces selected for the monitoring and control of those activities that pose a risk to the health and safety of plant floor personnel. A secure operating platform provides appropriate control over access to the critical information resources that run the system. Again, use of properly matched evaluation assessment levels of security to the known security targets can be verified as part of the interoperability checking process. The use of the Common Criteria specification in ISO/IEC 15408 (2000) can provide a consistent scheme for evaluating vulnerabilities and assessing appropriate security measures.

3 INTEROPERABILITY SCENARIOS

A key to the practical use of the framework approach described above is the availability of standard interfaces that can be tailored to inter-operate. The choices for standard interfaces are based on these interfaces being configured to exchange materials and information at the desired rate and volume. For instance, the method of denoting process interfaces in ISO 15745, (2002) and the method of describing a process in ISO 18629, (2002) must have a shared ontology for use in the manufacturing process domain. The shared ontology may be derived from the definitions in IEC 62264, (2002), Enterprise-Control System Integration. In this effort, the processes and their component activities are described relative to an enterprise model view, generically defined in ISO 15704, (2002). In IEC 62264, (2002), the process boundaries are chosen based on the types of information exchanged between the production control system and the other enterprise business functions.

An example of a set of areas that need alignment and harmonization among the various standards is given in Table 1. The rows list the standards providing interface specifications while the columns list the standards providing integration and interoperability frameworks. Each table element identifies the set of definitions, in each standard, that have to be aligned in order for the two standards to be used consistently within the same application. To address different types of applications in various industry sectors will require a more comprehensive table involving other standards, showing the areas needing alignment and harmonization of definitions. Future standards can be included in these tables, indicating the needed interoperability with existing standards.

Table 1: Common Areas That Need to Be Aligned Between Standards

	ISO 15704 Enterprise Modeling	ISO 15745 Application Integration Framework	ISO 16100 Manufacturing Software Capability Profiling
IEC 62264: Enterprise-Control System Integration	Enterprise model	Integration models	Capability classes
ISO 18629: Process Specification Language	Process lifecycles	Process description	Process description
ISO 13374: Condition monitoring & diagnostics of machines	Asset management model views	Information exchange and resource models	Capability model
ISO 20242: Application Service Interface	Enterprise modeling	Device interoperability profiling	
Pre EN ISO 19439 & 19440: Enterprise modeling	Generic model views		
Device Profile Guideline (IEC 65A - 65/290/DC)		Device integration model	
ISO 10303: Application Protocols	Enterprise lifecycle modeling	Resource integration model	

There are additional considerations for harmonized definitions among the row elements. For example, activity descriptions and data models in IEC 62264, (2002) and ISO 13374, (2002); data definitions & structures in IEC 62264, (2002) and pre EN ISO 19440, (2002); process descriptions in ISO 18629, (2002) and ISO 10303, (1999); resource views in ISO 15745, (2002) and ISO 15704, (2001); function views in ISO 16100, (2002) and ISO 15704, (2001); profile templates in ISO 16100, (2002) and ISO 15745, (2002).

The IEC 62264, (2002) standard also provides definitions of data units exchanged between a subset of functional activities that comprise the various enterprise processes. These activities are organized and performed in a particular sequence when a specific product is taken through its lifecycle – planning, design, implementation, production, distribution, and disposition. By correlating the activity definitions specified in the ISO 10303, (1999) STEP application protocols (AP) with the activity definitions defined in IEC 62264, (2002), the associated data structures being exchanged can be compared with respect to syntax, semantics, and synchronization.

4 SUMMARY

Use of an application integration framework facilitates the construction of a set of manufacturing application system models and the compilation of the standard interfaces, identified within the models, as required by the system resources that enable the manufacturing processes.

Using the combined integration model views for the process, resource, and information exchanges, a list of interfaces required to support the flows can be identified. The desired or as-built interface configurations can be described in a set of XML files, using the XML schemas defined in the ISO 15745, (2002) application integration framework standard. For each configured interface, an XML file contains the particular settings or values assigned to the interface parameters. The suite of XML files and the corresponding schemas can be labeled as the interoperability profile for a particular manufacturing application. The specific rules used to select the interface parameter values, based on the application's needs, can also be documented in a complementary set of XML files and schemas (XML, 2000/2001). The conditions applied to the integration model views can be described in terms of constraint statements attached to the UML diagrams for the various integration model views.

The models and the associated interfaces capture the control and coordination mechanisms needed to support the interoperability of the set of processes within the application system. The parameters and options detailed in the set of standard specifications associated with the required interfaces can be selected and verified to provide a set of compatible configurations. These compatible configurations of interfaces can represent a system solution that meets the requirements of a manufacturing application system. Similar manufacturing applications within an enterprise, a supply chain, or an industry can use a set of compatible interface configurations as a starting basis for developing more specific solutions that support system interoperability.

The statement of the interoperability of a set of standards used in an application can only be realized if the standards have been harmonized. A more comprehensive table of interoperable standards, similar to the chart in Section 3, needs to be developed. The elements in the charts have to be harmonized to support application integration. The critical effort for standards committees is to collaborate and to coordinate their results to achieve the interoperability of the standards, plus targeted industry dialogues. More efficient methods are needed to enable the collaboration and coordination, including web-based meetings, common repositories, and joint efforts.

5 REFERENCES

IEC 61158, (2000), *Industrial Process Control and Measurement Systems - Digital Communications: Fieldbus.*

IEC 61512, (1997), *Batch Control.*

IEC 62264/FDIS, (2002), *Enterprise-Control System Integration.*

IEC 65A-65/290/DC, (2002), *Device Profile Guideline.*

ISO 10303, (1999), *Product data representation and exchange.*

ISO 13374/FDIS, (2002), *Condition monitoring and diagnostics of machines - Data processing, communications and presentation.*

ISO 15704, (2001), *Requirements for Enterprise Reference Architectures and Methodologies.*

ISO 15745/FDIS, (2002), *Industrial automation systems and integration - Open systems application integration framework.*

ISO 16100/FDIS, (2002), *Industrial automation systems and integration - Manufacturing software capability profiling for interoperability.*

ISO 18629/DIS, (2002), *Industrial automation system and integration - Process specification language.*

ISO 20242, (WD/2002), *Industrial automation systems and integration -Application Service Interface.*

Pr EN ISO 19439, (2002), *Enterprise Integration – Framework for Enterprise Modelling.*

Pr EN ISO 19440, (2002), *Language Constructs for Enterprise Modelling.*

ISO/IEC 15408, (2000), *Information technology- Security techniques -Evaluation criteria for IT security.*

ISO/IEC 19501, (2001), *Information technology – Unified Modeling Language (UML).*

OPC DA, (2000), *Open Process Control: Data Access Specifications.*

XML, REC-xml-20001006, (2000), *Extensible Markup Language (XML)* 1.0 Second Edition – W3C Recommendation 6 October.

XML, REC-xmlschema-1/2-20010502, (2001), *XML Schema Part 1/2: Structures/ Data types* – W3C Recommendation 02 May.

MultiView Program Status: Data Standards for the Integrated Digital Environment

Richard L. Engwall[1], and John W. Reber[2]
[1]*RLEngwall & Associates, USA,* [2]*Trident Systems Inc, USA,* rlengwall@aol.com

Abstract: The MultiView (MV) Program is a USA Research & Development project, which focuses on developing data standards for the integrated digital environment. The aim is to achieve a high degree of interoperability of the Information Technology (IT) systems for complex engineer-to-order systems, products and processes over their life cycle.

Project objectives are to use available software and standards, provide a single schema for seamless integration of the data sets and to develop a framework for data access and communication over the life cycle of the products as well as the systems.

1 INTRODUCTION

The MultiView (MV) Program is a multi-year USA congressionally mandated Research & Development project, managed by the U.S. Army Logistics Integration Agency, developed by Trident Systems Inc. as prime contractor and Concurrent Technologies Corporation and RLEngwall & Associates as subcontractors. MultiView's focus is on developing data standards for the integrated digital environment. The program start was July 2000; program completion is dependent upon obtaining year-to-year available funding.

Enterprises that acquire and sustain complex modern weapons systems face unprecedented challenges in containing costs while taking their systems through concept, design, development, deployment, and retirement. The schema and associated data set required for specifying, developing, operating, maintaining, and disposing of such systems is extremely large and involves myriad subtle relationships among seemingly disparate domains. Added to this complex reality, contractors and program offices must develop, deliver, and manage systems meeting aggressive readiness requirements and shifting mission objectives within stringent budget constraints. Affordability has become as important as mission performance when developing and sustaining such systems.

Similarly, product customization to satisfy specialized customer requirements, time-to-market, and affordability have become as important to the industrial world as product performance is for commercial complex systems. Commercial companies are increasingly operating in a virtual extended enterprise environment striving to share selective information with their myriad of disparate distributed stakeholders. Therefore the industrial approach to solving these problems through standards and partnering activities can be used as a model for the Department of Defense (DOD) and the MV Program.

There is significant commonality of interest between DOD and U.S. industry in trying to achieve a high degree of interoperability of their Information Technology (IT) systems for complex engineer-to-order systems, products and processes over their life cycle. Success in so doing minimizes the number and cost of transactions and results in a lean more affordable operating mode for all involved. The path to this objective (and the specific objective) of the MV program involves the use to maximum extent possible of:

- Existing commercial enterprise software and standards
- A single schema for seamless integration of broad and varied data sets
- A framework or architecture for the communication and access to this data, over the life cycle of the system or product involved

2 DESCRIPTION OF PROBLEM

Military weapon system program offices must be able to define technical or fiscal metrics to assess the Total Ownership Cost (TOC) of their systems. Traditional system acquisition and life cycle management practices include the use of automated tools for modeling and simulation, configuration management, and supply support systems to create and manage technical data for systems both in development and in the field. However, each tool uses its own data representation and storage mechanism causing major problems in communicating between systems.

With a few exceptions, mostly in the commercial industrial world, no real interoperability exists among tools, even for exchanging data concerning the same technical area of the system. Human operators most often re-enter data for each tool employed in the process. Interoperability problems grow with automated support. As program management offices use advanced process modeling and planning techniques, and work with complex sets of data across multiple databases, as shown in Fig. 1 the interoperability problem multiplies. This presents a significant and growing challenge to programs to effectively integrate complex system data.

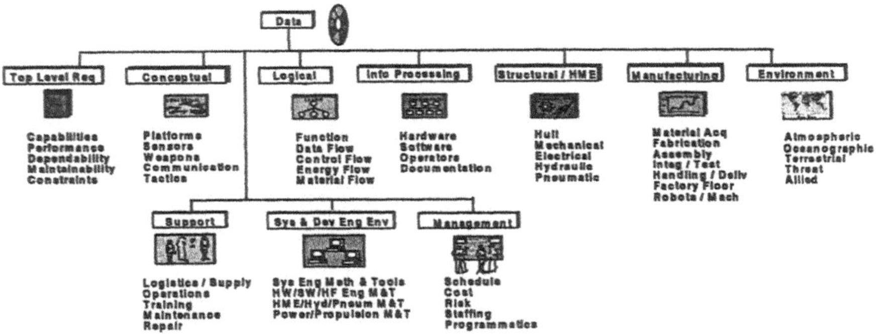

Figure 1 Product Life Cycle Multiple Domain Information Needs

In the commercial industrial world "best of breed" enterprise integration interoperability solutions are based on achieving as much integration as possible through use of:

- Applying Business-Process Re-engineering (BPR) to simplify business processes;
- Tailoring business streams to separate basic/core, options, and unique processes;
- Developing a structured information systems framework/application integration middleware solution focusing on a single call data communication employing a common objective model using standards accepted by the international community or de facto standards such as OMG-CORBA-2, (1999), CORBA-bus, to abstract data at the data interface business object level not the database/schema level;
- Deploying selective use of a portfolio of international information systems standards such as STEP, (ISO IS 10303, 1999), Web-based XML (2000, 2001), and ANSI X12 EDI data transfer/presentation solutions (2000);
- Deploying Commercial-off-the-Shelf (COTS) application solutions for production and management processes such as: ERP, PDM, MES, SCM, CRM, etc.

Application of these individual company COTS tools solutions just further exacerbates the overall enterprise interoperability problem. Most of the present enterprise interoperability successes have been with management-related textual business processes and not with the technical processes involving systems engineering, product and factory modeling and simulations, CAE, CAD, CAM and product life cycle support processes. The MV Program's prime focus is to address these technical business process areas and to integrate the technical information with the management business process areas.

3 MV APPROACH TO PROBLEM SOLUTION

Meeting the challenge to effectively integrate complex system data is a key to ensuring that complex systems (such as the Abrams Tank, the Navy's 21st Century Destroyer, and the Joint Strike Fighter, and commercial complex systems such as wide-body airplanes, automobiles, trucks, ships, trains, offshore oil rig platforms, satellites, etc.) are both mission and performance effective and affordable. The response to the challenge involves three principal parts:

- The first part is an organization of the system data through an integrated multi-domain data schema for representing system product and process data. This will be essential to developing and operating an advanced integrated environment.
- The second is an integrated environment that employs formal methods and automation to support the full range of data manipulation and communication required by complex system life cycle activities. This environment enables a broad spectrum of life cycle participants to evaluate alternatives in multiple domains simultaneously; provides a way for stakeholders to understand their needs in relation to the enterprise as a whole; and provides a continuous proactive means of identifying and successfully addressing key challenges for a complex system over time.
- The third is an evolved culture where enterprise-wide cooperation is the rule and individual contributions are encouraged and efficiently managed.

Working within an integrated environment based on these three elements will provide a common frame of reference in which sophisticated relationships among technical domains, and between these domains and a system's affordability, can be explicitly identified and analyzed. The key to realizing gains from the combined elements is the data schema, essential to integrating the disparate data sets in use by complex system program offices and other related enterprises.

The MV Program needs to stay abreast of evolving data transfer standards to gain leverage by building on their standardized business processes, data type schemas, data dictionaries, and publish-and-subscribe message techniques wherever feasible. The MV common data schema uses, rather than competes with, these other data transfer and data storage means of achieving interoperability.

The MV response to the challenge involves the analysis and assessment of two primary concepts, each to be investigated against an evolving set of applicable schema requirements employing the primary systems engineering process described in ANSI/EIA 632, (1999). These two concepts are:

1. An integrated multi-domain data schema for representing system product and process data. In this concept "data schema" means a depiction of the structure and constraints of the contents of an information system: a data model, which defines a vocabulary of terms, their properties and relationships, and how the information to which they refer must be organized.
2. Enhancement of COTS integrated ERP (Enterprise Resource Planning) systems already in use in multiple Aerospace and Defense companies and planned for several DOD military services and the Defense Logistics Agency (DLA).

Concept (1) requires the development of a data schema constructed from existing or evolving standards that can completely represent all required product and process data. Such standards include the IGES and STEP standards as examples of a partially complete standard set. Concept (2) currently requires the migration of legacy databases to an applicable Meta database in accordance with the database format selected by each ERP provider.

The MV approach contrasts with the current object-oriented interoperability approach employing middleware that does not necessitate the use of the data schema concept (1) but which requires an integration schema for integration of the various databases to an Application Architectural Framework.

All of the above concepts must be traded off against an evolving set of MV program requirements, particularly those of a derived nature. These are evolving because the cost and schedule of the ultimate solution must be timely and affordable and as such, each requirement must be examined for its meaningful application to the MV program.

An extensive initial set of requirements have been evolved starting with the MV Statement of Work, with additional requirements derived from the MV Concept of Operations, decomposition of associated "Use Cases," user interviews, DOD Instruction 5000.2, development of existing relevant standards activities and initiatives, and commercial enterprise resource planning and management systems. The requirements defined to date address the schema integration Framework, Schema, and Life Cycle User Information System Requirements. The first demonstration of the MV schema was successfully conducted in July 2001 in the PM Abrams pilot program.

In order to reduce the time necessary to develop the schema, avoid duplication of effort, and enhance industry acceptance and use, the MV team agreed on a strategy of reusing existing schemas that cover some portions of the MV scope rather than developing the MV data schema from scratch. Such reuse will take two forms: integration into the actual MV schema, and reference from the MV schema.

Promulgation of the MV schema as a standard is essential for its acceptance and adoption for use by both industry and the DOD. A MV Schema Transition Strategy was developed in February 2002 defining how the MV schema can be transitioned to a formal living standard that is sought-after, endorsed, broadly accepted, and maintained and used by industry and DOD.

4 CURRENT STATUS

Any Government or commercial organization faces a dilemma in determining the degree of support to which one commits to achieve their organization's interoperability objectives. The complexity of the organization, complexity of product and/or services, present degree of product and process information systems, subsystems, and life cycle domains interoperability all dictate a different interoperability strategy and business case. What is important though is to deploy a structured business case methodology to guide one to make the most cost effective interoperability implementation decisions.

The primary issue is that there are many focused application solutions that are addressing a piece of the total interoperability need. These efforts provide partial solutions to pieces of the virtual enterprise-wide interoperability objective for particular areas of the enterprise. These activities do not provide the integration or framework for the entire enterprise life cycle. The MV Program is the first top down systems engineering approach that views all of the technical and business complex systems life cycle data requirements in a virtual enterprise-wide environment. MV provides the framework that builds on many of today's partial integration solutions. The MV Program integrates data schemas of these partial solutions with a MV top-level schema. The MV Program technical approach is object-flavored, modeled in EXPRESS, utilizes the ISO 15926, (2001) as the integration model, builds on ISO 10303, (1999) STEP APs, and has integrated with a portion of the U.S. Department of Defense Data Architecture. Two of the most significant integration activities in the ISO standards community are the IIDEAS (Integration of Industrial Data for Exchange, Access, and Sharing) project sponsored by the ISO TC184 SC4 activity (ISO IS 18876, 2000) and Product Life Cycle Support (PLCS, 1999). IIDEAS provides the underpinnings of the integration approach that the MV Program has developed. The MV Program is working with the STEP and PLCS standards activities to incorporate their work into the MV schema.

In addition, the commercial world is moving towards implementation of XML (2000, 2001), SOAP (2000), and other standards for World Wide Web data transfer and application integration architecture middleware, as well as deploying extensive ERP systems. MV will augment their successful implementation.

In the first year, the MV Program built an Interoperability Testbed to validate the schema, framework, and translators for the Abrams Tank pilot. This effort will be followed with additional DOD complex systems pilot application validations that will be selected by the MV stakeholders. A draft ANSI/EIA standard will be prepared and implemented that will subsequently be transitioned to an ISO standard. On 5/12/02 the Government Electronics & Information Technology Association (GEIA) issued a press release announcing the formation of a new standardization project based on the MV Program. The project, designated as GEIA/EIA-927 Common Data Schema for Complex Systems, will involve both industry and Government experts. In subsequent years a not-for-profit consortium will be formed to promulgate and maintain the use of the MV schema standard.

5 REFERENCES

ANSA/EIA 632, (1999), *Processes for Engineering a System*, http://global.ihs.com/

ANSI X12 EDI, (2000), *Electronic Data Interchange.*

ISO IS 10303, (1999), *Product data representation and exchange STEP*, TC 184 SC4.

ISO 15926 EPISTLE Core Model 3.1, (2001), *Integration of Life-Cycle Data for Oil & Gas production facilities.*

ISO IS 18876, (2000), *Integration of Industrial Data for Exchange, Access, and Sharing*, TC 184 SC4.

OMG-CORBA-2, (1999), *Common Object Request Broker Architecture- Interoperability.*

PLCS, (1999), *Product Life Cycle Support*, planned submission to ISO TC184 SC4 in 2002.

SOAP 1.1, (2000), *Simple Object Access Protocol.*

XML, REC-xml-20001006, (2000), *Extensible Markup Language (XML)*, 1.0 Second Edition – W3C Recommendation 6 October.

XML, REC-xmlschema-1/2-20010502, (2001), *XML Schema Part 1/2: Structures/ Data types*, – W3C Recommendation 02 May.

Workflow Quality of Service

Jorge Cardoso, Amit Sheth, and John Miller
LSDIS Lab, University of Georgia, USA, amit@cs.uga.edu

Abstract: Workflow management systems (WfMSs) have been used to support various types of business processes for more than a decade now. In e-commerce processes, suppliers and customers define a binding agreement or contract between the two parties, specifying quality of service (QoS) items such as products or services to be delivered, deadlines, quality of products, and cost of service. Organizations operating in modern markets require an excellent degree of quality of service management. A good management leads to the creation of quality products and services, which in turn fulfills customer expectations and achieves customer satisfaction. Therefore, when services or products are created or managed using workflow processes, the underlying WfMS must accept the specification, be able to predict, monitor, and control the QoS rendered to customers. To achieve these objectives the first step is to develop an adequate QoS model for workflow processes and develop methods to compute QoS.

1 INTRODUCTION

Organizations are constantly seeking new and innovative information systems to better fulfill their mission and strategic goals. In the past decade, Workflow Management Systems (WfMSs) have been distinguished due to their significance and impact on organizations. WfMSs allow organizations to streamline and automate business processes, reengineer their structure, as well as, increase efficiency and reduce costs.

Our experience with real world enactment services (Miller, et al, 1998, Kochut, et al, 1999) and applications made us aware of the importance of Quality of Service (QoS) management for workflow systems. For organizations, being able to characterize workflows based on their QoS has three direct advantages. First, it allows organizations to translate more efficiently their vision into their business processes, since workflow can be designed according to QoS metrics. Second, it allows the selection and execution of workflows based on their QoS to better fulfill customers expectations. Third, it also makes possible the monitoring of workflows based on QoS, allowing managers to set compensation strategies when undesired metrics are identified.

Our goal is to develop a workflow QoS specification and methods to predict, analyze and monitor QoS. We start by investigating the relevant quality of service dimensions, which are necessary to correctly characterize workflows. Once the QoS dimensions are identified, it is necessary to devise methodologies to estimates QoS for tasks. Finally, algorithms and methods need to be developed to compute workflow QoS. In workflows, quality metrics estimates are associated with tasks and tasks compose workflows. The computation of workflow QoS is done based on the QoS of the tasks that compose a workflow.

This paper is structured as follows. Section 2 introduces our workflow QoS model and describes each of its dimensions. In section 3, we describe how QoS estimates for tasks are created. Section 4 discusses two techniques to compute workflow QoS from task QoS. Section 5 discusses related work in this area and section 6 presents our conclusions.

2 WORKFLOW QUALITY OF SERVICE

For us, workflow QoS represents *the quantitative and qualitative characteristics of a workflow application necessary to achieve a set of initial requirements*. Workflow QoS addresses the non-functional issues of workflows, rather than workflow process operations. Quantitative characteristics can be evaluated in terms of concrete measures such as workflow execution time, cost, *etc*. Qualitative characteristics specify the expected services offered by the system such as security and fault-tolerance mechanisms.

Workflow QoS is composed of different dimensions that are used to characterize workflow schema and instances. To our knowledge most of the research carried out to extend workflow systems capabilities, in the context of QoS, has only been done for the time dimension (Bussler, 1998, Dadam, et al 2000, Eder, et al, 1999, Kao, Garcia Molina, 1993, Marjanovic, et al 1999, Sadig, et al, 2000, Son, et al, 2001), which is only one of the dimensions under the workflow QoS umbrella. Even though some WfMSs currently offer time management support, the technology available is rudimentary (Eder, et al, 1999). The Crossflow project (Damen, et al, 2000, Grefen, et al, 2000, Klingemann, et al, 1999) is the one that most closely relates to our work, for which workflow cost is also considered.

Quality of service can be characterized along to various dimensions. We have investigated related work to decide which dimensions would be relevant to compose our QoS model. Our research targeted two distinct areas: operation management for organizations and quality of service for software systems (which include networking, middleware areas, and real-time applications.) The study of those two areas is important, since workflow systems are widely used to model organizational business processes, and workflow systems are themselves software systems.

2.1 QoS Model

Based on previous studies and our experience in the workflow domain, we construct a QoS model composed of four dimensions: *time*, *cost*, *fidelity*, and *reliability*.

Time (T) is a common and universal measure of performance. For workflow systems, it can be defined as the total time needed by an instance to transform a set of inputs into outputs. Task response time (T) corresponds to the time an instance spends to be processed by a task. The *task response time* can be broken down into two major components: *delay time* and *process time*. Delay time (DT) refer to non-value-add time needed in order for an instance to be processed by a task. While, those two metrics are part of the task operation, they do not add any value to it. Process time (PT) is the time a workflow instance spends at a task while being processed, in other words it corresponds to the time a task needs to process an instance.

Cost (C) represents the cost associated with the execution of workflow tasks. During workflow design, prior to workflow instantiation and during workflow execution it is necessary to estimate the cost of its execution to guarantee that financial plans are followed. Task cost is the cost incurred when a task *t* is executed, and can be broken down into two major components: *enactment cost* and *realization cost*. The enactment cost (EC) is the cost associated with the management of the workflow system and workflow instances monitoring. The realization cost (RC) is the cost associated with the runtime execution of the task.

We view **Fidelity** (F) as a function of effective design and refer to an intrinsic property or characteristic of a good produced or service rendered. Fidelity is often difficult to define and measure because it is subjective to judgments and perceptions. Nevertheless, the fidelity of workflows must be predicted, when possible, and carefully controlled when needed. Workflow tasks have a fidelity vector dimension composed by a set of fidelity attributes ($F(t).a_i$), to reflect and quantify task operations. Each fidelity attribute refers to a property or characteristic of the product being created, transformed, or analyzed. Fidelity attributes are used by the workflow system to compute how well workflows, instances, and tasks are meeting user specifications.

Task **Reliability** (R) corresponds to the likelihood that the components will perform when the user demands it and it is a function of the failure rate. This QoS dimension provides information concerning the relationship between the number of times a task reaches the state done or committed, and the number of times it reaches the failed/aborted state.

2.2 QoS and Production Workflows

One of the most popular workflow classifications distinguishes between *ad hoc* workflows, administrative workflows, and production workflows. This classification was first mentioned by (McCready, 1992). The main differences between these types include structure, repetitiveness, predictability, complexity, and degree of automation.

The QoS model presented is better suited for production workflows (McCready, 1992) since they are more structured, predictable, and repetitive. Production workflows involve complex and highly structured processes, whose execution requires a high number of transaction accessing different information systems. These characteristics allow the construction of adequate QoS models for workflow tasks. In the case of *ad hoc* workflows, the information, the behavior, and the timing of tasks are largely unstructured, which makes the procedure of constructing a good QoS model more difficult and complex.

3 CREATION OF QOS ESTIMATES

Determining useful estimates for the QoS properties of a task can be challenging. A combination of *a priori* estimates from designers as well as estimates computed from prior executions will be used, with the historical data playing a larger role as more data is collected. Additional complexities are due to the fact that QoS is parametric. For example, the response time of a service that takes an XML document as input will depend on the size of the document. Estimates for workflows can be developed in two ways: (a) estimates for the entire workflow can be created just like they are for ordinary/atomic services (*i.e.*, *a priori* estimates refined as execution monitoring data is collected), (b) the QoS properties can be synthesized from the QoS properties of the tasks making up the workflow. Synthesizing aggregate estimates requires several problems to be solved, among them (1) determination of transitions probabilities from transitions conditions and (2) dealing with correlation between individual tasks.

In order to facilitate the analysis of workflow QoS, it is necessary to initialize task QoS metrics and also initialize stochastic information indicating the probability of transitions being fired at runtime. Once tasks and transitions have their estimates set, algorithms and mechanisms such as simulation can be applied to compute overall workflow QoS.

3.1 QoS for Tasks

Task QoS is initialized at design time and re-computed at runtime when tasks are executed. During the graphical construction of a workflow process, each task receives information estimating its quality of service behavior at runtime. The re-computation of QoS task metrics is based on data coming from the user specifications and from the workflow system log.

3.2 QoS for Transitions

The same methodology used to estimate task QoS, is also used to estimate workflow transitions probabilities. The user initializes the transitions probabilities at design time. At runtime the probabilities are re-computed. If a workflow has never been executed, the values for the transitions are obviously taken from initial user specifications. If instances of a workflow *w* have already been executed, then the data used to re-compute the probabilities comes from initial user specifications for workflow *w* and from completed instances.

4 QOS COMPUTATION

Once QoS estimates for tasks and for transitions are determined we can compute overall workflow QoS. We describe two methods that can be used to compute QoS metrics for a given workflow process: analysis and simulation. The selection of one of the methods is based on a tradeoff between time and accuracy of results. The analytic method is computationally faster, but yields results, which may not be as accurate as the ones obtained with simulation.

4.1 Analytic Models

Comprehensive solutions to the difficult problems encountered in synthesizing QoS for composite services are discussed in detail (Cardoso, et al, 2002). This work presents a stochastic workflow reduction algorithm (SWR)

(Cardoso, 2002) for computing aggregate QoS properties step-by-step. At each step a reduction rule is applied to shrink the network. At each step the response time (RT), processing time (PT), delay time (DT), cost (C) and reliability (R) of the tasks involved is computed. Additional task metrics can also be computed, such as task enactment time and setup time. This is continued until only one atomic task (Kochut, et al, 1999) is left in the network. When this state is reached, the remaining task contains the QoS metrics corresponding to the workflow under analysis. The set of reduction rules that can be applied to a composite service (network) corresponds to the set of inverse operation that can be used to construct a network of services. We have decided to only allow the construction of workflows based on a set of predefined construction rules to protect users from designing invalid workflows. The algorithm uses a set of six distinct reduction rules: (1) sequential, (2) parallel, (3) conditional, (4) fault-tolerant, (5) loop, and (6) network.

4.2 Simulation Models

While analytical methods can be effectively used, another alternative is to utilize simulation analysis (Miller, et al, 2002). Simulation can play an important role in tuning quality of service metrics of workflows by exploring "what-if" questions. When the need to adapt or to change a workflow is detected, deciding what changes to carry out can be very difficult. Before a change is actually made, its possible effects can be explored with simulation. To facilitate rapid feedback, the workflow system and simulation system need to interoperate. In particular, workflow specification documents need to be translated into simulation model specification documents so that the new model can be executed/animated on the fly.

In our project, these capabilities involve a loosely coupled integration between the WfMS (METEOR, Miller, et al, 1998, Kochut, et al, 1999) and the simulation system (JSIM, (Miller, et al, 2002). Workflow is concerned with scheduling and transformations that take place in tasks, while simulation is mainly concerned with system performance. For modeling purposes, a workflow can be abstractly represented by using directed graphs (*e.g.*, one for control flow and one for data flow, or one for both). Since both models are represented as directed graphs interoperation is facilitated. In order to carry out a simulation, the appropriate workflow model is retrieved from the repository and translated into a JSIM simulation model specification. The simulation model is displayed graphically and then executed/animated. Statistical results are collected and displayed which indicate workflows QoS.

5 RELATED WORK

The work found in the literature on quality of service for WfMS is limited. The Crossflow project (Damen, et al, 2000, Grefen, et al, 2000, Klingemann, et al, 1999) has given a major contribution. In their approach, a continuous-time Markov chain (CTMC) is used to calculate the time and cost associated with workflow executions. While the research on quality of service for WfMS is limited, the research on time management, which is under the umbrella of workflow QoS, has been more active and productive. Eder (1999) and Pozewaunig (1997) present an extension to CMP and PERT by annotating workflow graphs with time. At process build-time, instantiation-time, and runtime the annotations are used to check the validity of time constraints. The major limitation of their approach is that only direct acyclic graphs (DAG) can be modeled. This is a significant limitation since the majority of workflows have cyclic graphs. Cycles are in general used to represent rework actions or repetitive activities within a workflow. Marjanovic, Orlowska, (1999) describe a workflow model enriched with modeling constructs and algorithms for checking the consistency of workflow temporal constraints. Son, (2001) present a solution for the deadline allocation problem based on queuing networks.

Recently, in the area of Web services, researchers have also manifested an interest for QoS. The DAML-S, (2001) specification allows the semantic description of business processes. The specification includes constructs to specify quality of service parameters, such as quality guarantees, quality rating, and degree of quality. While DAML-S has identified specification for Web service and business processes as a key specification component, the QoS model adopted should be significantly improved to supply a realistic solution to its users. One current limitation of DAML-S' QoS model is that it does not provide a detailed set of classes and properties to represent quality of service metrics. The QoS model needs to be extended to allow a precise characterization of each dimension.

6 CONCLUSIONS

We have shown the importance of quality of service management of workflow and introduced the concept of workflow quality of service (QoS). While QoS management has a high importance for organizations, current WfMSs and workflow applications do not provide full solutions to support QoS. Research is necessary in four areas: *specification, analysis algorithms and methods*, *monitoring tools*, and *mechanisms to control* the quality of service. In this paper, we focus on workflow QoS specification and the development of algorithms and methods to calculate QoS. Based on the reviewed literature on quality of service in other areas, and accounting for the particularities of workflow systems and applications, we define a workflow QoS model, which includes four dimensions: time, cost, fidelity, and reliability. The use of QoS increases the added value of workflow systems to organizations, since non-functional aspects of workflows can be described. The specification of QoS involves fundamentally the use of an adequate model and the creation of realistic QoS estimates for workflow tasks. Once tasks have their QoS estimated, QoS metrics can be compute for workflows. Since this computation needs to be automatic we describe two methods for workflow QoS computation: analysis and simulation.

7 REFERENCES

Ankolekar, A., et al, (2001), *DAML-S: Semantic Markup for Web Services,* in Proceedings of the International Semantic Web Working Symposium (SWWS), pp. 39-54.

Bussler, C. (1998), *Workflow Instance Scheduling with Project Management Tools,* in 9th Workshop on Database and Expert Systems *Applications DEXA'98,* Vienna, Austria: IEEE Computer Society Press, pp. 753-758.

Cardoso, J. (2002), *Stochastic Workflow Reduction Algorithm.* LSDIS Lab, Department of Computer Science, University of Georgia. http://lsdis.cs.uga.edu/proj/meteor/QoS/SWR_Algorithm.htm. 2002.

Cardoso, J. (2002), *Workflow Quality of Service and Semantic Workflow Composition.* Ph.D. Dissertation. Department of Computer Science, University of Georgia, Athens, GA.

Dadam, P. Reichert, M. Kuhn, K. (2000), *Clinical workflows the killer application for process oriented information systems,* in 4th International Conference on Business Information System (BIS 2000), Poznan, Poland pp. 36-59.

Damen, Z. et al, (2000), *Business-to-business E-Commerce in a Logistics Domain,* in The CAiSE*00 Workshop on Infrastructures for Dynamic Business-to-Business Service Outsourcing; Stockholm, Sweden.

DAML-S, (2001), *Technical Overview - a white paper describing the key elements of DAML-S.*

Eder, J. et al, (1999), *Time Management in Workflow Systems,* in BIS'99 3rd International Conference on Business Information Systems, Poznan, Polen: Springer-Verlag, pp. 265-280.

Grefen, P. et al, (2000), *CrossFlow: Cross-Organizational Workflow Management in Dynamic Virtual Enterprises.* International Journal of Computer Systems Science & Engineering, 15(5): p. 227-290.

Kao, B. Garcia Molina. H. (1993), *Deadline assignment in a distributed soft real-time system,* in Proceedings of the 13th International Conference on Distributed Computing Systems, pp. 428-437.

Klingemann, J. Wäsch, J. Aberer, K. (1999), *Deriving Service Models in Cross-Organizational Workflows,* in Proceedings of RIDE - Information Technology for Virtual Enterprises (RIDE-VE '99), Sydney, Australia, pp. 100-107.

Kochut, K.J., Sheth, A.P. Miller, J.A. (1999), *ORBWork: A CORBA-Based Fully Distributed, Scalable and Dynamic Workflow Enactment Service for METEOR.* Large Scale Distributed Information Systems Lab, Department of Computer Science, University of Georgia: Athens, GA, USA.

Marjanovic, O. Orlowska, M. (1999), *On modeling and verification of temporal constraints in production workflows*. Knowledge and Information Systems, 1(2).

Marjanovic, O. Orlowska, M. (1999), *On modeling and verification of temporal constraints in production workflows*. Knowledge and Information Systems, 1(2), pp. 157-192.

McCready, S., *There is more than one kind of workflow software*. Computerworld, 1992. November 2:86-90.

Miller, J.A. Cardoso, J.S. Silver. G. (2002), *Using Simulation to Facilitate Effective Workflow Adaptation,* in Proceedings of the 35th Annual Simulation Symposium (ANSS'02). San Diego, California, pp. 177-181.

Miller, J.A. Seila, A.F. Xiang, X. (2000), *The JSIM Web-Based Simulation Environment.* Future Generation Computer Systems: Special Issue on Web-Based Modeling and Simulation, 17(2): pp. 119-133.

Miller, J.A., et al, (1998), *WebWork: METEOR2's Web-based Workflow Management System.* Journal of Intelligence Information Management Systems: Integrating Artificial Intelligence and Database Technologies (JIIS), 10(2): pp. 185-215.

Pozewaunig, H. Eder, J. Liebhart, W. (1997), *ePERT: Extending PERT for workflow management systems,* in *First European Symposium in Advances in Databases and Information Systems (ADBIS)*. St. Petersburg, Russia, pp. 217-224.

Sadiq, S. Marjanovic, O. Orlowska, M.E. (2000), *Managing Change and Time in Dynamic Workflow Processes.* The International Journal of Cooperative Information Systems, 9(1, 2), pp. 93-116.

Son, J.H. Kim, J.H. Kim, M.H. (2001), *Deadline Allocation in a Time-Constrained Workflow.* International Journal of Cooperative Information Systems (IJCIS), 10(4), pp. 509-530.

Improving PDM Systems Integration Using Software Agents

Yinsheng Li[1], Weiming Shen[1], and Hamada H. Ghenniwa[2]
[1]National Research Council Canada, [2]University of Western Ontario, Canada
weiming.shen@nrc.ca

Abstract This paper addresses challenges in integrating distributed and heterogeneous Product Data Management (PDM) systems, introduces a virtual enterprise oriented agents-based integration solution, and presents an infrastructure with interactive and communicative agents as community and domain services to support distributed management of documents, BOMs, workflows and engineering knowledge, and secure communication in a collaborative product development environment. Considering the standardized Web-based agents and fundamental product data managed by PDM systems, the proposed paradigm is scalable and can also be used to integrate business systems in manufacturing enterprises.

1 INTRODUCTION

With the globalization of economy, an industrial organization should be able to react quickly and correctly to external changes and proactively manage the internal changes. Therefore, establishing an effective and efficient collaborative product development environment within an organization and with its partners becomes not only a requirement, but also an essential feature of its business model.

To that end, several tools and design paradigms have been proposed, such as Product Data Management (PDM). PDM is becoming an efficient facility for collaboration within and across industrial organizations. A PDM system takes product databases as basic information infrastructures; takes product structures (generally bills of materials) as a core organizing backbone; combines a series of engineering data and documents with each other, and allows for effective access, integration, management of product data and business workflows. Moreover, it supports integration between different applications, and facilitates collaborative activities among designers. A designer, with development workspaces in a PDM system, is able to take advantages of all factors related to a product lifecycle in a way that improves the quality of the end products, reduces dramatically trials, errors, costs and time-to-market of new products (Interleaf, 1998; ENOVIA, 1998; SDRC, 1996; PTC, 2002).

A number of challenges exist in designing and implementing PDM systems especially in cross-organization collaborative product development environments, such as virtual enterprises (VE), where there are different proprietary PDM systems (e.g., ProductManager by IBM (ENOVIA, 1998), IMAN by EDS (Interleaf, 1998), Metaphase by SDRC (SDRC, 1996), and Windchill by PTC (PTC, 2002)) preferred by their member companies due to certain historical and economical reasons. Therefore, providing integrity and consistency services for such collaborative environments with multiple PDM systems becomes very important. This integration will enable participants of a collaborative product development environment within and across organizations to access different product data seamlessly; extend their workflows to run through across the distributed collaborative workspaces; and come into being an integrated, efficient and unified product development platform for multidisciplinary collaborations (Peltonen et al, 1996; Jasnoch, Haas, 1996).

Although that some of the existing PDM systems have few successes in collaborating with one another, they still require tons of customization and high maintenance cost. This might be attributed to the complexity of the environment and to several key issues that need to be addressed in a comprehensive analysis and design approach, such as:

- Distributed management of documents, bills of materials (BOM), and workflows;
- Security issue on communication over the Internet within a single or multiple PDM systems (e.g., ProductManager transfers data over the Internet using built-in FTP with ASCII format);
- Security issue on access authorization based on distributed multidisciplinary teams;
- Privacy issue when sharing information.

In a previous project, a Web and coordination services-based integration paradigm for distributed PDM systems has been proposed and implemented (Li, 2000). Furthermore in this paper, we propose an agent-based approach to facilitate the integration in multiple PDM systems. With agents as interaction services layer, PDM systems get harmonized and combined with each other, especially in dealing with data consistency, communication security, and business privacy. Moreover, this approach provides a "virtually" homogenous environment for the essential PDM services such as document management, product structure management and workflow management across multiple organization platforms.

The paper is organized as follows: Section 2 presents our agent-based PDM systems integration approach; Section 3 describes distributed transactions under such integration; and Section 4 gives some conclusions.

2 AGENT-BASED PDM SYSTEMS INTEGRATION

2.1 Integration system upon domain and community services

Previously in (Li, 2000), we proposed two types of integration facilities for multiple PDM systems, namely, domain services and community services. Domain services collectively act as brokers between local and remote PDM systems, in the sense to make communication transparent between them. Community services, like in public human services, are managed by a group with representatives from all partners, and provide common services across the PDM systems community. While both the domain and community services provide the integration services across multiple PDM systems, the underlying components of each PDM system, manage all their product structures, workflows and applications in local development workspaces as depicted in Fig. 1.

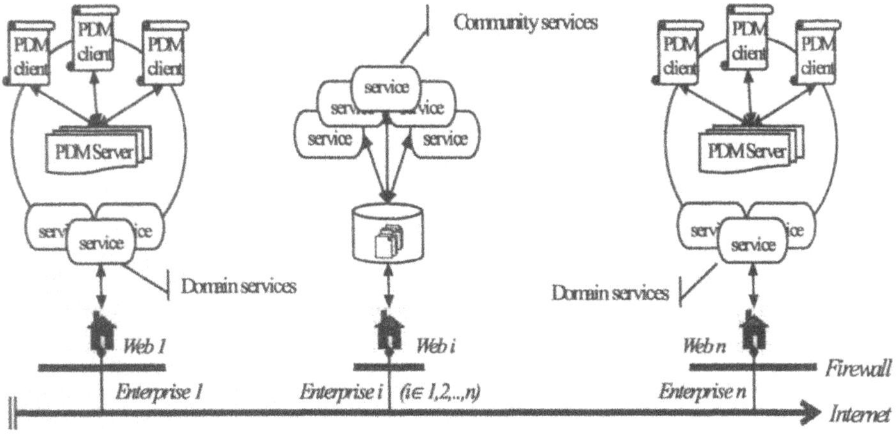

Figure 1: Integration schema based on community services

2.2 Agent-orientation and system integration

In order for PDM systems to enable and support collaborative product development, both community services and domain services require a technology that will package them transparently, and support interaction among them. The criteria for evaluating this technology include degree of automation, intelligence, independence and robustness. Especially, for community services, which require to be fair and impartial to get all partners' commitments to a public facility that is autonomous and neutral. To deal with these issues we propose an agent-based design for domain and community services. Agent-Orientation (AO) is a very promising design paradigm with a growing appeal in the distributed systems community. It is an emerging software engineering technology, which provides an adequate solution for modeling and designing collaborative distributed PDM systems. Particularly, AO is rich in providing several design features that are required for the integration of multiple PDM systems in a virtual enterprise or a supply chain, such as autonomy, proactiveness, flexibility, reconfigurability, etc. (Shen, et al., 2001; Ghenniwa, Kamel, 2000).

2.3 Integration architecture

As illustrated in Fig. 2, the proposed web- and agent-based architecture for integrating PDM systems is logically viewed as six layers, with each layer built on the next one:

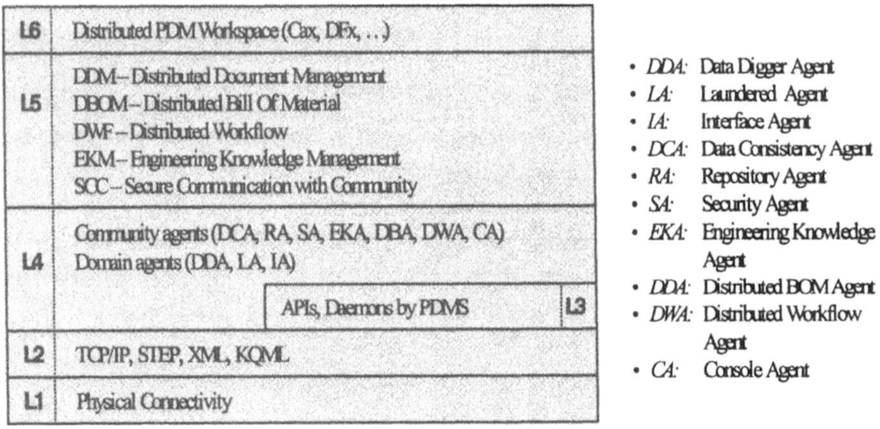

Figure 2: PDM systems integration architecture

1. Physical layer, the underlying network system connectivity with Intranet/Internet;
2. Communication protocols layer, including exchange protocols and specifications, such as TCP/IP for Intranet/Internet, STEP for inter-application product data exchange, XML for distributed content exchange, KQML for interaction between agent-based subsystems (Finin et al., 1993);
3. Resources layer, including programming resources in the product development environment, especially those professional APIs, Daemons provided by PDM systems vendors;
4. Integration services layer, i.e., agents layer, including community services and domain services;
5. Distributed transactions layer, including distributed management of documents, product structures (usually bills of materials), workflows and engineering knowledge, and secure communication within collaborative communities;
6. Applications layer, accommodating applications integrated into PDM system platforms such as CAD, CAM, DFM, DFA, and other application tools.

2.4 Agent identifications

This work focuses on the integration services layer. This layer provides services that seamlessly integrate PDM applications with physical connectivity resources, distributed communication protocols and application developing resources. Also, it supports the integration among the PDM applications at runtime. We categorize the roles of the agents based on the type of the service they participate in, either community service or domain service. Community agents usually locate at a Web server to provide public services and expertise across the community, such as virtual enterprise. These agents include:

1. Data Consistency Agent (DCA)
2. Repository Agent (RA)
3. Security Agent (SA)
4. Engineering Knowledge Agent (EKA)
5. Distributed BOM Agent (DBA)
6. Distributed Workflow Agent (DWA)
7. Console Agent (CA)

Domain agents usually locate at the partner enterprises and work together with other agents for system integration and information sharing. These agents include:

1. Data Digger Agents (DDA)
2. Laundered Agents (LA)
3. Interface Agents (IA)

The deployment scenario for these agents is shown in Fig. 3. The following paragraphs provide a brief description of the agents' roles.

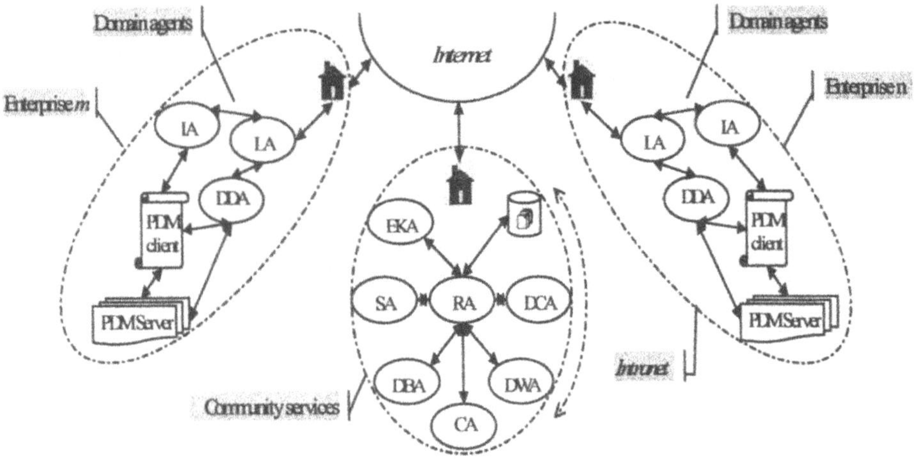

Figure 3: Integrating agents and their deployment

Data Digger Agents (DDAs) detect changes in product data and workflows, provide inputs of local partner PDM systems, and provide information to IAs. Almost all distributed PDM systems transactions involve DDAs. They run locally with PDM systems sides, get different knowledge and strategies to abstract, analyze and store the data about changes from specific local applications (e.g., PDM system clients, CAD, CAM, ...).

Laundered Agents (LAs) provide data format translation and standardization services, and facilitate intelligent communication between domain agents and community agents. As services on local ontology, they manage all information flow in and out of the environment.

Interface Agents (IAs) proactively assist their human-users in navigating the distributed PDM environment and identifying people and items that interest its human-users. IAs provide a notebook client for engineers, it can not only automatically capture engineers' design history and record them for future reuse with DDA's cooperation, but also allow engineers interactively write down design notes for reuse or reference by collaborators, and support online discussion with their remote associates efficiently and with a desired level of security.

Repository Agent (RA) is a foundational facility which maintains a complete meta-data model of the integration environment, including:

- Properties and status of agents, involved PDM systems and applications integrated into them;
- Member enterprises models built on personnel, activities, product information, and privileges;
- Standard elements and templates of distributed workflows and product structures (BOMs);
- Trigger rules and conflict resolution policies.

Data Consistency Agent (DCA) identifies any potential for data inconsistency in the distributed PDM environment. Then, it collaborates with DDAs and the RA to analyze situations, to determine the required update policies and procedures, and to execute the appropriate process.

Distributed BOM Agent (DBA) is responsible for BOMs decomposition and composition.

Distributed Workflow Agent (DWA) is responsible for workflow decomposition and composition.

Security Agent (SA) is a security service for agents' communication all over the community, similar to the Internet certificate authority.

Engineering Knowledge Agent (EKA) analyzes, translates and maintains design notes and other knowledge derived from DDAs and IAs at partners' domains, supplies engineers with these design experiences and history knowledge in the scope of the community, and allows engineering knowledge reuse and exchange. Engineers access these design knowledge and history by notebook-like IAs as if they were working in a single studio.

Console Agent (CA) acts as an administration interface console to all agents. By this interactive administration interface, administrators can initiate their specific environments when setting up, and allows further adjusting environment parameters and managing the system at runtime. All these administrations are through interactions between this agent and other agents in the community.

3 DISTRIBUTED TRANSACTIONS

3.1 Distributed documents management

In the current paradigm, the main concern for distributed documents management is data consistency, i.e., distributed version management and update. DCA, RA, DDA, and LA work together to keep documents consistent everywhere in the current distributed PDM community:

- The RA's database stores global descriptions about subject products, i.e., documents (in a PDM system, all files are documents), and their distribution across the distributed PDM community.

- When initiating the system, DDA in local web servers of member enterprises works together with PDM system services, retrieves the collaboration-related documents meta data (e.g., locations, and some other features) of product data from local PDM databases according to predefined mapping lists, which are retrieved by interacting RA.
- Before DDA interacts with outside, it firstly requests LA to translates and supplements submitted data into XML files with conventional schemas that receiver agents can recognize.
- DDA interacts with DCA in regular or real-time basis, and DCA uses the latest document meta data from RA to determine if subject documents have changed.
- As soon as a change is detected, DCA interacts with RA, provides the document with a new version based on conventional version policies, stores new meta data into the RA, finds out the impacted data objects, their locations and their DDAs by consulting product structure trees, bills of materials, and organization-organization relations lists, and then determines an updating program based on data consistency rules.
- Next, DCA contacts all the related DDAs in local web servers, notifies them to change the correlated data in their PDM databases by PDM system daemons and change mechanism, to keep them consistent with each other.

3.2 Distributed BOMs management

There are usually different BOM views in different life phases of a product. Moreover, in different PDM systems, there are different representations of BOMs even in the same phase. As a result, when integrating distributed PDM systems by agents, what distributed BOMs management concerns is transforming BOMs between different PDM systems. To this end, it is essential to define a global sharing BOM model, individual BOM models corresponding to the different product phrases and PDM systems, and furthermore work out the translating methods between the global BOM and individual BOMs, as an essential part of enterprise model in the RA.

DBA, RA, DDA, and LA work together to translate product structures in the community:

- When a BOM releases to another PDM system, the action is intercepted by DDA.
- DDA interacts with DBA through LA's standardization service.
- DBA interacts with RA, finds out the corresponding global BOM, individual BOMs and their mapping relationship, and executes the translation.

- The translated BOM is transferred to receiver DDA via LA, which forwards it to the goal PDM system (technically, usually by utilizing the exchange directory of the PDM systems and giving a fake notifying message to it, which responds the message to receive the data).

3.3 Distributed workflows management

Like BOMs, there are different representations of workflows in different PDM systems in a collaborative engineering environment. Therefore, when we get down to interlink these distributed, especially heterogeneous PDM systems, it is important to understand processes and their basic elements in different PDM systems, analyze distributed and collaborative design and engineering processes. And then, based on the understanding of them, a global workflow and a collection of translating relationships from the global workflow to specific workflows in related PDM systems are built as a main activity object of a common model of collaborative action in the RA. With above common object and translation lists, a distributed workflow can be executed remotely:

- When a distributed workflow, whose destination involves remote developer, is activated, DDA can find it by interacting with the PDM system at certain messages directory.
- DDA submits the workflow meta data to the DWA through LA.
- After interacting RA and getting the related global workflow template object and translating relationships, DWA analyzes, decomposes, and translates it into a specific common workflow as executing subject.
- Then, the DWA translates its to-be-executed element into a specific task that the predefined destination PDM system can understand, submits it to the PDM system to execute by LA/DDA.
- DWA holds and monitors the workflow execution. When a step is finished, DWA gets response from DDA/LA, and then gets to the next one.

3.4 Collaborative engineering knowledge management

The goal of the collaborative engineering knowledge management system is to combine distributed design workspaces into a virtual studio, allow engineers to collaborate and communicate their tips and decision-making evidences as convenient as they were in a same office. In this paper, all of its data detecting and collecting clients, end-user interfaces, and a centralized management console, have been improved as three standard agents: DDA, IA and EKA:

- DDAs detect predefined data changes and designers' operations, analyze and determine what should be submitted, recorded, and useful for reuse and communication.
- DDAs interact and submit captured information to IAs for users to confirm or modify. IAs display predefined items and provide engineers with interactive interfaces.
- DDAs get records and notes from IAs, whether they are input by engineers or edited content based on their submission.
- Periodically or activated by engineers' release, DDAs submit such data to EKA through LA's standardization.
- The formatted XML schema files are transferred to EKA through KQML. EKA checks, allocates and maintains them in a secure, orderly and classified manner. Sometimes, EKA interacts with RA to search related privileges and product features, and determine maintenance structure and access control list.
- With maintained information growing, a valuable facility for design knowledge referencing, reusing, and communicating is created. Now engineers can look up what have been recorded by all of them, what tips their co-workers have left, what is happening on their collaborators' sides, and with IAs supporting multimedia discussions, they can communicate with each other as they were working in the same office. Note that all above actions happen only they have valid priorities.

3.5 SA-based communication security between sites

Currently, even in a distributed collaborative product development environment using multiple PDM systems of the same type for product data sharing, there are few professional safeguarding considerations for site-site communications, without mentioning those happening between different types of PDM systems. The proposed paradigm provides a potential way to secure and maintain remote communications between PDM systems. With the SA, an Internet-like certificate authority can be built up by developing a special authenticating service within a community. Thereafter, the certificate authority can authenticate communicators, and with a public-key encryption solution (e.g., a symmetrical method, an unsymmetrical, or their integration), and offer them a certificate, a public key and digital signature. Finally, a set of secure infrastructures comes into being, and data security and integrity are subsequently guaranteed across all over the community.

4 CONCLUSIONS

In this paper, an agent-based solution for integrating distributed and heterogeneous PDM systems is proposed, and the details about the integration architecture, agent identifications and integrated mechanisms are presented. This is an attempt at applying agent technology to enhance the interoperability among existing PDM system entities. A promising collaboration structure as featured by community and domain agents has been applied to meet the requirements of a virtual enterprise. Due to the space limit, more details, particularly related to the system implementation issues, could not be included in this paper.

The proposed paradigm is an enhanced and systematic improvement of the previous research presented in (Li, 2000). Major related implementations were done during our previous research and the feasibility of multiple PDM systems integration based on community and domain services had been proved. As a result of software agents being good at wrapping up legacy systems as a methodology, the proposed paradigm is feasible in such an environment, by showing more autonomous, flexible, intelligent, and scalable features as various types of its agents come up with inherently. Nevertheless, a lot of research and implementation efforts are still necessary. Some of those identified include: determining the nature and model for each of community and domain agents; working out well-accepted implementation methods for each of distributed transactions that involves multiple even heterogeneous PDM systems; building data and security infrastructure and facilities for the agent-integrated layer; researching global product structure, workflow/processes; analyzing concurrent accesses and conflicts resolution strategies and facilities; selecting implementation technologies; and developing prototype systems.

Noted that the current architecture and system is towards integrating a product development environment based on PDM systems. However, a further extension can be expected. For example, with the similar methodology, many other enterprise business systems such as MRP II, ERP, SCM, and CRM etc., can be interlinked despite they are distributed and heterogeneous. At the same time when they are integrated and abstracted, with their agents having similar architecture and interaction protocols, a complete paradigm for virtual enterprise and supply chain can be expected.

5 REFERENCES

ENOVIA Corp. (1998), *ProductManager: Advanced Customization Environment,* Script, Reference and Operations Manual, Version 3.2.0, IBM Corporation.

Finin, T. Fritzon, R. McKay, D. McEntire, R. (1993), *KQML - A Language and Protocol for Knowledge and Information Exchange.* Tech. Report, University of Maryland, Baltimore, MD.

Ghenniwa, H. Kamel, M. (2000), *Interaction Devices for Coordinating Cooperative Distributed Systems,* Automation and Soft Computing, Vol.6, No.2, pp.173-184.

Interleaf Inc. (1998) *IMAN Online Help 2.1.0*/sunos5, Interleaf Inc.

Jasnoch, U. Haas, S. (1996), *A collaborative environment based on distributed Object-oriented databases,* Computers in Industry, Vol. 29, pp. 51-61.

Li, Y. (2000), *Integrated Infrastructures for Collaborative Development in a Virtual Enterprise, PhD Thesis,* Tsinghua University, Beijing, China.

Peltonen, H. Pitkänen, O. Sulonen, R. (1996), *Process-based view of product data management,* Computers in Industry Vol. 31, pp. 193-203.

PTC, (2002), http://www.ptc.com/products/windchill/

SDRC Corp. (1996), *Online Books for Metaphase 2.3,* SDRC Corp.

Shen, W. Norrie, D.H. Barthes, J.-P. (2001), *Multi-Agent Systems for Concurrent Intelligent Design and Manufacturing*, Taylor & Francis, London, UK.

Ontologies for Semantically Interoperable Electronic Commerce

Leo Obrst[1], Howard Liu[2], Robert Wray[3], and Lori Wilson[2]
[1]*MITRE, USA,* [2]*self employed* [2]*Soar Technology, USA,* lobrst@mitre.org

Abstract: In this paper we discuss the use of ontologies to support semantically interoperable B2B electronic commerce. First, we describe the nature of B2B and the kinds of applications used. Second, we present arguments towards why B2B needs ontologies and the nature of the problems faced. Finally, we discuss the interaction of ontologists and domain experts in the building of ontologies for business, and some of the tools available for developing ontologies.

1 INTRODUCTION

We are interested in ontologies in the product and service space to support semantically interoperable Business-to-Business (B2B) electronic commerce (e-commerce). Ontologies in this space include domain ontologies (lower ontologies), an upper ontology and upper model, and shared middle ontologies, as in Fig. 1. We are assisted by subject matter (domain) experts who know various technical product areas. Although difficult problems remain to be solved (Obrst, et al, 2001), we firmly believe that ontologies remain the best solution for robust, semantically interoperable electronic commerce, as they do for many other applications.

An ontology defines the common words and concepts (meanings) used to describe and represent an area of knowledge. Ontologies are used by people, databases, and applications that need to share domain information (a domain is just a specific subject area or area of knowledge, like medicine, automobile repair, banking, tool manufacturing, etc.) Ontologies include computer-usable definitions of basic concepts in the domain and the relationships among them. They encode knowledge in a domain and also knowledge that spans domains. They represent shared conceptualizations.

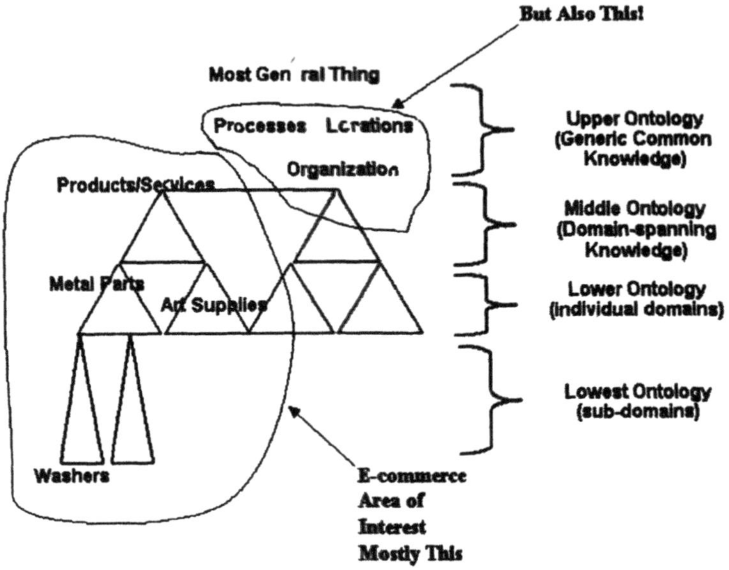

Figure 1: Upper, Middle, and Lower Ontology

2 THE NATURE OF THE BUSINESS-TO-BUSINESS (B2B) ENTERPRISE

B2B electronic commerce is everything that land commerce is, plus more: automated support for information and transaction flow for vertical and horizontal commercial interoperability. Given this definition, B2B provides marketplace platforms on the Internet that support the following:

- Multiple trading models. Trading models include auctions, reverse auctions, exchanges, Request-For-Proposal/Request-For-Quote, bookstores, trading hubs, etc. The platforms are used for and by commercial organizations.
- Rich information content on products and services for both buyers and sellers. This content might be realized in the form of catalogs, product guides, market and domain editorial content, news, advertising. The information content may also be annotated in machine understandable form, as well as organized and tailored for human use.

- Support for buying and selling. B2B transactions often require much more than the simple exchange of goods for money. Support functions include financing, privacy/security, payment processing, order management, profiling/personalization, product configuration, planning/scheduling and forecasting, product life cycle and inventory management, business processes, workflow, and rules, logistics, distribution, and delivery.

3 WHY ONTOLOGIES ARE NEEDED FOR B2B

B2B e-commerce needs ontologies to support semantic interoperability. Consider two critical motivations for ontologies in B2B e-commerce. First, there is an *informational* use. An ontology is a structured conceptual model of an e-commerce domain. This structuring of the information space supports parametric search and navigation using product and service knowledge by prospective buyers to discover what to buy, and subsequently to determine pricing and availability. In this case, fairly static knowledge embodied in the ontology (e.g., this retailer sells ball peen hammers) maps to the dynamic data of the vendors (model numbers of the hammers of manufacturers, distributors carrying specific models, selling prices, etc.). Furthermore, an ontology can model not only product and service knowledge, but also knowledge about buyers and sellers, i.e., users. By employing user role

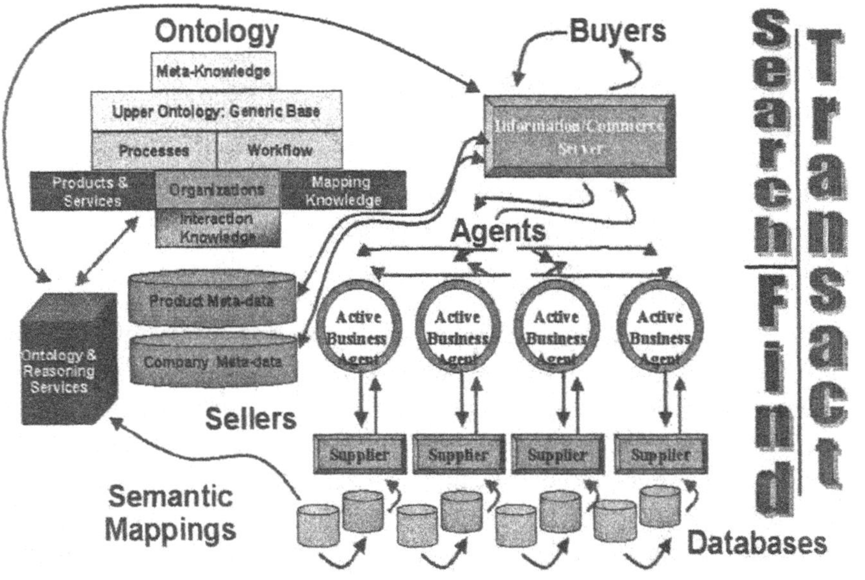

Figure 2: An E-Commerce Application Using Ontologies

knowledge (sometimes called user profiling or personalization), for example, queries can be customized relative to that user's experience and interests.

E-commerce also needs ontologies for *transactional* purposes. Knowledge of a company's organizational structure, workflow, processes, and products/services can be used to assist directly in buying and selling. For example, Fig. 2 depicts one view of an architecture and flow of knowledge within a prospective ontology-driven B2B marketplace infrastructure, linking buyers to semantically mapped suppliers via software agents or web-based service-oriented applications for both informational and transactional purposes. In this framework, multiple heterogeneous databases map to a common ontology that thus enables a meaningful comparative view to be displayed to a prospective buyer.

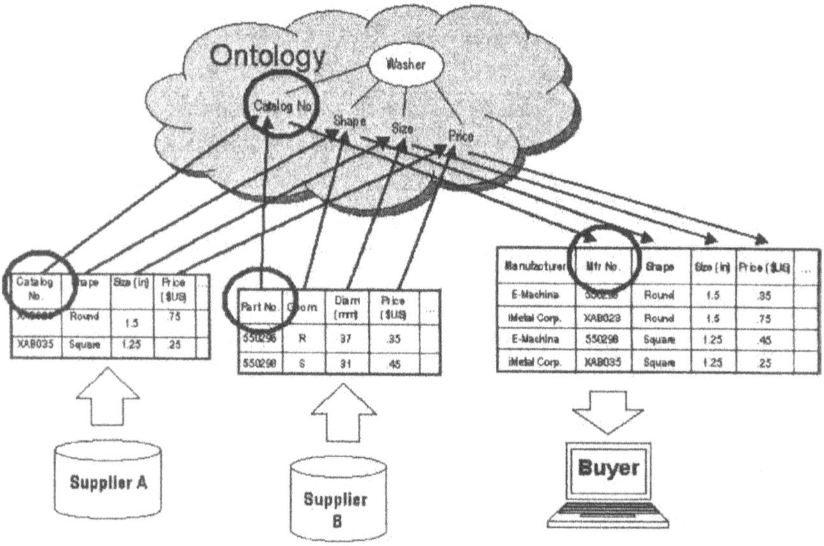

Figure 3: Buyers and Sellers Linked by Ontology

The use of ontologies in e-commerce thus goes a long way towards solving two unsolved obstacles to successful B2B E-Commerce that involve semantic interoperability. *The heterogeneous vendor database problem* results from distributors, manufacturers, service providers with databases that differ significantly in format, structure, and meaning. In Fig. 3, different suppliers use different fields for conceptually similar products. The buyer should not have to refer to "Part Number" to look at one vendors product, and "Catalog Number" in another. As the figure shows, the ontology provides a consistent representation for the heterogeneous databases, with mappings from the ontology representation to the specific databases.

The second problem ontologies address *is the standards and common vertical conceptual model problem.* What is the meaning of the terminology employed in the product and service space and the relationships between terms? In Fig. 3, what is a washer? Obvious choices include a part used with nuts and bolts and a colloquialism for a household appliance. Which is meant? Ontologies can provide formalizations of the concepts underpinning common business terminology, and this formalized meaning can be made sound, consistent, extensible, reusable, modular, and logical.

Ontologies must be built to support the representation requirements of many IT applications, most of which will presume some form of classification of products and services. However, many business classification systems are ad hoc, inconsistent, and non-integrated, with little association between classification systems. In order to address this apparent incompatibility, we suggest a distinction be made between *representation* and *presentation.* Representation is the underlying structure and codification of the product and service knowledge space to be supplied by the eventually developed ontologies. This representation is semantically sound, consistent (though incomplete in the sense that additional refinement could always be made), controlled, modular, reusable, and provides some support for application presentation needs.

Presentation remains largely the responsibility of the application. Applications could choose to use their own terminology and classification display, as long as that terminology and structure had linkages to the underlying representation. An application intended for a buyer, for example, might display a different structure and terminology from that of an application intended for a seller. Furthermore, even within a buyer application, the terminology and displayed structure could be different based on the role of the prospective buyer/user. For example, a technically savvy engineer using a catalog search application would typically employ search terminology (or equivalently, navigation through a classification system/taxonomy) based on technical specifications. Within the ontology, this terminology would be *entity-centric* and use entity-centric concepts. An entity in this usage is typically a *thing,* i.e., a product object. But a non-technical purchasing analyst would typically employ search terminology based on his/her own company's environment or processes. Within the ontology, the terminology would thus use *process- or function-centric* representation). In either case, however, the navigational path employed (via relational links or inference) should arrive at the identical, parameterized product or service. Ontologies would thus at least partially support multiple ways of presenting the product and service information.

The vision, of course, for using a common representation is to enable a consistent *ontological or conceptual search* across data and applications, so that semantically meaningful documents and data concerning products and services are returned to the user. This search by definition includes the notion of *parametric search*, which is related to the notion of *product configuration,* that is, a search informed by ontological and other properties of the searched-for product/service. Correspondingly, a common representation supports *ontological classification* of products and services: search assisting primarily buyers, product classification assisting primarily sellers.

Many emerging Web-based standards and languages implicitly define ontologies (example: Unified Business Language (UBL, http://)), and draw upon the experience and expertise of B2B companies (Ariba, CommerceOne, VerticalNet) and electronic commerce consortia (RosettaNet, X12 Electronic Data Interchange, etc.)

4 ONTOLOGISTS AND DOMAIN EXPERTS

Domain experts provide knowledge for the ontologies. Ontologists determine how to represent that knowledge. Ontologists usually teach domain experts some of the fundamental concepts about ontologies and ontological engineering, along with how to use ontology editing tools. As soon as domain experts begin creating domain ontologies unassisted, ontologists can shift their responsibilities to formulating designs for an overall knowledge architecture, to set guidelines for building ontologies, to integrate the ontologies with applications, and to enhance the ontology building environment to make the ontologies more expressive and more maintainable.

5 ONTOLOGY TOOLS

Ontology development tools are now entering the market. Most of the tools until recently were research tools originally funded by programs of the US Defense Advanced Research Projects Agency (DARPA) (Patil, et al, 1992; Cohen, et al, 1998) such as Ontolingua/Chimaera (McGuinness, et al, 2000) and Protégé (Noy, et al, 2000). Both of these tools use frame-based knowledge representation languages developed for artificial intelligence (AI), such as the Open Knowledge Base Connectivity (OKBC) language (Chaudry, et al, 1998). In contrast, Cyc (Guha & Lenat, 1990), which has been a commercial product for a number of years uses a first order logic (FOL) based language. One advantage of Cyc is that it provides a freely available *upper ontology*. An upper ontology (or more appropriately, a set of integrated ontologies) attempts to characterize basic, commonsense knowledge notions that humans know so well that we typically don't know we know them: that is, distinctions between kinds of objects in the world (a tangible product vs. an intangible process), events and processes, how parts constitute a whole and what that means, and general notions of time and space.

Other newer tools for creating ontologies include the commercially available:

OntoEdit, (http://) and the research tool OilEd, (http://).

Both of these tools use knowledge representation languages such as:

RDF, (http://) and DAML+OIL, (http://) (itself a fusion of DAML, (http://) and OIL, (http://).

RDF, DAML+OIL, and the newly emerging Ontology Web Language (OWL) are being developed as standards under the W3C, (http://) to support the so-called *Semantic Web,* (http://), (Berners-Lee, et al, 2001).

Other, more generic tools, which can help build an infrastructure for ontologies, include both Java and Common Lisp (e.g., Allegro Common Lisp, (http://)).

6 CONCLUSIONS

Electronic commerce, in general, and B2B e-commerce, in particular, needs ontologies, to ensure semantic interoperability. Ultimately, ontologies provide a *lingua franca* or *trading language,* to transact the business of buying and selling. Mature ontologies will serve as an intermediate commercial vocabulary with a common set of meanings or concepts for commercial products and services to which individual, disparate product catalogs, commercial databases, and marketplace applications can map. Given M buyers and N sellers, without the use of an intermediate ontology, the problem of integrating between commercial systems and databases is very complex: M x N integrations (each buyer to each seller) have to be performed. With the use of ontologies, only M + N integrations have to be performed (each buyer or seller to the ontology). That is a huge savings in time, effort, and money for any business. The bottom line is: ontologies save you money, and that's always a competitive advantage.

7 REFERENCES

Allegro Common Lisp, http://franz.com/

Berners-Lee, T. Hendler, J. Lassila, O. (2001), *The Semantic Web*. In The Scientific American, May, 2001. http://www.scientificamerican.com/2001/0501issue/0501berners-lee.html

Chaudri, V. Farquhar, A. Fikes, R. Karp, P. D. Rice, J. P. (1998), *Open knowledge base connectivity specification*. Specification V. 2.0.31, SRI and Knowledge Systems Laboratory, Stanford University.

Cohen, P. Schrag, R. Jones, E. Pease, A. Lin, A. Starr, B. Easter, D. Gunning D. Burke, M. (1998), *The DARPA High Performance Knowledge Bases Project*. In Artificial Intelligence Magazine. 19 (4), pp.25-49.

DAML, http://www.daml.org/]

DAML+OIL, http://www.daml.org/2001/03/daml+oil-walkthru.html

Guha, R.V. Lenat. D. (1990), *Cyc: A mid-term report*. Microelectronics Technology and Computer Corporation (MCC), Austin, TX, Technical Report ACT-CYC-134-90.

McGuinness, D. L. Fikes, R. Rice, J. Wilder, S. (2000), *An Environment for Merging and Testing Large Ontologies*. Proceedings of the Seventh International Conference on Principles of Knowledge Representation and Reasoning (KR2000), Breckenridge, Colorado, USA, April 12-15. Cohn, A.G. Giunchiglia, F. Selman, B. (Eds.), San Francisco, CA: Morgan Kaufman.

Noy, N. F. Fergerson, R. W. Musen, M. A. (2000), *The knowledge model of Protege-2000: Combining interoperability and flexibility*. In 2nd Intl. Conf. on Knowledge Engineering and Knowledge Management (EKAW'2000).

Obrst, L. Wray, R. Liu, H. (2001), *Ontological Engineering for E-Commerce: a Real B2B Example,* in Proc. of the International Conference on Formal Ontology in Information Systems (FOIS-2001), Oct. 17-10. http://www.fois.org/fois-2001/index.html

OIL, http://www.ontoknowledge.org/oil/.

OilEd, http://img.cs.man.ac.uk/oil/

OntoEdit, http://ontoserver.aifb.uni-karlsruhe.de/ontoedit/

Patil, R. Fikes, R. Patel-Schneider, D. Mckay, D. Finin, T. Gruber, T. Neches, R. (1992), *The DARPA Knowledge Sharing Effort*: Progress Report. In Proc. of Knowledge Representation and Reasoning Conference (KR-92).

RDF, http://www.w3.org/RDF/

Semantic Web, http://www.w3.org/2001/sw/

UBL, http://www.oasis-open.org/committees/ubl/

W3C, http://www.w3.org/]

8 TERMINOLOGY

Term	Definition
DAML+OIL	DARPA (Defense Advanced Research Projects Agency) Agent Markup Language-Ontology Inference Layer: these are two XML- and Web-based languages to support the Semantic Web, which have recently fused. DAML originated from a US DARPA-sponsored program; OIL originated from a European Union-sponsored program. Together they constitute the most semantically expressive language available for WWW documents. The combined language is now supported by the W3C web standards consortium, and is soon to be superseded by the Ontology Web Language (OWL).
Frame-based Knowledge Representation Language	A knowledge representation language or language for expressing ontological information derived originally from the artificial intelligence (AI) language called KL-ONE, which itself is one of the earliest formalizations of the notion of semantic network. The notion of a frame comes from the early LISP programming language terminology used by early KR languages. In frame terminology, a concept is a class, and a relation is a slot. Attributes (sometimes called properties) are just slots defined on a domain (a specific class subtree) or one of its subdomains (a subclass of a domain class).
OKBC	Open Knowledge Base Connectivity language. This is a language for knowledge access and interchange (an API) derived from the Generic Frame Protocol, developed in the early 1990s by knowledge representation technologists under the support of the DARPA Knowledge Sharing Effort (CITATION). This protocol became the OKBC under the support of the DARPA High Performance Knowledge Base (HPKB) program, 1996-1999 0.
Ontology	An ontology models the meaning of domains of interest: the objects (things) in domains, the relationships among those things, the properties, functions and processes involving those things, and constraints on and rules about those things.
RDF/S	Resource Definition Framework/Schema. These are two languages. The first (RDF) expresses instance-level semantic relations phrased in terms of a triple: <subject, verb, object>, i.e., <object1, relation1, object2>. The second (RDFS) expresses class level relations describing acceptable instance level relations.
Semantic Web	"*The Semantic Web is an extension of the current web in which information is given well-defined meaning, better enabling computers and people to work in cooperation.*" (Berners-Lee, et al, 2001)

9 ACKNOWLEDGMENTS

The views expressed in this paper are those of the authors alone and do not reflect the official policy or position of The MITRE Corporation or any other company or individual. Finally, we wish to thank the anonymous reviewers for their cogent comments and suggestions.

PART 5. COMMON REPRESENTATION OF ENTERPRISE MODELS

Enterprise models are crucial for the success of the enterprise. This sentence was very clear for people involved in the workshop, but we are not sure that people in the enterprises have the same view; at least we are not sure about the word "crucial".

Many industrial users think of models as a blueprint of the enterprise. As this has been the case originally, it is not true any more. Enterprise models or business-process models nowadays not only provide an understanding about the enterprise operation, but also are actively used for knowledge management, decision support through simulation of operation alternatives and even for model-based operation control and monitoring.

It is very important that the user community is aware of this evolution and understands its implications. Whereas in the old days model creation was a skill left to experts, it will become a need for many people in the enterprise to be able evaluate process alternatives for decision support. That means we need executable models as well as a common representation of the models for the model users to enable understanding and easy manipulation of the models. However, common representation does not imply an Esperanto like language, but rather a set of dialects aimed at the different user groups, but based on a common set of modelling language constructs.

The workgroup reports address the user orientation (Kotsiopoulos) and discuss new support technologies for enterprise integration (Goranson). Critical issues discussed include the role of the user and his requirements in the modelling process. Especially emphasis was on the use of current process information needed to evaluate proposed solutions and the use of formal methods for semantic mappings between different tools and models (Kotsiopoulos). The working group explored methodologies needed to support user-enabled business process modelling for model based decision support.

The second workgroup focused on radical but practical strategies for greatly improving process modelling in an enterprise context (Goranson). The group's work centred on improving user benefits in the context of common models, enterprise context and enterprise views. Major problems addressed were: multi-world views, soft modelling and meta-modelling theories. Several discrete research projects are proposed.

In this section of the proceedings the papers presented address the subject of enterprise model representation from two different points of view: a) the development of an inter-lingua (UEML Unified Enterprise Modelling Language) among enterprise modelling tools and b) the description of related concepts. The paper by Petit compares the expected UEML development process with the problem of database integration. Methodological clues for the definition of the meta-model of a UEML are represented. Jochem in his paper complements Petit by defining requirements and an approach to support common representation by a UEML. Panetto in his paper describes the role of UML in enterprise modelling. It illustrates the semantics approach defined in the UML standard showing the UML semantics representation of some UEML constructs.

The paper by Kotsiopoulos is a response to the call for a common underlying domain theory to address the mismatch between syntax and semantics of enterprise modelling languages. It proposes categorical morphisms of object interactions as a strong candidate theory on which all modelling constructs can potentially be mapped. Semantics of particular modelling languages and architectures can be obtained as specialisations of the general theory. Basic features of CIMOSA are derived as an example.

Innovative concepts, which extend existing modelling language concepts for the modelling of distributed business processes are discussed in the paper by Grabowski. Emphasis is on the capability of assigning objects to different partners.

Hawa presents in his paper an analysis of the methodological aspects of selected methodologies for enterprise integration (CIMOSA, PERA, IE-GIP among others). A set of characteristics that must be provided by an enterprise integration methodology is defined.

The paper by Vallespier addresses the necessity and rationale to take enterprise control into account. Man-based decision-making oriented enterprise control is proposed as a complementary approach with respect to formalised views used in enterprise modelling and integration.

The Editors

Kurt Kosanke
CIMOSA Association Böblingen, Germany

Angel Ortiz Bas
Polytechnic University of Valencia, Spain

Steps in Enterprise Modelling

a Roadmap

Ioannis L. Kotsiopoulos[1], (Ed.), Torsten Engel[2], Frank-Walter Jaekel[3], Kurt Kosanke[4], Juan Carlos Mendez Barreiro[5], Angel Ortiz Bas[6], Michael Petit[7], and Patrik Raynaud[8]

[1]Zenon S.A., Greece, [2]Fztr PDE, Germany, [3]FhG-IPK, Germany, [4]CIMOSA Association, Germany, [5]AdN Internacional, S.A. de C.V., Mexico, [6]Universidad Politecnica de Valencia, Spain, [7]Univ. Notre-Dame de la Paix, Namur, Belgium, [8]PSA, France, ikotsiopoulos@zenon.gr

Abstract: see Quad Chart on page 2

1 INTRODUCTION

Advances in Information Technology have made Enterprise Modelling possible for many enterprises of today. A variety of software tools has appeared in the market, processing power has dramatically increased, modelling architectures have evolved and even matured. Despite such advances however, widespread use of models, as a strategic decision support tool encompassing large industrial sectors, remains unattainable. The working group analysed the current situation, identified major problems and issues as causes and suggested a roadmap for the next steps in Enterprise Modelling.

The following Quad-Chart (Table 1) summarises the work of the group that addressed those requirements. It identifies the approach taken to resolve the issues and proposes a project and ideas for future work for testing and enhancing the proposed solutions.

Table 1: Working Group Quad-Chart

EI3-IC Workshop 4 Common Representation of Enterprise Models	***Workgroup 1: Steps in Enterprise Modelling: a Roadmap***	***2002-02-20/22 IPK, Germany***

Abstract:
Critical issues, which will affect the near future of Enterprise Modelling, include the identification of the modelled enterprise, the role of the user and his requirements in the modelling process and the use of formal methods for semantic mappings between different tools and models. Particular emphasis is placed on the relationship of the user to the model life cycle: the user should be enabled to use current process information in order to evaluate proposed solutions.
The working group explored methodologies needed to support user-enabled business process modelling for model based decision support.
	Major problems and issues:
	– How to convince users of the value of EM?
	– How to reduce the gap between user expectations towards EM and modelling expert results?
	– How to enhance the faithfulness of models to the reality and the maintainability of models?
	– How to enhance EM to enable model based decision support?
	  – How to guide the user in modelling and evaluating process alternatives?
	  – How to link business-process models to the actual operational data bases of the enterprise?

Approach: – Re-define the role of business-process models in the enterprise – Identify the needs of the user for business-process model based decision support – Identify mechanism for mapping different user representations to the underlying common business-process model – Identify mechanism to link the business-process model to the operational data of the enterprise – Discuss the needs for formal underpinning of the common business-process model	**Results:** – Established the need for a public view of the business-process model to be the blueprint of the enterprise – Identified a project on user enabled business process modelling directed towards model based decision support aimed to develop the necessary methodologies, user guidance and ICT with focus on system consistency assurance and adaptation of model representation to the users way of thinking **Future work:** – Identify the common set of modelling language constructs (e.g. UEML) from which the representations needed by the different users can be derived – Develop the methodologies to support the users in modelling and evaluating alternatives to the current way of doing business

2 CURRENT SITUATION

Models provide structure to information generated and manipulated by the enterprise. This structure (model) is, in turn, provided by another more generic structure (language, meta-model) and this again by another, even more generic, based on definition formalisms. Each of these levels of increasing genericity is executed by humans (actors) to whom specific roles are assigned as parts of the model creation process. To understand the issues affecting modelling today one must address not only the model itself but the actors and the roles (e.g. the process) they play in its creation.

- The lowest level is that of occurrences, themselves divided in two types: those of fully tangible artefacts and happenings (products, performed processes, resources, etc.) and those of less tangible informational artefacts (e.g. data objects in databases).
- The next upper level is that of models. It is usually populated by classes of objects in order to describe the commonalities of a set of occurrences at the level below. Different models can be created and may describe partially overlapping sets of occurrences.
- The next upper level is that of languages: their constructs are instantiated into models lying on the next lower level. Different languages may provide similar but still slightly distinct constructs.
- The uppermost level is generic: it contains definitions of formalisms used to describe the elements of the next lower level, e.g. languages.

Referring to Fig. 1, the actors of the process are shown on the left side of each level. They create, use and modify information according to their roles:

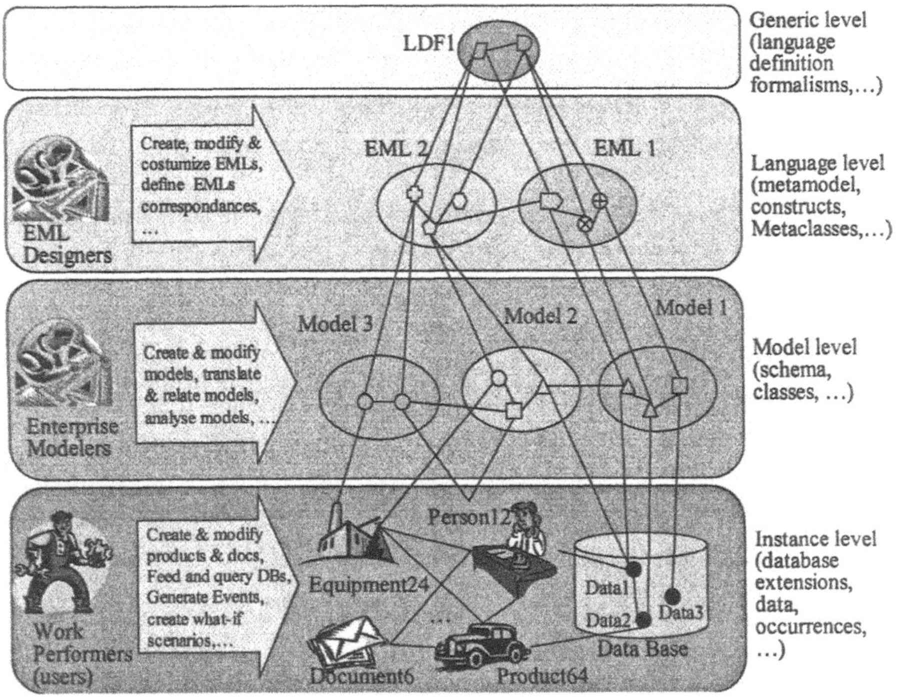

Figure 1: The enterprise model creation process

- Work performers (users): they perform the processes of the enterprise by generating, modifying and destroying instances of products, events, occurrences of business processes, objects such as documents, data through databases etc.
- Enterprise modellers: they analyse, create and modify models and links between them
- Enterprise Modelling Languages (EML) designers: they define syntax, semantics and correspondences between languages.

Modern practice in the creation of enterprise models distinguishes three roles within the above process, performed by different people-actors, namely work performers, modellers and language designers.

Modellers are not work performers, however, meaning that the knowledge of the enterprise they have is rarely as deep as that of work performers. Moreover, it is this knowledge that has to be extracted and validated in order to produce a model. On the other hand, work performers usually have no explicit mental model of their work and therefore no understanding of the consequences of their actions on other parts of the enterprise. A similar, rigid separation of roles exists between language (EML) designers and modellers.

3 MAJOR PROBLEMS AND ISSUES

The result of the allocation of roles being as described in the previous section amounts to modelling errors, imprecise abstractions and model maintenance difficulties. The situation worsens in proportion to the size of the model.

Corresponding problems arise at the language level. Because EML designers are not enterprise modellers, languages often are:

- Too generic for the purpose, causing unnecessary modelling effort
- Insufficient or not adapted to the situation at hand
- Semantically imprecise resulting in differences in understanding between EML designers and enterprise modellers.

From an outside observer's point of view, the model creation processes of today are characterised by a marked deviation between user expectations and actual results produced by the modelling experts.

It is therefore not surprising that users still have to be convinced about the value of enterprise models. The working group considers this to be the central issue facing enterprise modelling today and, to this end, identifies the following steps as being critical to its solution.

- Enhancement of models so that they are faithful to reality and easy to maintain. For this to happen roles in the model creation process must

be brought together: work performers, model experts and EML designers.
- Enhancement of the entire model creation process so as to enable model-based decision support, which relies on actual enterprise data. This calls for competencies on the user side (inclusion and evaluation of process alternatives), as well as the model side, which should be linked to the operational enterprise data bases.

4 APPROACH

The working group believes that the role of business-process models within the enterprise life cycle should be seen under a new light. To draw a parallel, any form of management by humans is based on a mental model of the situation at hand. Our brain can take decisions only when a model of reality (built according to perception, logic, data etc.) is available.

We envisage a similar role for our models of the enterprise: they should not be seen only as a description of some activities but as a true blueprint of the architecture of the entire enterprise in operation; a means of making this architecture explicit and transparent to the users.

4.1 User-enabled modelling

One significant problem area identified so far is the distinct allocation of roles to different actors participating in the model creation process (Fig. 1). Instead, they should be interlinked and partly overlapping.

Starting with the lowest level, there are significant advantages in putting the users in charge of their own model, which can then be a reflection of their superior process knowledge. What needs to be overcome is the alienation caused when they are faced by large complex models.

Although help by experts will always be required, especially at the first stages of drawing a model, users should eventually be able to have full ownership of their modelling domain and full access to its decision support capabilities.

To achieve this, a new model creation process paradigm is required with features, largely not available today, such as:
- A user-defined modelling universe customised to his own domain of interest, such that:
  - It does not limit the expression of his needs
  - It does not lead to inconsistencies and semantic conflict with other users (possibly implemented through sub-typing out of some underlying semantic domain as described in section 4.3 below)

  - It supplies a limited set of clearly defined constructs implemented with a similarly equipped modelling language
  - It supports user-created new object-types, through a registration and administration process with the underlying semantic domain, thus safeguarding semantic uniformity throughout the entire model
  - It adheres to a commonly agreed enterprise ontology to which all semantics and objects used by the various models are anchored (for example, the same person can be used as a resource on one model and an organisation unit on another)
- A set of derivable views on a larger model, adapted to the needs of specific users
- A consistency preservation mechanism between model and enterprise data bases
- A flexible, customisable software support tool

A model creation process with those characteristics does indeed bring together all roles: work performers, modellers and language designers. Users do modelling and language customisation tuned to their needs, thus transversing all three levels of Fig. 1. We use the term *user-enabled modelling* to refer to the new model creation process paradigm.

Finally, although there are significant advantages in bringing power to the users, it must be acknowledged that unification and integration for different models and languages is likely to encounter partly redundant and/or conflicting items.

4.2 Requirements on support technology

User enabled modelling cannot be effective without the support of software technology. We shall briefly refer to the needs of each role in the process in relation to existing products and tools:

- Work performers: tools for creation of occurrences and what-if scenarios. Examples of existing such tools include various DBMS, systems, process control and enactment tools, simulation tools, workflow management systems, etc. A new class of needed tools would be interfaces between models and enterprise data bases so that decision support is based on real operational data.
- Enterprise modellers: tools for model creation, modification and integration. Existing categories of tools, in diminishing order of effectiveness, take care of: enterprise modelling (various in the market), model verification (syntax and consistency checks), CASE, database integration and executable code-generation. New needed support includes tools for: modelling adapted for particular users (i.e. providing an interface tailored to the language chosen by the user), exchange of

models through some import/export facility, model linking and integration platforms and model translators
- EML designers: tools for creation and modification of languages and constructs. The capabilities of existing tools lie in:
  - EML definition and adaptation (generic syntax definition and checking tools such as Lex, Yacc, (Levine, et al, 1992) and (Arabatzis, et al, 1995), where an early application of the tools to a model execution prototype is reported
  - Meta-CASE (see Engelbert, 2000 for a survey)
  - Modelling which enables language customisation or extension (see MO²GO, http://, Merge, http://, Popkin, http://)
  - Knowledge modelling environments (e.g. Metis, http://)
  - Multi-level modelling environments enabling the definition of meta-models, models and instances in a single setting (e.g. Conceptbase, http://)
  - Integration and linking of other modelling tools or CASE tools (e.g. Pohl, 1999)

New capabilities would comprise: generation of dedicated modelling tools and translators for customised EMLs, language integration and linking (consistency checking), meta-models integration, semantics definition and mapping between different EMLs.

4.3 Formal semantics

The semantics of Enterprise Modelling Languages (EMLs) are pinned to real world objects and features; therefore they exist by definition, even in an intuitive form (referred to as "designations" by Jackson (1995), for example). Despite this, the arrival of computing power and modelling tools has pushed for more machine-automated features such as error checking, code generation, simulation, animation etc. To support those, formalisation of semantics is imperative, something which, however, has not been fully achieved, despite the calls from theoreticians. More precisely, although research effort has been present, fragmentation and lack of co-ordination have so far resulted in no widely acceptable definition, unification and resolution mechanisms for the semantics of modelling languages.

Given this state of affairs, the new model creation process paradigm put forward in section 4.1 seems not feasible: interoperability between all types of user-defined modelling universe is a key feature. The problem now appears to be even more complex than before: the notion of semantics is divorced from the notion of language, as each modelling universe may use a subset of the same language but different semantics. Fig. 2 shows a semantics mapping mechanism (to be formally defined) such that:

- An underlying semantic domain exists as a universal enterprise-wide "receptor" pegged to the enterprise ontology
- A semantics composition/decomposition operation is available
- It appears that assigning unique formal semantics to a certain language is rather undesirable. One could assign multiple compatible formal semantics for an EML, thus accommodating different analysis needs. Exchange of models is then taken care of by the Underlying Semantic Domain (Fig. 2).

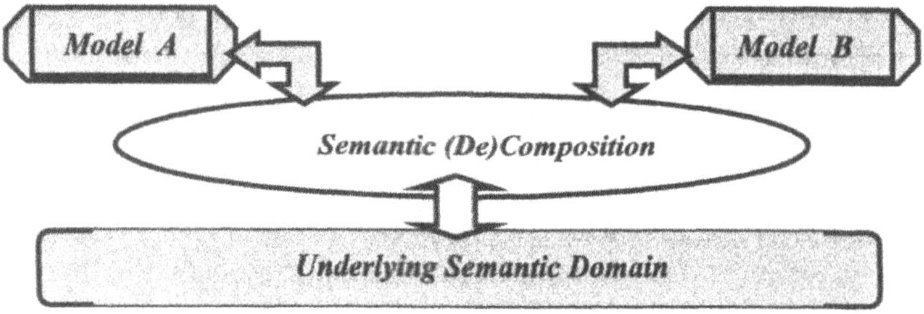

Figure 2: A semantics mapping mechanism

Finally, an Underlying Semantic Domain cannot be constructed without the existence of a corresponding Underlying Domain Theory. This could provide mapping mechanisms to a corresponding underlying language such as the proposed Universal Enterprise Modelling Language (UEML) (Petit, et al, 1997) and also be used as a representation of the enterprise ontology. Capabilities on abstraction, unification and universality should be present in the mathematical framework of such a theory (Kotsiopoulos, 2002).

5 CONCLUSIONS AND FUTURE WORK

The working group has reviewed the enterprise modelling scene of the last five years and has identified the need for a public view of the business-process model as a blueprint of the enterprise. The new model creation process paradigm needed is epitomised in the following call for research and development.

5.1 Project Proposal: User Enabled Process Modelling

Rational: The need for model based operational decision support requires process models to be modified for representation of potential solutions to the problem to be solved. The optimal source for such potential solution is the person responsible for the business process under consideration. Therefore, he has to be enabled to do the required model modifications himself.

However, such decisions can only be made if the model represents reality not only by representing the current structure of the process, but also by using current process information in the evaluation of the proposed solutions. The model has to be linked to the data bases of the process.

Objectives: To develop the necessary methodologies, user guidance and ICT support for user enabled process modelling with focus on system consistency, assurance and adaptation of model representation to the users way of thinking.

Future directions along this line of work have been identified as:

- The establishment of a common set of modelling language constructs (e.g. such as those to be provided by UEML, (Jochem, 2002)) and semantics from which the representations needed by the different users can be derived
- The development of methodologies to support users in modelling and evaluating alternatives to the current way of doing business

6 REFERENCES

Arabatzis, T. Papaioannou, D. Didic, M. Neuscheller F. (1995), *"Elval pilot. Aluminium casting traceability supported by CIMOSA"*, Computers in Industry 27.

Conceptbase web site: http://www-i5.informatik.rwth-aachen.de/CBdoc/

Englebert V. (2000), *"A Smart Meta-CASE Towards an Integrated Solution"*, PhD Thesis, University of Namur, Computer Science Department.

Jackson M. (1995), *"Software Requirements and Specifications: A lexicon of practice, principles and prejudices"*. Addison-Wesley.

Jochem R. (2002), *"Common Representation through UEML - Requirements and Approach"*, this conference.

Kotsiopoulos I.L. (2002), "Language Semantics: Towards a Common Underlying Domain Theory", this conference.

Levine J.R. Mason T. Brown D. (1992), *"Lex & Yacc, 2nd/updated edition"*, O'Reilly & Associates, October.

Merge web site: http://www.fzi.de/cad/projekte/merge/index.html

METIS web site: http://www.metis.no/

MO^{2}GO web site: http://www.um.ipk.fhg.de/mogo/bprhome.htm

Petit M (Ed.) Goossenaerts J. Gruninger M. Nell J.G. Vernadat F.B. (1997), *"Formal Semantics of Enterprise Models – Workshop 4, Working Group 2"*, in Enterprise Engineering and Integration, Proceedings of ICEIMT'97, Springer-Verlag.

Pohl K. Weidenhaupt K. Dömges R. Haumer P. Jarke M. Klamma R. (1999), *"Process-Integrated (Modelling) Environments (PRIME): Foundation and Implementation Framework"*, ACM Transactions on Software Engineering and Methodology (TOSEM), Vol. 8(4).

Popkin Software $ Systems Inc., System Architect tool, http://www.popkin.com/

New Support Technologies for Enterprise Integration

H. Ted Goranson[1], (Ed.), Roland Jochem[2], James G. Nell[3], Hervé Panetto[4], Chris Partridge[5], Francesca Sempere Ripoll[6], David Shorter[7], Peter Webb[8], and Martin Zelm[9]

[1]Old Dominion University, USA; [2]FhG-IPK, Germany; [3]National Institute of Standards and Technology, USA; [4]CRAN - Research Center For Automatic Control, France; [5]LADSEB-CNR-BORO, Italy; [6]Universidad Politecnica de Valencia, Spain; [7]IT Focus, UK; [8]BMT Defence Services Ltd, UK; [9]CIMOSA Association, Germany, tedg@sirius-beta.com

Abstract: see Quad Chart on page 2

1 INTRODUCTION

The following Quad-Chart (Table 1) summarizes the work of the group. It identifies the approach taken to resolve the issues and proposes several projects and ideas for future work for testing and enhancing the proposed solutions.

2 BACKGROUND

Enterprise modeling is a strange beast, because by its nature it synthesizes two major types of views: local views and a global one. Historically, modeling was a local activity, focused on individual processes and individual domains (activity, information, role, etc.). These process models, suitably constrained, were then aggregated in an ordered fashion to form a "model" of the whole enterprise, used for insight, analysis and optimization.

Table 1: Working Group Quad-Chart

<table>
<tr><td>EI3-IC Workshop 4
Common Representation of Enterprise Models</td><td>Workgroup 2
New Support Technologies for Enterprise Integration</td><td>2002-February-20/22
Berlin, Germany</td></tr>
<tr><td colspan="2">Abstract:
The workgroup focused on radical but practical strategies for greatly improving enterprise modeling and process modeling in an enterprise context. The group's work centered on improving user benefits in the context of common models, enterprise context and enterprise views. Major problems addressed were: multi-world views, soft modeling and meta-modeling theories. Several discrete research projects were proposed.</td><td>Major problems and issues:
– Enterprise modeling encompasses two major types of views, local and global
– These views tend to use different terminology, modeling methods, and ontologies; and occupy different space in virtual enterprises
– Globally modeled things tend to be "soft" and non-deterministic
– To engineer enterprises needs methodology to merge the soft and deterministic aspects
– The infinity of tacit knowledge needs to be classifiable into apropos chunks</td></tr>
<tr><td colspan="2">Approach:
– Use the GERAM as reference and extend where necessary
– Design focused, professionally facilitated workshops for a selected mix of experts
– Seek participants with expertise in logic; situation, type, and graph theories; knowledge representation; agent systems; and international standards
– Try to exploit prior research in relevant areas not normally encountered in the enterprise-modeling community</td><td>Results and Future work:
Five to six projects were identified:
– Soft modeling: such as, non-determinism, uncertainty, social and cultural dynamics, and tacit knowledge
– Introspective modeling: so that models can control other models
– Multi-world modeling: Domains such as legal, financial, and production contain conceptual discontinuities.
– Multi-level modeling: Improve the interaction between process models and enterprise models
– Meta-tools for modeling: to better accomplish multi-level and multi-world models
– Additive abstraction: model the processes directly into the enterprise context</td></tr>
</table>

As time progressed, the discipline allowed for actual modeling of enterprises as a congruent activity with the aggregation of process models. Many in the field believe that the future of the discipline is in this power to explore both ways: tweak the enterprise model and see the implications at the process level, and the other way around.

But with this vision, one inherits a collection of superficially related problems associated with the different sorts of heterogeneity one finds in an enterprise. Some of these have been widely discussed and possible solutions are forthcoming:

- Different modeling methods
- Different ontologies
- Different corporate "zones" in a virtual enterprise (A "zone" may be a function like marketing, or a profit center or supplier.)
- To some extent, different roles and goals.

But there are a host of additional problems that real enterprises present, which are apparently hard to address but which would yield immense benefit if solved in some way. These are briefly discussed below. Suggested projects to address these problems follow.

2.1 Soft Modeling

A previous ICEIMT workshop addressed this set of problems. ["Ontologies as a New Cost Factor in Enterprise Integration" this volume] but substantially more about the problem is understood now.

Several kinds of "softness" exist in an enterprise, and must be accommodated in some way at the enterprise level. All of these share the quality of not being representable by deterministic models based on first order logic.

Nondeterminism (including nonlinearity): Traditionally, a process is modeled precisely because it is deterministic, and one wishes to understand and engineer the revealed mechanics. But many typical enterprises are sufficiently complex that they include non-deterministic processes, or would like to either include such processes or model them if they could be managed in some way. The problem of modeling Nondeterminism has several dimensions that can be collected under the notion of "apparent determinism." In this notion, some key elements of the process appear deterministic at the enterprise level (for instance cost, quality or time constraints) but the details of the process get more unclear as one zooms down. This flies in the face of current approaches, which depend on ordered aggregation strategies. At some point in the decomposition, an explicit, semantically rich "place holder" of the non-deterministic causal dynamics needs to enter the picture. Such a placeholder would represent that you know something but not everything about an element, relationship or mechanism.

Uncertainty: Quite apart from the softness of nondeterminism (which concerns "how"), there is a an issue of uncertainty concerning "what." This appears at all levels, but often at the process level it is handleable by recourse to probabilities. This is because processes are usually repeatable and often have histories that can be incorporated statistically. But at the enterprise level, there are uncertainties of a more profound, disruptive and unpredictable nature, such as natural disasters. Most of the important of these cannot be usefully represented statistically. There needs to be a way to represent and reason about objects or elements that are partially undefined or largely speculative.

Social/cultural dynamics: This is a broad class that appears in several forms and appears at all levels of the enterprise. It encompasses the phenomenon covered by the "soft" sciences: sociology, anthropology, psychology and the like. At all levels one often needs to understand and to some extent to engineer collaborative dynamics. At the process level this is usually seen as team dynamics: at the enterprise level, corporate culture. Also included are personnel qualification and certification within the collaborative fabric of the enterprise. But there is more. Increasingly, the enterprise is engaged in providing customer satisfaction through values that themselves are soft, such as product styling, lifestyle branding or direct social uplifting in a service or product interaction. These soft dynamics of customer motivations are a higher order of softness that needs careful consideration. Relating these motives to product features and process elements is a real challenge.

Tacit knowledge: The tacit knowledge problem is the most recognized of this class, which makes the lack of support at this late stage rather frustrating. Tacit knowledge is all the implied knowledge that parties share when they collaborate. It is a major bugaboo in modeling because the tacit knowledge is largely unrecognized and extraordinarily expensive to make explicit. Lack of doing so makes models and model transactions only an incomplete shadow of the real world. In many cases, the limits are fatal. (Note that current methods of managing the non-determinism involve the correct use of tacit knowledge to "convert" non-determinism into determinism.)

Effective softness through combinatorics or other modeling "holes": Tacit knowledge is one kind of information that is usually not captured by models, leaving them incomplete. But there is a whole class of other incompleteness that is a reality in modeling large enterprises. Unlike the three items above, the information is missing not because it is unrepresentable. But it just fell through the cracks, was overlooked, or some models contain errors, or some integration or translation process was misapplied, or the model is simply out of date. Or, even when the modeling operation is perfect, sometimes the combinatorials of the situation make it impractical to carry everything, so pruning occurs as a matter of expediency.

Whatever the cause, there are almost certainly large holes in the enterprise model of any reasonably interesting enterprise. One would hope that in the spirit of making things explicit, one could represent these "holes" in some way and reason over them. Perhaps the only utility is a fidelity metric measuring the accuracy and completeness of the model. But perhaps a greater facility for self-correction and guided completion might be affected.

2.2 Introspective Modeling

This is one area that is not well supported because of accidents in the history of how enterprise modeling developed. Process models were focused on specific tasks, usually tasks directly associated with the basic work of the enterprise. Such processes usually have "second order" processes associated with them, processes associated with monitoring and adapting the basic process for example. These can be likewise modeled and managed on a case-by-case basis.

But enterprise models combine all these. It is desirable to treat them all the same from one perspective just for consistency in aggregation and management. But they are of a different nature: basic processes do the work of the enterprise; while second order processes produce better first order processes. In fact, essentially all of the analyses applied through enterprise modeling are of this second order type. Their existence is the justification of doing enterprise modeling in the first place. And it is often the case that 25% or more of the processes in the enterprise are of this second order type.

This condition where they are treated the same but are different requires the quality of "introspection." Introspective models have some way of "understanding" other models and therefore themselves. Despite the basic requirement for introspection, virtually no major model techniques support it. This needs to be remedied.

Introspection is required for system optimization that is managed just like the basic (first order) processes are. It is required for well-behaved state and configuration management. It is probably required for scalable approaches to trusted, secure system. It is certainly required for any measure of autonomous self-correction.

2.3 Multiworld Modeling

The multiworlds problem is also new to enterprise modeling, and also is overlooked because of its origin in focuses on local domains. When one models a shop floor, one can assume that all the behavior conforms to the same fundamental laws. For instance, physics applies. The same basic corporate policies and external regulations apply. There will still be lots of "soft" stuff missing, but it will all be missing from the same world.

Enterprises are not so well behaved. Enterprises are not logical machines; they have large components from different domains that interact as a system, and not all these domains exist in the same "world." Some of those that differ actually have different causal mechanics, different "physics" if you will.

For example, in the legal world, truth operates by different rules. In that world, many "truths" vary from fact-based reality and much of the activity of those people in the enterprise is to tolerably maintain that distance for the benefit of the enterprise. (The workgroup did not have time to harmonize the diverse definitions of the use of the word "truth," so it is in quotes here. The different opinions concerned long-standing arguments over whether truth is based in nature or is a fabricated quality. That the truth about "truth" was unattainable is submitted as an example of the problem phenomenon. For readability, the term should be understood as quoted below.)

Here is another example, the so-called "black pot" defense. A man borrows a pot from his neighbor and returns it broken. His neighbor sues. The lawyer argues 1) my client never borrowed the pot; 2) it was broken when my client got it; 3) it was not broken when my client returned it. In many legal systems, these assertions can be made in a non-exclusive manner. (Incidentally, this is a primary reason why expert systems have not been applied in the legal domain, as they have, say in medical diagnosis.)

In addition to parallel truths and truths that are non-fact-based but based on constrained proofs, one has other peculiarities of the legal world. There are lots of artificial concepts, actors and relationships that have no "real world" counterpart. These include ownership, intellectual property, rights, liabilities and value. One needs to model not only these notions, but also the processes we use to know and use these notions.

All of these are accommodated to a crude extent in ordinary process models, appearing as constraints and attributes. But that is not how they are managed in the legal world. And to model the enterprise in a truly holistic way, that world must be included. And the legal world is just the tip of the other worlds iceberg. Everyone has a favorite horror story of something that made sense in the financial corner of the enterprise, but when "integrated" into operations required processes to act in a way that was obviously harmful, or counterproductive or downright stupid.

Marketing and human resources have some of these otherworlds distinctions. And in their cases, it is complicated by also involving soft elements as well.

Clearly, there are some solid overlaps of worlds: human resources and legal worlds share the same rich causal notions of rights. Legal and financial worlds share the same rich causal notions of ownership value. But there are other areas of mismatches. These are not a matter of different modeling methods, nor a simple matter of semantic confusion. These are "discontinuities" in the basic causal structure of the worlds, meaning that there cannot be a "smooth," coherent system of models that covers all spaces. Obviously, these discontinuities can make big trouble when introspective control is applied: for instance a financial collection of processes changing physical operations and missing some clear "reality" of those processes resulting in the horror stories noted above.

Most of the group recognizes that this is a very hard problem. The intent in raising it is not to suggest a solution. No reasonably inexpensive or practical solution may exist. Rather, the group would like to suggest a lightweight mechanism that raises a flag when such causal discontinuities occur. The flagging of a problem would throw the issue to intervention by a human team.

The workgroup was not unanimous in understanding the problem this way as already noted. An alternative view suggested that there can only be one truth, one real world. The problem is not so much about the world, but about what we know or believe about the world. Engineering activities using traditional physics will probably assess the various possible ways in which the current state of affairs may have come to pass. There is only one way in which the world is – so we may have to hedge our bets as to which way that is and accommodate a range of possibilities.

In this perspective, the "black pot" example trades on the difference between what the law "believes" and what the lawyer believes. The notion of artificial concepts, actors and relationships opens one up to questions about how we come to know and use things that are not 'real' and so cannot have any causal relationship with the real world. So the multiworlds problem is that they seem like different worlds to the people 'living' in them. Further analysis may show that there is no consistent way of looking at them all.

In any case, there was unanimous agreement that despite the different definitions, the problem either does exist or appear to exist in a way that has the same effect.

2.4 Multilevel Modeling

There is another problem in the enterprise that superficially is related to the multiworlds problem. It is better behaved in internal "physics," and in fact relatively easy to describe, but it seems to be a deceptively thorny problem in aggregating models at various levels from bottom to top.

At present, enterprise modeling assumes that the world effectively consists of two levels: that of processes and that of process assemblies that at some point can be perceived as an enterprise. Strictly speaking, an enterprise can be seen as a large process, which can be successively subdivided into subprocesses until an intuitively atomic level is reached. So far as the other way, any reasonable collection of processes can be thought of as an enterprise. And these sub-enterprises can be combined. In fact, most approaches to virtual enterprises use this assumption.

Alas, the real world disappoints again. Actually, enterprises are not at all homogeneously behaved at all levels. The way that manufacturing cells are managed and measured is fundamentally different than the way a plant is.

The way that a design team is managed and measured is a different thing altogether from how a company is. This is partly an effect of scale, and partly a result of the forces that come to bear: for instance, a company is directly responsible to financiers, while the effect of financier metrics at the design team level is substantially filtered and transformed.

In reality, there are "levels" at which the dynamics and metrics of the system change in substantial ways from those below and above. These levels, their number and placement, will vary according to situation. Often, these differences are hidden in the tacit knowledge of company policies, which is why we are discovering the existence of these layers as virtual enterprises become more common.

The aggregation methods used in composing enterprise models have to be more specific than mere agglomeration. This is especially the case as virtual enterprise paradigms become more used, and as EVA (economic value added) metrics are applied at the lower levels turning them into selfish profit centers.

2.5 Meta-tools for Modeling

All of the above led the workgroup to consider the general class of meta-tools. This is a broad category, created only for the discussion. It includes:

- Tools to model the enterprise of enterprise modeling. The observation was made that if modeling is so good, why don't we use it ourselves?
- Tools that would become some of the reusable "second-order" tools noted above. Good candidates are automatic syntax analyzers and generators, tools to explore new model paradigms, and "interlingua" tools that would map between syntax and semantic variants.
- Formalisms and theories that are already mature but which have not been well applied to the modeling problems described above. Examples are situation, type and category theories.

Very near term actions were fleshed out for the use of an extended workgroup. These involved exploring tools which can model other tools, and using the CIMOSA web site to expose some of the relevant theories to the modeling community.

3 SUGGESTED ACTIONS AND PROJECTS

The discussion of the workgroup was centered on a number of high value problems. These were considered in turn, and specific actions identified for tests, research or further discussion. Each of these is presented below, in no particular order.

In each case, practical implementation was a guideline, with revolutionary capability as a goal. The group felt that rather than reinvent what exists, a new capability should be easily "insertable" into existing tools and methods. Many of the projects noted below address rather difficult problems.

The general feeling was that a complete near term solution to many of these problems is unachievable. But some tracking and notation of the inadequacy would be a real step forward. For example, the "many worlds" problem of below is certainly not practically solvable. It involves harmonizing incompatible causal mechanics. If a facility merely indicated that there was a problem with some indication of its nature, which would be of significant use.

3.1 Project 1: Soft Modeling

Problem: This project develops a roadmap for graceful implementation of soft modeling within existing and emerging modeling techniques. Both process and enterprise models will be addressed.

Approach: Projects 1 through 4 are proposed with the same approach, though they should be run independently, in parallel.

The project is suggested as a direct follow-on to the ICEIMT, using the same administrative processes. International experts will be identified for focused, facilitated workshops. These will be much longer (four–five days). The participants will be paid to produce position papers and attend. Professional facilitation will be provided.

In other words, for these projects, no new research is planned, merely the exploitation of prior research in relevant areas not normally encountered in the enterprise modeling community.

Disciplines of interest will be logicians, and theorists in the category, situation, type and graph theories, Experts in knowledge representation in agent systems will be included, as will leaders in the relevant standards efforts.

Expected Benefits: Major national research establishments are aware of the problem, but are unsure of the expected benefits and promising approaches. The benefit of a well-structured research roadmap should mobilize pent up interest and produce some early, leverageable results.

3.2 Project 2: Introspective Modeling

Problem: This project addresses the problem of models which monitor, measure and change models in the enterprise modeling context. It focuses on developing a set of priorities for implementation, an ordered list of the expected difficulties and benefits, and an indication of leverageable technologies.

Approach: (See Project 1)

Expected Benefits: As with project 1, the benefits of a research roadmap should bring light to a cloudy area. A well-defined roadmap is probably too much to expect from this project, given the diversity of communities involved. But there are likely many ready solutions that might be adapted, once the proto-roadmap is produced.

3.3 Project 3: Multiworld Modeling

Problem: This project concerns identifying the basics of the multiworld problem as described above. No immediate solution is expected, instead a thorough elucidation of the problem.

Approach: (See Project 1)

Expected Benefits: The benefits of this project are patterned after the two above. Of the three, this will be least mature in terms of a well ordered research plan. But immense value is expected from a clear understanding of the problem itself. As with modeling in general, a model of the problem will produce immediate insight into pitfalls for the entire community of enterprise modelers, giving them a virtual "placeholder" for problem areas that will require intervention by human teams.

3.4 Project 4: Multilevel Modeling

Problem: This project begins progress toward an ordered understanding of the multilevel problem described above. The group supposes that an understanding of this problem is a necessary step in applying enterprise modeling in interesting virtual enterprise cases.

Approach: (See Project 1)

Expected Benefits: The project is expected to result in a demonstration agent system (perhaps open source) that exhibits multilevel evolution. This system will emulate levels seen in common types of virtual enterprises. This problem is expected to one, which many clever enterprise modeling centers will jump on once first results are demonstrated. Quite possibly, the solutions that emerge will be in reaction to the result of this project instead of directly building on it.

3.5 Project 5: Additive "Abstraction"

Problem: This project addresses a need not described above.

The basic problem is that enterprise modeling at present is a two-step process: processes are modeled, and then they are aggregated. Each is a lossy step, involving abstraction.

The workgroup believes that these two steps produce a rather profound loss of relevant behavior because abstracting for effective individual processes probably is in a different "direction" than for enterprise registration.

The project intends to produce a roadmap toward a process humorously called "additive abstraction," or contextualization. In this approach, processes are modeled directly into the enterprise context, immediately inheriting the information needed for the local process owner to see relevant characteristics of the enterprise.

This will certainly result in more complex process models than the "direct" process models of today. But the expectation is that they will be more useful in the enterprise context.

Approach: Of the projects in this report, this is the only one suggested as a "traditional" project, with a dedicated project team chartered with producing a prototype system.

International involvement is suggested, in part to tap the large but distributed knowledge base. But the international scope is also intended to bypass the long, languid process usually resulting from normal national and European Union channels. Academics will be involved for their expertise, but the prime mover is suggested to be a lean and mean commercial prototyper.

The open systems process is suggested for maximum technology transfer and visibility. The project will explore four inter-related concepts: speech acts, metaphor, narrative and "explanation."

Speech acts are a formal parsing and ordering of the information exchanged between active entities. This is a well-studied area when the information is well-behaved and the entities "understand" each other well, but the approach will be extended for the four cases addressed in the projects above.

The resulting speech act performatives will guide the exploration into metaphor. The use of metaphor is expected to be an aid in contextualizing the information. The focus of the exploration into metaphor will be on notations that convey certain contextual patterns while making that context clear to non-specialists.

Both the performatives and metaphoric context will suggest means for building structured narrative. This narrative is expected to provide the structure for an annotation strategy that will provide the explanations desired.

The target application is as a structured annotation language to add to existing modeling environments. The notion is to allow existing methods to support what they already do, and revert to attached annotations for those areas (soft, multiworld...) which they do not. The project will focus on this annotation language to incrementally improve the information and context captured in a structured computable way.

Multimedia delivery and presentation methods will be explored.

Expected Benefits: The direct benefit of this project will be to define a new approach to modeling: for enterprise models rather than composable process models. This is expected to hit a sweet spot in the cleverness of the community and produce a rash of new modeling approaches. Direct input to the developing Process Specification Language and Unified Enterprise Modeling Language programs is expected.

3.6 Project 6: Definitions

Problem: This project works toward definitions of enterprise and behavior in the expanded context identified by the workgroup. The impetus for the project grew out of an unexpected difficulty the group had in closing on such definitions. The fundamental problem seems to be a matter of the simultaneity of purposes: enterprise view to optimize the system through processes, and process view of enterprise characteristics to improve the enterprise through incremental action.

Approach: This is proposed as a virtual, web-based project. It will leverage the many people already funded to work in this general area. The only costs will be professional facilitation, which need be substantial because of the continuing need for focusing and refinement.

Draft, web-based results and email-based discussion will feed the process. Total convergence on consensus positions is not expected. Consensus will be encouraged where possible. Otherwise, a crisp description of contrary positions is sought, together with underlying analysis of the philosophies behind the differences.

Expected Benefits: The result of this project will be directly fed to standards efforts, producing and expected greater understanding of the unique challenges of common enterprise model representation.

Some Methodological Clues for Defining a Unified Enterprise Modelling Language

Michaël Petit
University of Namur, Belgium, mpe@info.fundp.ac.be

Abstract The need for a Unified Enterprise Modelling Language (UEML) that would be used as in inter-lingua among enterprise modelling software tools has been established. The process of defining this UEML goes through the elaboration of a precise meta-model describing the constructs of the language. A possible and reasonable approach for the definition of this meta-model is to integrate (parts of) meta-models of existing enterprise modelling languages. This approach has similarities with the well studied problem of databases integration in which the models of several databases have to be integrated into a single one. In this paper, we make an analogy between the two problems and review a state of the art methodology proposed in database integration to derive methodological clues for the definition of the meta-model of a UEML.

1 TOWARDS A UNIFIED EML

Enterprise modelling (EM) has long been recognised as a valuable activity (Vernadat, 1996). However, the current situation in this domain prevents us from getting the most benefits from EM. Many enterprise modelling languages (EMLs) exist and offer similar but slightly different constructs for modelling. A model written in a particular language can rarely be understood by people not familiar with this language. The different language supporting tools offer different interesting functionalities, but the absence of common understanding of the models by the different tools prevents the exchange of models created by each tool, disabling the user from using these functionalities in a model without rewriting it completely in another language.

As an enterprise is a complex entity and because modelling needs are diverse, several models are often produced with different languages but no integrated model is available that could provide a coherent and complete view of the enterprise.

This situation has lead to the identification of the need for an inter-lingua for enterprise modelling (Goossenaerts, *et al*, 1997). Such a UEML (Unified Enterprise Modelling Language) could be used to exchange models among tools and would constitute a base to commonly understand models written in different languages, providing the possibility for an integrated model of the enterprise. Some efforts have already been made to define such a language and have resulted in the European pre-standard ENV 12204 (CEN, 1996). Such a need was also identified by other projects such as PSL and projects on ontologies (Schlenhof, *et al*, 2000, Fox, 1992, Gruber, 1993).

One of the important steps in defining a language is to elaborate a meta-model. The meta-model corresponds to an abstract syntax and describes the constructs of the language, their properties and restrictions on the way their instances they can be combined to form models. It usually serves directly to define a concrete syntax (textual or graphical) and is the basis for implementing repositories of software tools.

This paper proposes methodological guidelines for the definition of the meta-model of a UEML. In particular, some research work from the field of database integration is proposed as an inspiration for defining methodological clues. As some issues in this field are similar to the ones to be solved to define UEML, we propose that lessons learned and tools in the database integration field are reused. This paper is an attempt to make an analogy between these two research fields.

2 THE DATABASE INTEGRATION PROBLEM

The database integration problem arises when several databases exist and contain overlapping or related data, potentially implemented in heterogeneous environments. The problem in these situations is to provide a mechanism for accessing these databases that hides the location of data and provides a seamless access to logically related data present in more than one database. Parent and Spaccapietra, (2000) define database integration as "the process which:

- Takes as input a set of databases (schema and population), and
- Produces as output a single unified description of the input schemas (the integrated schema) and the associated mapping information supporting integrated access to existing data through the integrated schema."

As an example, consider a situation where some information about persons (such as e.g. name and birth date) is stored in a database made of Cobol files and other information (such as name and address) is stored in another SQL database. Integrating these databases would be necessary to answer queries such as "list persons aged more than 50 living in Brussels".

3 AN ANALOGY BETWEEN UEML META-MODELLING AND THE DATABASE INTEGRATION PROBLEM

UEML should be defined on the basis of the set of existing EMLs. Therefore, it should include constructs that relate as closely as possible to those of the majority of currently used EMLs. UEML should thus be an integration of most existing EMLs. The definition of UEML should therefore be made by examining different EMLs and systematically considering the inclusion of their constructs in UEML. The definition of a meta-model for UEML can be in some respects compared to a database integration problem in which:

- The database schemas to be integrated are the meta-models of a set of candidate EMLs;
- The data on which an integrated vision is desired are the various enterprise models created in these different EMLs.

In Parent and Spaccapietra, (2000) a general methodology for database integration is proposed on the basis of a review of current approaches and solutions to the database integration problem. It consists of three major steps:

1. Preparation for integration
2. Investigation and definition of correspondences
3. Integration

In the sequel of this section, we describe the activities that have to be performed in each of these steps, make an analogy with the UEML definition problem and show the possible implications of the methodology for the definition of UEML.

3.1 Step 1: Preparation for integration

3.1.1 Preparing for the integration of databases

When facing a database integration problem, a natural first activity consists of collecting information about the databases to be integrated. One of the most important pieces of information about a database is the description of its content in terms of classes of data that it may contain. This description is referred to as the database schema (or database model). The schemas of all databases to be integrated therefore have to be collected. If they are not available, they have to be defined. If only a technological description of the database is available, a reverse engineering process is necessary to obtain a conceptual schema useful for integration at the conceptual level. An example of a software tool that supports this reengineering activity is described in (Hainaut, *et al.*, 2000).

When integrating heterogeneous databases (databases built using different technologies such as files, SQL databases, Object-oriented databases, ...), the schemas are usually of different nature or quality. Additional treatment of the schemas is therefore often necessary to reduce discrepancies among them so that the schemas can be integrated more easily. Three kinds of modifications to the schemas are described by Parent and Spaccapietra, (2000): syntactic rewriting, semantic enrichment and representation normalisation.

First, a *syntactic rewriting* of the schemas might be necessary if the schemas are initially expressed in different languages. If one schema is expressed e.g. using an entity-relationship notation and another is expressed using UML class diagrams, the comparison of the elements of the schemas will be more difficult than if a single language was used. This requires choosing a common language to express the schemas. According to Parent and Spaccapietra, (2000) the chosen language must be rich enough to allow the expression of all information relevant to the different schemas, but must not be too rich to avoid too man modelling alternatives when defining the schema. The later problem, known as semantics relativism, calls for languages with minimal semantics embedded (no complex modelling elements).

Second, the schemas to be integrated might require *semantic enrichment*. This means adding semantic information to the schema such as initially unstated constraints, implicit assumptions, ...

Finally, though the choice of an adequate common representation language reduces the semantics relativism problem, it is probably not possible to avoid it completely. Therefore, the schemas might require a *representation normalisation* that imposes the use of consistent representational strategies within the schema when choices are possible. For example, if an attribute of a class in a schema may have many values for each object of that class, such a strategy might be to represent this attribute as a separate class linked to the initial class with a relationship.

3.1.2 Preparing for the definition of UEML

Similarly to the database integration problem, the UEML definition requires the collection and definition of the schemas to be integrated. In this case, the schemas are the meta-models of a set of candidate enterprise modelling languages considered for "integration" in UEML. Meta-models of EMLs are usually not directly available. Most EMLs are described in terms of their syntax, but do not always make explicit all relationships among language constructs and constraints that apply to obtain valid models written in those languages. A first necessary exercise is therefore to define precisely the meta-models of these candidate languages. In some cases, the meta-model can be obtained by reverse engineering the meta-models implemented in supporting software tools that use database technologies to store enterprise models in a repository. In other cases, the meta-models have to be elaborated by hand on the basis of the literature describing the EML.

In previous work, we have done the exercise of defining such meta-models. To express these meta-models we used a common language called Telos (Mylopoulos, *et al.*, 1990). The choice of Telos is justified by its formality (mathematical foundation), expressiveness (especially for expressing constraints) and limited number of concepts (preventing too many choices of representation). The first meta-model we have defined is the one of CIMOSA, a well-known EML, which is the result of a European project (AMICE, 1993). The complete meta-model can be found in (Petit, 1999). The second one is the meta-model on the ENV-12204 pre-standard (CEN, 1996). This pre-standard could be considered as an initial attempt of a UEML definition. The definition of the meta-model on the basis of the published standard raised a number of problems and open issue that are reported in (Férier, *et al.*, 2000). In both cases, we have applied implicitly semantic enrichment and representation normalisation. Based on our understanding of the studied documents, we have added constraints and resolved perceived inconsistencies. Representation strategies, while not explicit, were usually applied because the models were elaborated by a small number of people.

Nowadays, class diagrams from the UML (Booch, *et al.*, 1999) are often proposed for defining meta-models of languages. Compared to Telos, UML offers more representational choices and may therefore make meta-models more difficult to compare and integrate. However, it might be more intuitive because it is less formal. Note however that deriving a meta-model expressed in UML from the Telos descriptions is quite easy. We are currently performing some preliminary work on the definition of a meta-model expressed in UML for the Workflow Process Description Language (WPDL) defined by the Workflow Management Coalition, (1999).

For the definition of UEML, a more systematic way of working would be needed by:

- Making explicit the representation strategies used to define the meta-model, and systematically apply these strategies;
- Validating the meta-model through interaction with the language designers or owners and validating the semantic enrichments applied.

3.2 Step 2: Investigation and definition of correspondences

3.2.1 Investigation and definition of correspondences in databases

The next step in the methodology consists in establishing what is common in the databases that are candidates for integration. This amounts for investigating and establishing correspondences among these databases. Parent and Spaccapietra, (2000) explains that the correspondences have to be defined at two levels. *At the data level* (population of the database), the correspondences among instances of classes present in one database and the ones present in another have to be identified. To establish these correspondences, the semantics of the instances have to match. Two instances are considered to have the same semantics if they describe the same real world element. *At the schema level*, correspondences among classes are established if the correspondences among instances apply to a significant set of instances of these classes. The correspondence is thus generalised at the class (schema) level. The correspondence may be further characterised as an equivalence (if both classes have sets of instances that represent exactly the same set of real world elements), as an inclusion (if all instances of one class have a corresponding instance in the set of instances of the other), as an intersection (if there exists correspondences among instances but no equivalence and no inclusion), ... Furthermore, a notation for describing these correspondences is proposed.

3.2.2 Investigation and definition of correspondences between EMLs

For the definition of UEML, we need to establish correspondences among the classes defined in the different meta-models of the EMLs candidate for being integrated in UEML. But as Parent and Spaccapietra, (2000) suggests, we may only define correspondences at this level if the correspondences among instances of these classes can be generalised. This means that we first have to investigate correspondences among models elements created with the different EMLs and then generalise them if these model elements have the same semantics. In our case, the semantics of model elements is more complicated than the one of database instances because enterprise models usually represent sets of elements or happenings of the real world rather than individual elements. For example, a model element such as an object class actually represents a whole population of a database, whose elements themselves have a correspondence in the real world. A correspondence among languages can therefore only be established if model elements created with these languages represent the same set of elements from the real world. The same principle applies to the comparison of the semantics of behavioural models elements such as processes. In this case, the semantics is even more complex to compare since processes have dynamics semantics, potentially representing infinite sets of behaviours. The semantics comparison must in this case make sure that the sets of process behaviours described by both models are corresponding. This comparison can become possible and be computer-assisted if the semantics of process models is defined formally. Further research would however be needed to allow this kind of automation.

A consequence for the UEML definition process is that to establish correspondences among language constructs (classes of EMLs meta-models), we need models created using these languages. Therefore, case studies have to be carried out in which a single reality is modelled with these different languages. Then correspondences among the obtained models have to be established by comparing the semantics of the obtained model elements in terms of the sets of real world elements or behaviours they represent. On the basis of these correspondences, tentative correspondences may be defined at the language level, among elements of the meta-models. These correspondences can then be validated or infirmed on the basis of further case studies.

A notation for explicitly defining correspondences both at the model and language levels is therefore necessary. In (Petit, 1999), we have proposed a structure corresponding to this notation for a framework made of several languages for Manufacturing Systems modelling. This structure, formalised in Telos, is based on the meta-models of the languages of the framework. Potential correspondences at the language level are defined as "mapping rules" among language constructs, whereas correspondences at the model level are seen as applications (instances) of the mapping rules and establish correspondences among model elements. It should be noted that in our case, the mapping rules were not meant to be general nor automatic, so that the models creator can decide if the rule applies or not on particular elements of the models. This allows for the definition of different kinds of correspondences as described in (Parent, Spaccapietra, 2000) (equivalence, inclusion, intersection, …).

3.3 Step 3: Integration

3.3.1 Integration of the databases schemas

The third and last step is to define the integrated schema of the database by considering the inclusion of elements from the original schemas in the integrated schema. This process treats systematically each correspondence identified in step 2 and decides which elements to include in the integrated schema. Potential conflicts have to be resolved at this level. A simple example of conflict is a "description conflict" which occurs when two corresponding classes have different sets of attributes. In this case the conflict has to be resolved by deciding which of these attributes have to be included in the integrated schema. Parent and Spaccapietra, (2000) provides a list of possible conflicts and references to literature that propose systematic ways of solving theses conflicts. When solving conflicts, different strategies can be adopted depending on the objective followed when doing the integration. For example, some strategies may seek completeness of the integrated schema while others may seek simplicity. Parent and Spaccapietra, (2000) therefore insists on the importance of defining the objective of the integration beforehand and adopting an adequate strategy.

3.3.2 Integration of EML meta-models into one UEML meta-model

The meta-model of UEML should be defined by considering the integration of a number of relevant EMLs meta-models. In some respects, the ENV12204 is already such integrated meta-model. It was defined mainly on the basis of CIMOSA and IEM (Mertins, Jochem, 1999). However, the meta-models of these languages were not made explicit in a single representation language and the correspondences among these meta-models were not described explicitly. In this process, the conflicts that arose were solved intuitively without being made explicit. This has the drawback that no explicit trace has been kept of the relationship existing between the constructs of the original languages and those present in the integrated ENV12204 meta-model.

The strategy for resolving conflicts during integration has to be defined and depends on the objective of UEML. A reasonable objective of UEML could be the interoperability of a set of existing enterprise modelling tools. In this case, an adequate global strategy would be to only include a construct in the UEML meta-model if there exist corresponding constructs in the meta-models of at least two languages to be integrated. Hence, if a construct or attribute were specific to a tool, it would not be useful to make it available to the other tools, since it would not make any sense for the other tool. If the objective is rather to obtain a logically integrated model of the enterprise, a strategy where more completeness of the UEML meta-model is sought would be more appropriate.

In any case, an explicit definition of the correspondences, both among constructs of the original languages meta-models and between these constructs and the corresponding constructs in the UEML meta-model is useful. The approach of mapping rules proposed above seems adequate for this. The explicit definition of the correspondences is important not only for explaining the constructs of the UEML by making reference to the constructs of other existing and known languages, but also because they are the basis for e.g. a specification of model exchange and translation mechanisms to be implemented in EM tools (in the case of a "model exchange among tools" scenario) or specification of mechanisms for query processing on a logically integrated enterprise model (in the case of an "integrated enterprise model" scenario).

4 ADDITIONAL METHODOLOGICAL HINTS FOR THE DEFINITION OF UEML

A very important step, as discussed in section 3.3 is the definition of the objective of UEML at the very beginning. The meta-models integration strategy and the content of the UEML itself will depend on the settled objective.

A second step is the identification of candidate languages to be "integrated" into UEML. A strong candidate is the ENV12204 pre-standard. This language is currently being substantially revised by CEN TC310 WG1. One of the changes is an explicit and better definition of its meta-model. This new meta-model could serve as a base for the UEML meta-model and be augmented or improved by considering other languages.

A good way of working could be to work with languages on a pair wise basis, rather than considering them altogether.

To remove complexity, the integration process could be first performed on subsets of the considered languages. These subsets could be the core of the languages (set of simple or atomic constructs). After integrating these core constructs, additional composite or complex constructs could be considered for integration.

Software tools should be used whenever possible to support the definition of the UEML meta-model. Some of the tools used in the database engineering and database integration area seem appropriate for this.

5 CONCLUSION

In this paper we have investigated the analogy between the problem of defining the meta-model of a Unified Enterprise Modelling Language (UEML) and the problem of database integration. Some commonalities exist between the two problems. Based on a general methodology proposed by Parent and Spaccapietra, (2000), some methodological hints for the definition of the meta-model of UEML are identified. Some initial work of the authors is also cited.

As the database integration problem has been studied for some time now, a large body of literature describing the issues to be resolved, possible solutions to them and supporting tools to solve them is available. Many of them can be reused within the context of the UEML meta-model definition. This paper is only a preliminary study of the link between these two problems. Additional work is obviously needed to better identify relevant solutions and tools that can be reused for UEML meta-modelling.

6 ACKNOWLEDGEMENTS

The author would like to thank Jean-Luc Hainaut and Philippe Thiran for enlightening discussions on the database integration problem and its similarity to meta-models integration. We also thank the anonymous reviewers for their appropriate suggestions for improvement of the paper and Gaëtan Delannay for initial proof-reading.

7 REFERENCES

AMICE, (1993), *CIMOSA: Open System Architecture for CIM*, Springer-Verlag.

Booch *et al*, (1999), *The Unified Modeling Language User Guide*, Addison-Wesley.

CEN, (1996), ENV 12204: *Advanced Manufacturing Technology - Systems Architecture – Constructs for enterprise modelling*, TC310 WG1 (currently under revision).

Férier, L. Heymans, P. Petit, M. (2000), *Some Hints for a Clarification of CEN ENV 12204*, Invited paper at the Workshop on Evolution in Enterprise Engineering and Integration, Berlin, May 24-26.

Fox. M.S. (1992), *The TOVE project: Towards a common-sense model of the enterprise*. In Proceedings of the International Conference on Object Oriented Manufacturing Systems, Calgary Alberta.

Goossenaerts, J. Gruninger, M. Nell, J.G. Petit, M. Vernadat, F.B. (1997), *Formal Semantics of Enterprise Models*, in K. Kosanke and J.G. Nell (Eds.), Proc. of ICEIMT'97, International Conference on Enterprise Integration an Modeling Technology, Springer-Verlag.

Gruber, T.R. (1993), *A translation approach to portable ontology specifications*, Knowledge Acquisition, 5(2).

Hainaut, J-L. Henrard, J. Hick, J-M. Roland, D. Engelbert, V. (2000), *The Nature of Data Reverse Engineering*, In Proc. of Data Reverse Engineering Workshop, March 2, as part of Reengineering Week 2000, Zurich, Switzerland, available at http://www.info.fundp.ac.be/~dbm/publication/2000/dre2000_jlh.pdf

Mertins, K. Jochem, R. (1999), *Quality-Oriented Design of Business Processes*, Kluwer.

Mylopoulos, J. Boride, A. Jarke, M. Koubarakis, M. (1990), *Telos: A Language for Representing Knowledge about Information Systems*, ACM Transansaction on Information Systems, 8(4).

Parent C. Spaccapietra, S. (2000), *Database Integration: the key to data interoperability*, In Papazoglou, M.P. Spaccapietra, S. Tari, Z. (Eds.), Advances in Object-Oriented Data Modeling, MIT Press.

Petit, M. (1999), *Formal Requirements Engineering of Manufacturing Systems: a Multi-formalism and Component-based Approach*, PhD Thesis, Computer Science Department, University of Namur, Belgium.

Schlenoff, C. Gruninger, M. Tissot, F. Valois, J. Lubell, J. Lee, J. (2000), *The Process Specification Language (PSL) Overview and Version 1.0, Specification*, National Institute of Standards and Technology, Gaithersburg, MD, USA, available at http://www.mel.nist.gov/psl.

Vernadat, F.B. (1996), *Enterprise modeling and integration: principles and applications*, Chapman & Hall.

Workflow Management Coalition, (1999), *Interface 1: Process Definition Interchange Process Model*, Document Number WfMC TC-1016-P, Version 1.1, available at http://www.wfmc.org/

Common Representation through UEML – Requirements and Approach

Roland Jochem
FhG-IPK, Germany, Roland.jochem@ipk.fhg.de

Abstract: At the current state of technology, we can state that Enterprise Modeling (EM) is now a reality in many large companies. Enterprise Engineering practices are developing and enterprises are under pressure to adopt engineering procedures based on models. However, interoperability between Enterprise modelling methods and also modelling tools is still very weak compared to real needs. Although both EM and EI are nearly not introduced in SMEs. They are only exposed when they take part in supply chains. In this case, they are told what to do and which tools they have to use. It is the author's opinion that these technologies would better penetrate any kind of enterprises if there were a standard interface in the form of a unified enterprise modelling language (UEML), which is based on a consensus on all modelling tools available on the market. This paper presents requirements and an approach to support common representation by a unified enterprise modelling language (UEML).

1 INTRODUCTION

The manufacturing world is in permanent change. Nowadays, it is moving from an economy of scale to an economy of scope under a global economy for mass customisation. For many companies around the world, staying in business means:

- To meet customer requirements,
- To reduce the time-to-market of their products, and
- To manufacture products at low cost with increased quality.

Thus, there is a need for better *process management* and for more *integration* within decentralised and modular individual enterprises (e.g. most discrete parts manufacturing companies) as well as among enterprises belonging to the same group or co-operating on collaborative projects (e.g. AIBUS consortium). Integration aims at providing quickly ***the right information at the right place at the right time*** under the right format throughout the enterprise, is therefore evolving (Vernadat, 1996). Enterprise Integration concerns:

- Efficient business process integration and co-ordination;
- Support to teamwork or computer supported collaborative work (CSCW) for concurrent design and engineering activities;
- Increased flexibility throughout the company;
- Total quality deployment to be introduced as early as possible in the product life cycle; and
- Collaboration of IT solutions, systems and people to face environment variability in a cost-effective way.

Among all these issues, process integration and co-ordination remains the most challenging problem because of its knowledge intensive nature.

Furthermore, it must be stressed that *integration is a never-ending process.* First, because it is a goal. Second, because the enterprise is in a permanent process of change. Its introduction must be carefully planned and documented by a master plan, and once started, procedures for continuous process improvements must be put in place (Jochem, 2001).

In a more general definition Integration consists of putting components together to form a synergistic whole. Different types of integration can be listed. They include in terms of Enterprises:

- Horizontal versus vertical integration
- Intra-enterprise versus inter-enterprise integration (Fig. 1)

The major problems currently faced by industry, which significantly limits the wide use of enterprise modelling, and engineering techniques concern Interoperability, Enterprise Integration, and Sharable Enterprise knowledge.

Interoperability: Business companies must use various tools from different vendors for obvious reasons of vendor independence and need for many kinds of functionality. Unfortunately, each vendor system comes as a stand-alone tool with its own proprietary language (or interface) forcing users to learn different languages and to model several times the same concepts due to non interoperable, closed systems.

Enterprise Integration: Any time two companies go for partnership or merging, they have to at least connect but in many cases tightly integrate their information systems and co-ordinate their business processes. Enterprise modelling is usually recognised as a prerequisite phase to Enterprise Integration to build a common vision or consensus in terms of business operations.

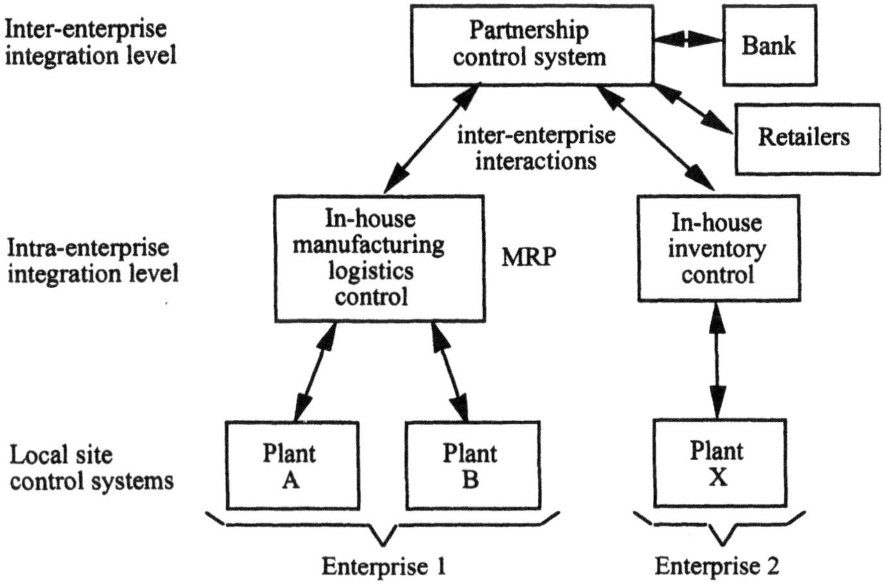

Figure 1: Inter-enterprise integration vs. intra-enterprise integration (Vernadat, 1999)

Sharable Enterprise Knowledge: Many business activities could be leveraged within single companies or within networks of enterprises (virtual enterprises, extended enterprises, large supply-chains) if they could share knowledge about the enterprises. This is not yet achievable at the level requested by industry because most tools encode fragmented knowledge in non-sharable way and do not access common repositories (IST-2001-34229, 2001).

This limitation has been recognised by standardisation bodies both at the European (CEN TC 310) and ISO (TC 184/SC5) levels. They have proposed guidelines for enterprise modelling/engineering (ENV 40003, 1993, ISO 15704, 1998) and modelling constructs (ENV 12204,1996).

Over the last decades, numerous efforts have been carried out in the field of Enterprise Modelling and Integration. With the application of modelling and integration principles in manufacturing (e.g. CAD/CAM integration, supply chain integration, CIM, Concurrent Engineering or just-in-time operations), significant improvements in terms of competitive advantages and industrial excellence are expected by means of better communication, co-ordination and co-operation between all levels and components of companies (IST-2001-34229, 2001).

Most major Projects (e.g. ESPRIT/CIMOSA, ICAM/IDEF, IPK/IEM, ESPRIT/CCE-CNMA, LUT/CIM-BIOSYS, PERA, GRAI/GIM, GERAM) have demonstrated the necessity of developing enterprise models to support analysis, design and management of the business processes that are executed in companies. These processes must be modelled from different points of view and at different levels for the purpose of building more integrated systems (stand-alone companies or networked enterprises). These ideas have even contributed to standardisation and unification efforts to harmonise concepts and terminology (CEN ENV 40003 and its reworked version, CEN ENV 12204 and its reworked version, ISO 15704, (1998), ISO/IEC 15414 (ODP), (2000), IFAC-IFIP (1997), OMG (2001), BPML (2001)).

Several commercial tools for enterprise engineering as well as process management are also being proposed based on these concepts (e.g. ARIS ToolSet, FirstSTEP, PrimeObjects, Bonapart, MOOGO, etc. in Mertins, 1998).

However, business users still face a variety of difficulties in their day-to-day work:

- Wide variety of available languages and technical approaches for modelling.
- Significant semantic gap between current different modelling languages.
- Poor interoperation capability of process modelling and management tools.
- Insufficient coverage by most languages of modelling views required by integrated engineering and management.
- Ignorance by most current enterprise modelling languages of aspects such as strategic goals, intentions, human roles and behaviour, know-how and other so-called "soft issues".
- Diversity of graphical modelling representations (diagrammatic or semi-formal notations) and multitude of meanings for similar concepts (inconsistent semantics), which make difficult common understanding of enterprise models.
- Lack of a common standard language and exchange format make model exchange from one tool to another nearly impossible.

Because of this Tower of Babel situation, it becomes a necessity to define a unified language for universal use by business users as well as within the enterprise modelling community and which would address these problems (Vernadat, 1999). We therefore propose the development of such a language, called Unified Enterprise Modelling Language (UEML), by analogy with UML devoted to conceptual systems modelling.

2 OBJECTIVES AND APPROACH OF UEML

The main objective of the UEML project is to define, to specify and to validate a set of core constructs and related services to support a Unified Language for Enterprise Modelling, named UEML, to serve as a basis for interoperability within a smart organisation or a network of enterprises (IST-2001-34229, 2001).

This UEML will:

- Provide the business community with a common user interface to be used on top of most commercial enterprise modelling and workflow software tools,
- Provide a standardised mechanism for exchange of enterprise models among these tools,
- Support the implementation of open and evolutionary Enterprise Model Repositories to leverage enterprise knowledge engineering capability.

Taking into consideration previous work and existing tools, the main business objective of the UEML Project is to provide industry with a unified and expandable modelling language. The language, to be used on top of existing systems, will be defined as an open approach (i.e., an expandable set of core constructs with formal and graphical specifications) to describe the structure, behaviour and organisation of enterprises (be they related to service or goods industry). The language will be applicable to all sectors of industry and services, all enterprise dimensions and various kinds of enterprise processes (IST-2001-34229, 2001).

Therefore, it is important to mention that the objective of the project is *not* to develop a new language, which will replace the ones already in use, but rather to propose a language serving as a *gateway* (i.e. a mediating facility) between existing EM or workflow tools. UEML will be used as a facilitator within business applications of one enterprise or of a network of enterprises (extended enterprise).

Fig. 2 provides a global framework enabled by UEML and shows the expected result of the project (IST-2001-34229, 2001).

The global structure of the project will be based on two main elements:

UEML language for Modelling/Engineering Application: construct definition and specification, and demonstration that it can be implemented as a common user-interface on various commercial tools and used for model exchange.

Enterprise Knowledge Management *based on a common Communication Layer and Repository services implementation* (construct classes and type hierarchy definition, UEML API's, repository classes and services) demonstrating how an open, expandable, vendor independent infrastructure can be developed to implement UEML constructs and build shared Enterprise Knowledge Repositories to be accessed by various heterogeneous application systems

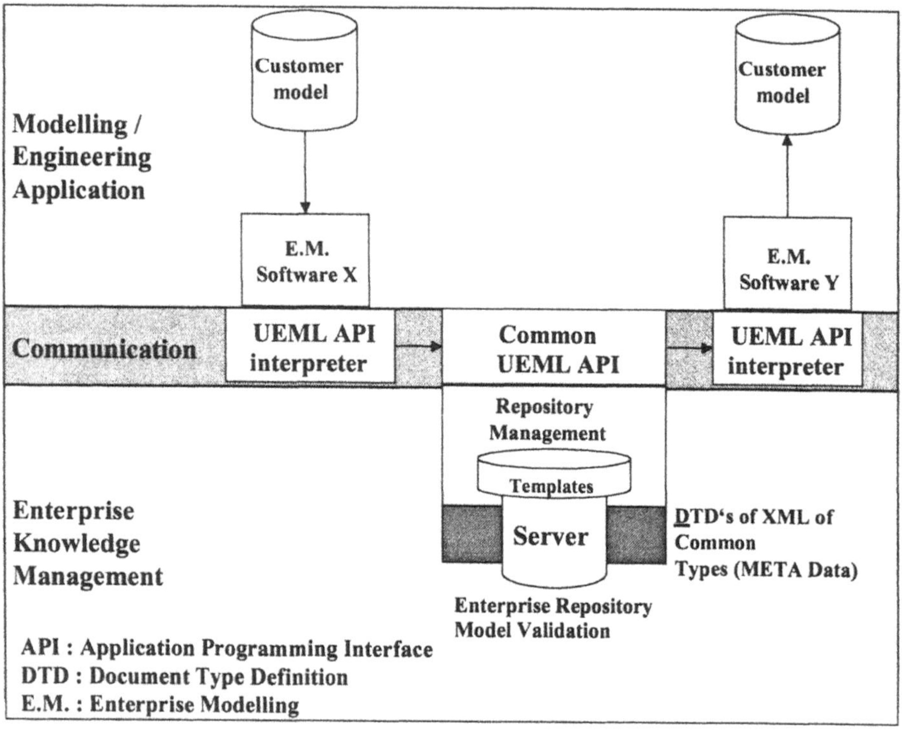

Figure 2: Expected results of the UEML Project

In the future, UEML should even become the basis for new system development. Assuming that the UEML language be unanimously accepted by the user community and strongly supported by standardisation organisations (especially, CEN and ISO), *this work could significantly influence the Enterprise Modelling/Extended Enterprise software development industry.* For instance, the UEML constructs could be the basis for the development of new generations of modelling and simulation tools to support enterprise analysis and engineering as well as inter-enterprise workflow management tools (ISI-2001-34229, 2001).

3 RELEVANCE AND BENEFITS

The relevance of UEML is clear because of its relevance to industrial companies as well as to consultants and software developers. First, a unified language that allows the development of exploitable models in an appropriate framework at different management levels of the company and at different levels of details would be welcome by the business user community.

Second, by learning UEML the business user will become able to interact with many tools without having to learn other dedicated languages, to move models from one tool to another more easily, and therefore be able to share or exchange models with his business partners. Third, tool developers will be provided with a clear and precise language definition to build a UEML interface for their proprietary tool. Last, UEML will bring some form of standardisation in the field of enterprise modelling by imposing a universal way of describing business processes and related elements, like UML currently does for conceptual modelling.

Furthermore, this common format and way of expressing enterprise models should make it easier to develop so-called reference models, i.e. partial, reusable models, which could be shared and commonly developed within a company or within a community of business partners (Mertins, 1999).

4 THE UEML PROJECT

A "UEML" project has been implemented as an IST Thematic Network project (IST-2001-34229, 2001) which sets up a feasibility study to analyse the market potential of such a visual enterprise languages, to accurately define the specifications of a core of such a language, to validate it, to demonstrate and to disseminate the concepts.

The UEML Working Group will be composed with two kinds of memberships (IST-2001-34229, 2001):

- A core membership of the eight initial partners of the UEML project consortium and
- A network membership consisting of none-consortium organisations

The members of the UEML consortium are:

GRAISOFT (associated to the LAP/GRAI-University Bordeaux 1 and LABRI-University Bordeaux 1, France), INRIA (associated with CRAN-University Henri Poincaré Nancy 1, France), COMPUTAS AS, Norway, CIMOSA Association, Germany, IPK/FhG Berlin, Germany, University of Torino, Italy, University of Namur, Belgium, University of Valencia, Spain.

All partners of this core membership have a strong background in the usage of one or more enterprise modelling languages and work in this research topic for a long time.

The UEML network will be established by the UEML consortium and will be composed of people from any kind of institution – industrial enterprise (end-users or provider of I.T. products), research, or standardisation bodies - who have an interest in the UEML core constructs definition and more generally in the complete UEML elaboration.

The UEML network will be supported by the UEML Working group for travelling expenses and also for assigned work.

Figure 3 below shows the relation between the two groups.

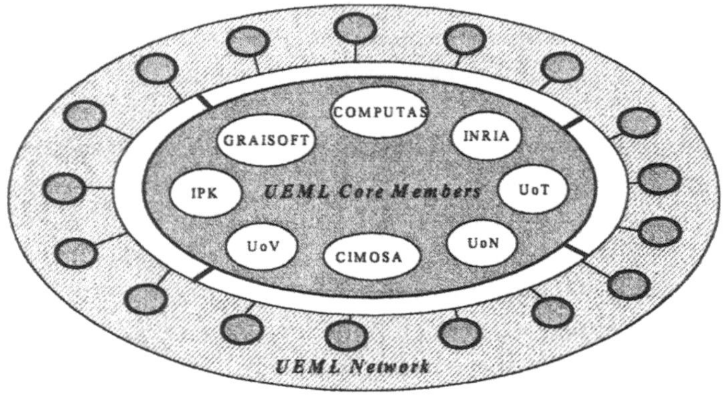

Figure 3: Relation between UEML Core Members and UEML

The UEML network is an integral part of the project. The network has the goal to reach a common understanding and an improved consensus on enterprise modelling language constructs in the industry and academic community. It serves two purposes: 1) to disseminate general knowledge about UEML and project results and 2) to gather user requirements. Co-operation in the network is performed by the participation to core membership meetings, working groups, and information exchange via Web portals, electronic mail and distributed papers.

The members of the UEML network will be recruited from academia and various industry segments, from SMEs and large enterprises, from associations, European projects, standardisation bodies and industry initiatives. We intend to co-operate with the leading organisations in this field.

5 CONCLUSION

It must be well understood that it is not the intention of this project to re-invent the wheel, i.e. to invent yet another language, but rather to consolidate the accumulated knowledge and experience in the field of enterprise modelling by proposing a universal user-oriented language, or common tool interface, by generalising existing ones to, avoid the Tower of Babel situation currently prevailing and limiting the wide use of enterprise modelling technology.

The first step in that direction is the new established IST-Thematic Network Project "UEML" that will prepare the launching of a development project to define and implement a 'complete' UEML.

The development of the UEML language will be accompanied by the development of Enterprise Ontologies (formal descriptions of entities as well as of their properties, relationships, constraints and behaviours). These Enterprise Ontologies will provide formal meta-models and micro-theories to enterprise modelling concepts.

In the future, this complete Unified Enterprise Modelling Language will allow to:

- Provide the business community with a common visual template based language to be used on top of most commercial enterprise modelling and workflow software tools,
- Provide standardised mechanisms for sharing knowledge models and exchanging enterprise models among projects, overcoming tool dependencies,
- Support the implementation of open and evolutionary enterprise model repositories to leverage enterprise knowledge engineering services and capabilities.

6 REFERENCES

BPML, (2001), *BPML – Business Process Modelling Language,* BPMI.org.

ENV 12204, (1996), *Constructs for Enterprise Modelling,* CEN TC 310 WG1.

ENV 40003, (1993), *Framework for Enterprise Modelling,* CEN TC 310 WG1.

IFAC-IFIP Task Force, (1997), *GERAM: Generalized Enterprise Reference Architecture and Methodology,* Version 1.5, IFAC-IFIP Task Force on Architecture for Enterprise Integration.

ISO 15704, (1998), *Requirements for Enterprise-Reference Architectures and Methodologies,* TC 184 SC5t WG1.

ISO/IEC 10746, (1992), *Information Technology - Open Distributed Processing - Basic Reference Model of Open Distributed Processing.*

ISO/IEC 15414 (ODP), (2000), *Information Technology - Open Distributed Processing - Reference Model - Enterprise Viewpoint.*

IST-2001-34229, (2001), *Unified Enterprise Modelling Language (UEML).* Description of Work. European Commission IST Project.

Jochem, R. (2001), *Integrierte Unternehmensplanung auf der Basis von Unternehmensmodellen.* Dissertation TU Berlin.

Mertins, K. Jochem, R. (1998), *MOOGO.* In: Bernus, P., Mertins, K.; Schmidt, G.: Handbook on Architectures of Information Systems, Springer-Verlag.

Mertins, K. Jochem, R. (1999), *Quality-Oriented Design of Business Processes.* Kluwer.

OMG, (2001), EDOC – *Enterprise Distributed Object Computing*

Vernadat, F.B. (1996), *Enterprise Modeling and Integration: Principles and Applications,* Chapman & Hall.

Vernadat, F.B. (1999), *Enterprise Modeling and Integration – Myth or Reality.* In: Proceedings of CARS&FOF 99 Conference, Aquas de Lindoia, Brazil.

UML Semantics Representation of Enterprise Modelling Constructs

Hervé Panetto
CRAN CNRS UMR 7039, France, herve.panetto@cran.u-nancy.fr

Abstract: Enterprise modelling contributes to understand enterprise structure by providing an explicit description of enterprise processes. Among many key issues in an engineering project, formalisation appears to be a suitable technique to check the global consistency between all the various specifications a system is intended to cover. This paper deals with the use of UML semantics representation by means of stereotypes and OCL invariant formalisation to cope with a global consistency of the UML definition.

1 INTRODUCTION

Enterprise modelling contributes to the understanding of enterprise structure by providing an explicit description of enterprise processes, which could help in performance measurement and improvement to make the best possible decision by the enterprise managers. Among many key issues in an engineering project, formalisation appears to be a suitable technique to check the global consistency between all the various specifications a system is intended to cover. Applying that within the enterprise modelling framework, leads to the formalisation of some existing enterprise standards such as CIMOSA (AMICE, 1993, Vernadat, 1998) in order to provide them with refutable foundations.

Our approach is based on the UML (2001) meta-modelling of CIMOSA constructs (Panetto, et al, 2000) and, more generally, of the European Pre-Standard ENV 12204 (CEN, 1995) constructs, in order to establish enterprise constructs described with a common language, UEML (Unified Enterprise Modelling Language) (Kosanke, 1999, UEML IST TN, 2002), which formalises, not only their definitions and their relationships, but also the constraints they have to meet in order to gain semantics. The first section outlines the formalisation requirements in enterprise modelling and illustrates the UML modelling of enterprise constructs. The second section illustrates the semantics approach defined in the UML standard. The third section shows the UML semantics representation of some UEML constructs. Conclusion and prospects are discussed in the last section.

2 UEML CONSTRUCTS

The European Pre-Standard ENV 12204 contains definitions, descriptions and detailed attributes of the common constructs (IFAC/IFIP, 2001) extracted from enterprise models such as CIMOSA (Fig. 1), GERAM, GRAI, ... and the relationships between these (Fig. 2).

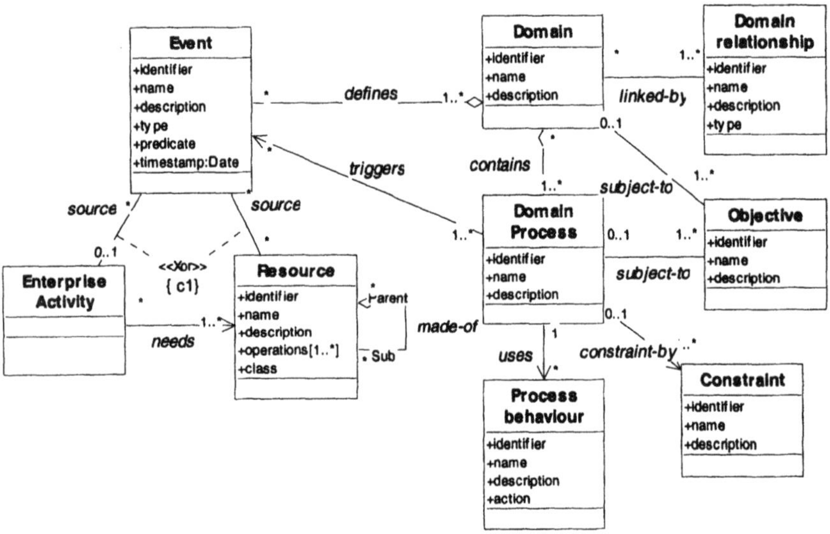

Figure 1: Part of the CIMOSA constructs (Panetto, et al, 2000)

These constructs to be standardised suffer from the lack of a semantic foundation for formally verifying their use in the scope a particular modelling process. Indeed, these constructs are promoted without any ability to check their conformance with the user's requirements. Moreover, their instantiation within a particular enterprise model is not guaranteed to respect some enterprise constraints and properties. Formalisation of these constructs (including enterprise properties) is expected to cope with these two key issues in enterprise models verification.

The practical issues in formalising UEML constructs aim to meta-model them using class diagrams from UML and to formalise constructs constraints and relationships using the OCL (Object Constraint Language) as defined in UML standard.

The formal quality of the system model can be reached by the quality of a formal modelling language. A model is a representation used to formalise a system with semantics while a meta-model is a model used to formalise another model with semantics. Indeed, Gödel's second incompleteness theorem states that any formal system that is interesting enough to formulate its own consistency can prove its own consistency if and only if it is inconsistent. This means that a model cannot be formalised by itself, but only by a higher-level meta-language.

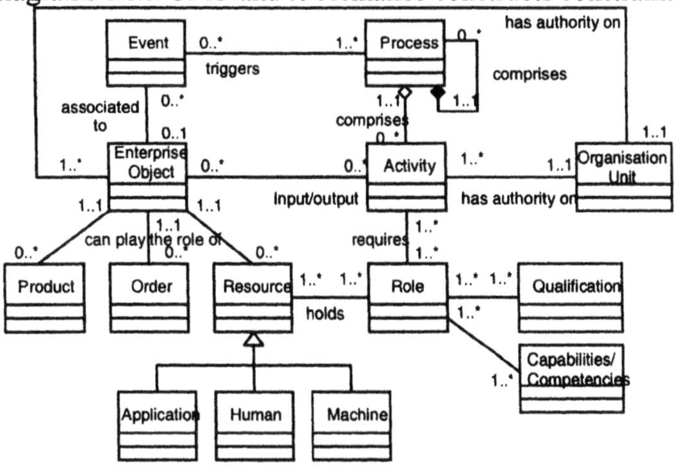

Figure 2: Selected constructs and concepts from ENV 12204:1995 (IFAC/IFIP, 2001)

Such meta-languages manipulate basic concepts of the formalised model to help its understanding. For example, the Fig.3 and Fig.4 represent meta-models of relational model and UML with respectively sNets (sNets Formalism, 1998) and MOF, (1997).

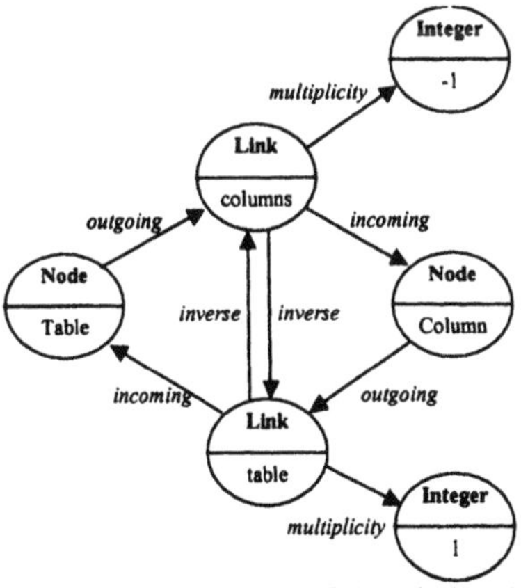

Figure 3: Meta-model of the relational model with sNets

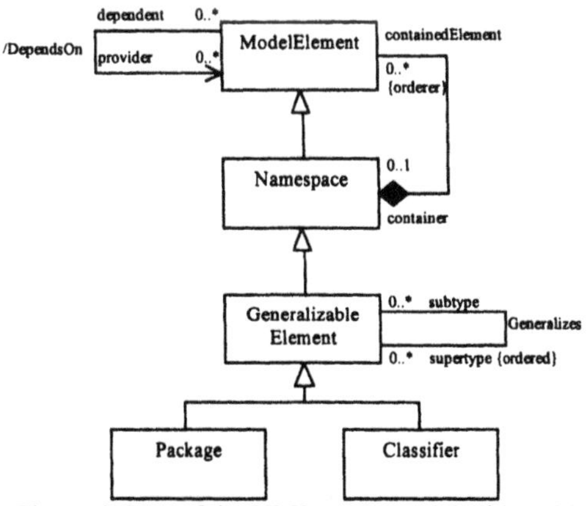

Figure 4: Part of the UML meta-model with MOF (Meta Object Facilities)

As has already been done for products data definitions in ISO STEP 10303 application protocols (ISO, 1994), constructs can be defined as object classes or template structures, which can be assembled to model a system. Due to the complexity and the variety of models, their coherent integration needs the definition of a limited set of constructs that can be applied for their formal representation. A construct is a generic object class or template structure, which models a basic concept independently of its use (ISO, 1994). As an example, the *"if-then-else"* control structure is a particular construct of programming languages.

The identification of constructs consists of meta-modelling the models using a formal meta-language to define the basic concepts that they use. Integration of different models is done by analysing their respective constructs by mean of their meta-model and their definitions, in order to build new constructs that merge their respective capabilities. The objective is *not here to build a new modelling language* but only to formalise constructs that *help to understand the common concepts* of different modelling languages, their relationships and constraints.

3 UML

The Unified Modeling Language (UML), an OMG standard, is a widely adopted and used modelling language. The UML emerged from the unification that occurred in the 1990s following the "method wars" of the 1970s and 1980s. Even though the UML evolved primarily from various second-generation object-oriented methods (at the notation level), the UML is not simply a third-generation object-oriented modelling language.

The UML is defined by nine languages. In this work, we use the Class Diagram, which defines objects with their attributes, their operations and the relationships between them. Research work in progress aims to use state-transition diagrams to describe the dynamic behaviour of operations. Moreover UML standard specifies the Object Constraint Language (OCL), an expression language that enables one to describe constraints on object-oriented models. OCL is a formal constraint language based on first order predicate logic. It formalises constraints, which are restrictions on a model or a system. Thus, a constraint states, «this should be so». Constraints are attached on every modelled item. This is called the *context* of the constraint.

There are three types of constraints:

- An *invariant* formalises a condition that must always be met by all instances of the class
- A *precondition* to an operation is a restriction that must be true at the moment that the operation is going to be executed.

- A *postcondition* to an operation is a restriction that must be true at the moment that the operation has just ended its execution.

As a modelling language, UML can be used to meta-model enterprise modelling standards to ensure their integration through unique and coherent definitions.

In order to extend its meta-model, UML provides an expendability mechanism through the definition of so called "Profiles". A profile contains one or more related extensions of standard UML semantics. These are normally intended to customise UML for a particular domain or purpose. They can also contain data types that are used by tag definitions for informally declaring the types of the values that can be associated with tag definitions. In effect, these extension mechanisms are a means for *refining* the standard semantics of UML and do not support arbitrary semantic extension. They allow the modeller to add new modelling elements to UML for use in creating UML models for process-specific domains such as enterprise models. *Constraints* can also be attached to any model element to refine its semantics.

4 CONSTRUCTS SEMANTICS

The construct semantics representation deals with defining an UML Profile using the extensibility mechanisms of UML, which allow modellers to customise UML for specific domains. Profiles are used for:

- Defining new meta-classes (stereotypes),
- Defining new meta-attributes (tagged values),
- Defining new meta-associations (tagged values, referencing to other model elements),
- Defining new constraints.

The UML standard already defines 8 profiles: Scheduling, performance and time, Enterprise Distributed Object Computing, CORBA, EJB, Software Process Engineering Management, EAI and QoS and fault tolerance. A profile defines a projection of a reference meta model and provides a mechanism to define facets that can be applied to model elements and combined.

Moreover, as the UML specification relies on the use of well-formedness rules to express constraints on model elements, this profile uses the same approach. The constraints applicable to the profile are added to the ones of the stereotyped base model elements, which cannot be changed. Constraints attached to a stereotype must be observed by all model elements branded by that stereotype. If the rules are specified formally in a profile (for example, by using OCL for the expression of constraints), then a modelling tool may be able to interpret the rules and aids the modeller in enforcing them when applying the profile.

As an example, the “Enterprise Object” construct is defined as an “Enterprise Object” stereotype, based on the UML “Class” meta class (Fig. 5).

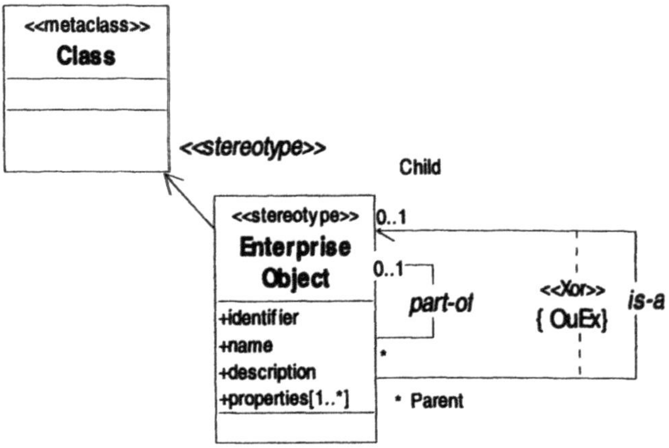

Figure 5 : Stereotype definition

That stereotype defines that an “Enterprise Object” could be “part-of” another “Enterprise Object” and that an “Enterprise Object” could be a subclass (“is-a” relationship) of another “Enterprise Object”. That stereotype defined, also, tagged values such as identifier, name, description and a set of properties.

An invariant constraint represented by well-formedness rules ensures the consistency in the relationships between modelled elements. Such a formal rule could be:

```
context EnterpriseObject
   inv : self.partOf->forall(p | p <> self)   (1)
   inv : self.properties->forall(p | p.stereotype.name = "Enterprise
Object"
 implies not self.partOf->exists(p | p = self) (2)
```

That invariants state that (1) a particular “Enterprise Object” could not be part of itself, and (2) a particular “Enterprise Object” could not be itself included in the set if its own properties.

Instantiation of that stereotype in a particular model aims at defining a stereotyped class that should meet the previous invariant formalisation. For example, Fig. 6 shows the “Client

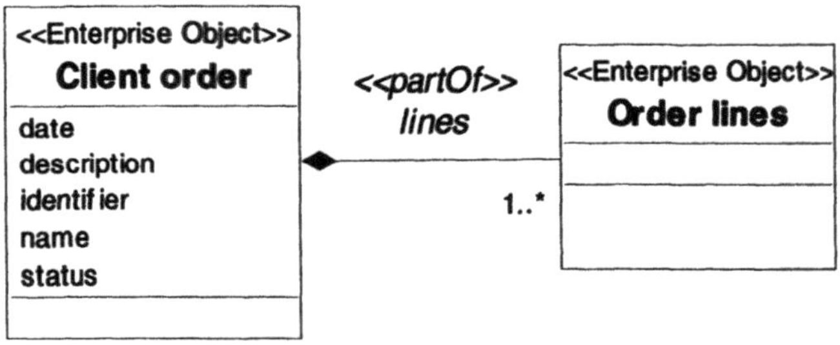

Figure 6: A Stereotype instantiation

order" object and the "Order lines" object as instances of the "Enterprise Object" stereotype. The "partOf" relationship between these two "Enterprise Objects" comes from the "part of" relationship defined in Fig. 5.

The same invariant as defined previously is applied to that model ensuring its consistency. In particular, that rules avoid the definition of a relationship between two "Client orders". The only authorised relationship is the "partOf" composition between a "Client order" and one or more "Order lines".

5 CONCLUSION

There is a need to provide a semantic foundation for formally verifying its use in the scope a particular modelling process. UML provides extensibility mechanisms able to formalise enterprise modelling constructs. Constraints are also expressed and could be used, by engineering tools, to aid the modeller in ensuring the global consistency of its model. These rules are expressed in the generic view of the model. There are tools that can interpret these rules using class instances values for particular models. In order to be able to verify them in partial model (for domain-based models), work is in progress to translate them into the B language (Abrial, 1996), which allows properties proofs, based on non refutable mathematical theories.

6 REFERENCES

Abrial J.R. (1996), *The B Book: Assigning Programs to Meanings*. Cambridge Univ. Press.

AMICE, (1993), *CIMOSA: Open System Architecture for CIM*, Springer-Verlag.

CEN, (1995), European Pre-Standard ENV 12204, *Advanced Manufacturing Technology, Systems Architecture, Constructs for Enterprise Modelling*, TC 310/WG1 (currently under revision).

IFAC/IFIP Task Force (2001), "*Architectures for Enterprise Integration*", UEML Interest Group.

ISO 10303, (1994), *STEP, Standard for the Exchange of Product Model data*, TC 184 SC4.

Kosanke, K. Vernadat F.B. Zelm, M. (1999), *CIMOSA enterprise engineering and integration*, Computers in Industry, Volume 40, Issues 2-3, Pages 83-97.

MOF Specifications, (1997), Joint Revised Submission, OMG Document ad

Panetto H. Mayer F. Lhoste P. (2000), *Unified Modeling Language for meta-modelling: towards constructs definitions*, Proceedings of ASI'2000, ISBN 960-530-050-8.

sNets Formalism, (1998), technical report, LRGS, Université de Nantes.

UML 1.4, (2001), *Unified Modeling Language*, Object Management Group standard.

Vernadat, F.B. (1998), *The CIMOSA languages*, Handbook of Information Systems. P. Bernus, K Mertins and G. Schmidt (Eds.), Springer-Verlag

Language Semantics

Towards a Common Underlying Domain Theory

Ioannis L. Kotsiopoulos
Zenon S.A., Athens, Greece, ikotsiopoulos@zenon.gr

Abstract: The paper is a response to the call for a common underlying domain theory to address the mismatch between syntax and semantics of enterprise modelling languages. To this end, we propose categorical morphisms of object interactions as a strong candidate theory on which all modelling constructs can potentially be mapped. Implications of such a framework are fundamental properties of models such as genericity of modelling, modelling environments and model correctness. Semantics of particular modelling languages and architectures can be obtained as specialisations of the general theory. Basic features of CIMOSA are derived as an example.

1 INTRODUCTION

The mismatch between syntax and semantics of languages for enterprise modelling has been identified by ICEIMT'97 as a serious shortcoming in the course of enterprise integration. This is the fundamental reason why models, even syntactically compatible ones, can neither be exchanged between tools, nor can they be federated leading to "agility" nor "level 4 and 5 integration in the enterprise", (Hollocks, et al, 1997). To this end, Working Group 2 at Workshop 4, (Petit, et al, 1997), identified the need for both a Unified Enterprise Modelling Language (UEML) and precise semantics for its constructs. To achieve the latter, the working group called for a single definition of the constructs in terms of a *common underlying domain theory*. Although theories such as situation calculus, state transition diagrams, temporal logic, process algebras etc (Petit, et al, 1997) have already been employed, fragmentation and lack of co-ordination of research effort has so far prevented an overall evaluation of their adequacy and suitability. This renders the domain open to competition from candidate theories, one of which is the subject of this paper.

We argue the case that category theory morphisms of object interactions is a strong candidate for an underlying domain theory within which:

- All generic enterprise modelling concepts and constructs can potentially be mapped so as to provide definition, unification and resolution mechanisms for semantics of modelling languages,
- Fundamental properties such as genericity of modelling, architecture completeness and model correctness can be addressed using the objects of the theory.

The reason for such confidence stems from the nature of category theory itself. It is essentially a meta-mathematical theory, designed to deal with mathematical objects at the most general and abstract level allowable by the axiom-driven deductive process of mathematics and logic. If we accept that this is the highest form of human reasoning known to us, models themselves are nothing but maps of selected aspects of the real world to mechanisms of this reasoning. Loosely speaking, the "modelling power" of a semantically rigorous modelling language appears analogous to the generality allowed by its underlying domain theory.

2 OBJECTS

We introduce objects themselves as categorical morphisms called "Observed Processes" and we extend this characterisation to interactions among them, (Ehrich, et al, 1990). The generality of category theory can ensure that neither "useful real objects" nor "useful interactions" are excluded from such a formalism.

Following (Ehrich, et al, 1990), an object is seen as a mapping device between events and values of certain types called *attribute-value pairs* or *observations*. Both attribute-values and events share a general property: they are *atoms of behaviour*, that is, they occur at some single point in time and they "express" themselves through a designated *alphabet*. A collection of such atoms with the same single time occurrence is appropriately called a *behaviour snapshot*.

Dynamic behaviour of an object in this setting results from the attachment of behaviour snapshots to points in time. Formally this is defined as a map $\lambda : t \to S, t \in TIME$, with *TIME* some time domain be it discrete or continuous. λ is called a *trajectory* over *S*, the set of behaviour snapshots, while a set of trajectories is called a *behaviour*.

An *object* is a mapping $ob : \Lambda \to \Omega$ between different behaviours, which, formally, can be given the structure of a morphism in the (complete) category of behaviours, (Ehrich, et al, 1990). Both Λ and Ω are snapshots on the time axis, where Λ is termed *process* part and Ω *observation* part.

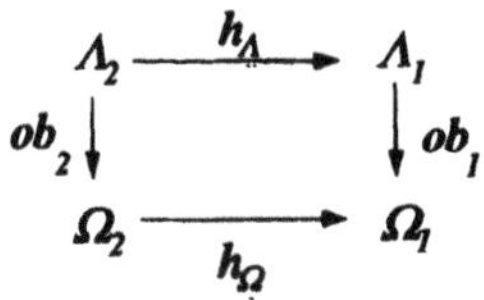

Remark. This notion of an object as a behaviour morphism allows complete freedom of choice for the process and observation parts. Moreover, an object triggers its own observations according to the process behaviour (the principle of encapsulation).

Interaction between objects can be expressed as operations (morphisms) on objects. An *object morphism*, symbolised by $ob_2 \rightarrow ob_1$ is a pair of behaviour morphisms $h_\Lambda : \Lambda_2 \rightarrow \Lambda_1$ and $h_\Omega : \Omega_2 \rightarrow \Omega_1$ on objects $ob_1 : \Lambda_1 \rightarrow \Omega_1$ and $ob_2 : \Lambda_2 \rightarrow \Omega_2$ so that the left diagram commutes.

New objects within this category can also be constructed via morphisms. Of particular importance to us are the abstractions of objects produced by restricting either the observation part of an object (object view) or the process part (object trigger). Combining both operations and under some technical conditions (Kotsiopoulos, 1993) a *partial observation morphism* on ob_2 can be defined and used as the main abstraction mechanism. The outcome of this morphism is an object ob_1, which selectively restrict both the process and the observation part of the original object ob_2.

We are in a position now to give substance to the relationship of the "real world", i.e. physical objects, such as an industry processed product under intermediate or final treatment (e.g. a semi-completed form, a large paper roll etc.), to models. We consider a (finite) set of primitive objects, which correspond to all the functions of the physical objects of the enterprise. To those we apply partial observation morphisms, in order to construct a new set of processable objects, which we term the *enterprise objects*. Interaction between them is represented as another set of object morphisms. In this way, what is commonly called "the enterprise and its objects" is nothing but a set of objects and object morphisms partially observed from a physical space and time ensemble and performing some target function.

3 ENVIRONMENTS

Objects are usually seen as operating within a larger set or a larger object, frequently called an "environment". Here we also define a similar concept, adapted to the requirements of modelling architectures, that is amenable to hierarchical control structures similar to the organisation structures of an enterprise (business-process models).

The general idea of an environment is shown in the left diagram. The grey areas are linked with time scale preserving behaviour morphisms to an object (left) and to an object morphism (right). The morphisms e_Λ and e_Ω representing the links to the environment are called *encoding* and *decoding* morphisms accordingly. The encoding morphism "translates" the commands of the environment into commands, which the process part of the object "understands", while the decoding one does the reverse. Should there be other behaviour morphisms with corresponding domains and encoding and decoding morphisms, a common environment for all of them can be set by suitably enlarging the original.

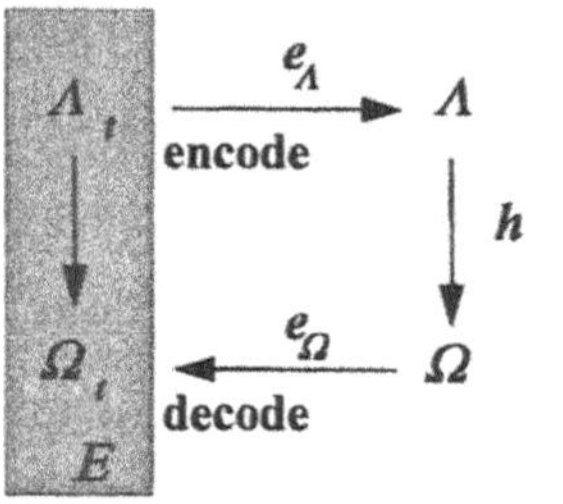

The idea of an environment here depicts it both as a supplier and a recipient of selected subsets of behaviour, without reference to the internal workings of the embedded objects or morphisms (events and/or observations), a notion close to the common concept of a physical environment. In (Ehrich, et al, 1990) this is closer to the definition of an implementation of an object over another. In (Kotsiopoulos, 1993) this corresponds to the Petri Net identification of the CIMOSA constructions.

4 IMPLICATIONS OF THE CATEGORICAL FOUNDATION

As mentioned in (Kotsiopoulos, et al, 2002) all enterprise modelling languages have intuitive semantics; they represent real world "objects", "processes", "actions". Although what we examine here is formal (mathematical) semantics (taking advantage of the reasoning power of an underlying domain theory), intuitive or "physical layer" semantics cannot be ignored. It is actually through them that such a language is judged by the user.

This puts additional requirements on our formal semantics framework: it must not only be embedded within a theory, but also be broad enough to allow mapping to real physical entities and their actions. The categorical framework we introduced views virtually all quantifiable or machine processable activity within the enterprise as an object and/or object interaction, that is morphisms of behaviour (objects) and morphisms of such morphisms (object morphisms). It is this ability provided by the categorical framework, which allows us to formalise the relationship between an enterprise model and reality.

Suppose we have a very competent observer, who can record all the states of all entities in an enterprise at any single moment in time. Provided physical objects of the enterprise have been identified, his observation will be just a set of cause and effect behaviour snapshots, representing the causality of enterprise processes. Using the framework of section 2, these can be readily seen as enterprise objects and object morphisms. Actually, the parallel composition of those into a single object morphism, representing the entire enterprise is also possible.

Let us think that a certain enterprise has been selected and that enterprise objects and morphisms between them have correctly been identified and abstracted from real objects and their interactions. To examine issues of integration and monitor overall performance, a more structured expression of the enterprise machinery is desired and therefore a "model" is employed. But what is a model and is there a way of separating good models from bad ones? The representation of objects as observed processes and of interactions as morphisms can be employed to characterise a "correct" model of an enterprise, with respect to the required abstraction.

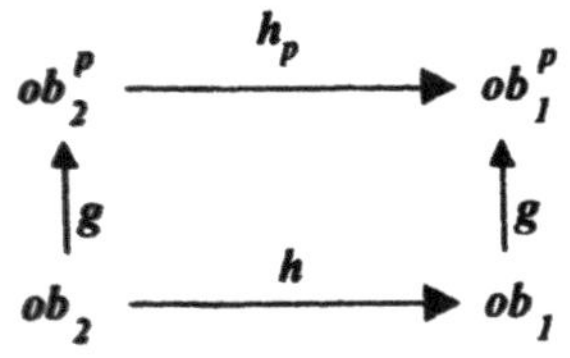

Definition. Let $h: ob_2 \rightarrow ob_1$ be a morphism between enterprise objects and $g: ob_2 \rightarrow ob_2^p$ be a partial observation morphism on ob_2. Then h_p belongs to a (correct) *model* of the enterprise if for every morphism *h* between enterprise objects the attached diagram exists and is commutative.

The images of the morphism g are the *objects of the model*, while a composition of those (and their morphisms) is the enterprise model itself.

Indeed, following (Ehrich, et al, 1990), the category *OB* of objects and their morphisms is cocomplete. Parallel composition of objects can therefore be semantically mapped as a colimit in *OB*. It follows that the entire enterprise, as well as its model, can be thought of as an interacting set of two composite objects (representing an input and an output object accordingly) connected via an object morphism.

An aggregation based on more permanent characteristics i.e. a type of those objects together with a corresponding type of morphisms is what we commonly call a *reference architecture* (for enterprise modelling).

Note that the constructs of a single reference architecture cannot always ensure model correctness; indeed, this is the very reason for which different architectures exist: to ensure model correctness, i.e. that each object morphism of the model, such as h_p, satisfies the previous definition. This is particularly important for model-enacted or regulated operation (Kotsiopoulos, 1999), where a correct model implies that:

- Not every action which takes place in reality is modelled
- No action inconsistent with the model does take place.

A related property of an architecture, *completeness*, concerns the suitability of a certain architecture for a purpose or a particular class of enterprises.

This has interesting implications on practical issues such as model re-use and software tool requirements for true support of the modeller. A mathematical logic representation of the criteria involved is possible through model theory and situation theory of meaning (Bernus, et al, 1996). Associating completeness criteria with our correctness principle in a uniform semantic and representational context (objects, morphisms, model theory and situation semantics) remains an interesting open question.

5 EXAMPLE. MAPPING OF ENV 12204 AND CIMOSA SEMANTICS

Both ENV 12204, (CEN, 1995), as well as CIMOSA, (1994), models are based on objects and their properties, themselves being abstractions of enterprise objects. Whether the abstraction is "good" or not is for the model correctness test to say. The formalism is built around object processing blocks, called "(Business/Domain) Processes" or "(Enterprise) Activities" accepting objects as "Function Inputs" and producing objects as "Function Outputs".

We shall realise this by posing additional structures to enterprise objects, thus abstracting them to ENV12204 (CIMOSA) objects. The same will be done with their corresponding morphisms and environments. In our view, *an ENV12204 (CIMOSA) compliant model is a set of category morphisms on ENV12204 (CIMOSA) objects, embedded in an ENV12204 (CIMOSA) environment.*

The method is of importance in its own right as it may be used for the construction or the extension of modelling architectures. We apply a bottom-up approach by starting from the smallest functional unit in ENV12204 the Enterprise Activity.

Consider an object morphism called an *Enterprise Activity (EA)* embedded in an environment called the *Business Environment.* Enterprise Activities act on objects in an input/output fashion and are deployed by the Business Environment, realised by the ENV and CIMOSA as a hierarchical structure performing sequencing operations on them. Although not currently present in the ENV the embedding contains:

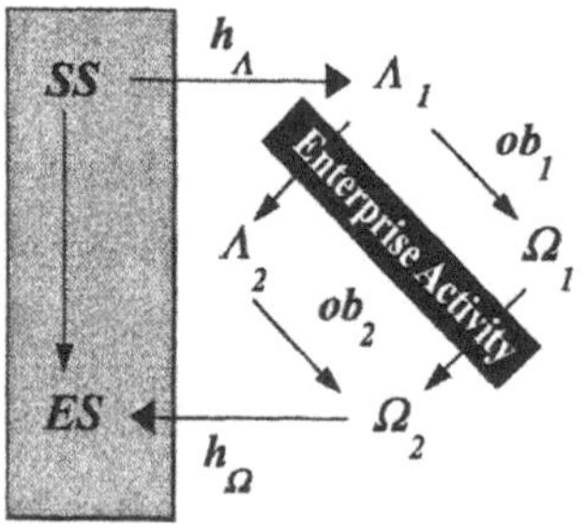

- Encoding morphisms "triggered" by start events in *SS*, the set of all *starting events* for the launch of all EA morphisms
- Decoding morphisms valued in *ES*, the set of all attribute-value pairs for the *ending statuses* of the EAs launched by a model

In this way, the Business Environment has a finite alphabet equal to $SS \cup ES$. ENV and CIMOSA map this to a simple discrete event calculus called Procedural Rules, (CIMOSA, 1994), which, in turn, can be mapped to Hoare's Communicating Sequential Processes, (Hoare, 1985) and is termed *Business Process*. Hierarchies can also be built in similar ways. As only events are present at this level, each Business Process is indeed a composite object morphism and its input and output objects are ENV12204 (CIMOSA) compliant objects (schematic diagram on the left).

A further level of decomposition is done by CIMOSA by decomposing Enterprise Activities into *Functional Operations*. Our framework allows this mapping too. We define an object ob_e and call it an *elementary object* such that: $ob_e : e \to \Omega$, where e is a singleton in the event space and Ω is a singleton in the space of observations consisting of n-tuple vectors with typed components. ob_e acts as an object analysis mechanism mapping objects into named entities with attributes. Indeed, an elementary object is uniquely identified by the name of a single event and a set of attribute-value pairs. Moreover, the attachment of the single event on the time axis is independent of the particular point of attachment.

By the same token, a morphism between elementary objects, called an *elementary morphism*, can be defined and mapped to a computable (in the sense of ENV 12204) structure similar to a callable function of a programming language, or, in CIMOSA terms, a *Functional Operation*. We refer to (Kotsiopoulos, 1995) for full technical details. An (CIMOSA) *Activity Environment* can also be constructed, such that the encoding morphism is a partial observation with identity in its process part and the decoding morphism is an inclusion. We endow this environment with all the control structure of the CIMOSA Activity Behaviour pseudo-code as described in the Technical Baseline (CIMOSA, 1994).

Remark 1. All elementary objects release their events and attribute values to the Activity environment. This is in accordance to the CIMOSA definition of a Functional Operation as a grain of the model which is either fully executed (i.e. its elementary objects released to the environment) or not at all.

Remark 2. As far as the (CIMOSA) Business Environment is concerned an Enterprise Activity is seen only as a starting event and an ending status, occurring at single moments in time. It follows that for the purposes of this environment, an Enterprise Activity is semantically mapped to an elementary morphism.

Remark 3. CIMOSA considers *"object views"* instead of *"objects"*. Since either of them is constructed through a partial observation morphism on some enterprise object, we chose not to rename here.

Remark 4. Both notions of an environment and an elementary morphism can accommodate in a natural way the call of ICEIMT'97 for level 4 and 5 integration, (Hollocks, et al, 1997). New concepts such as that of an agent, identified by the working group of the 1st workshop of ICEIMT'02 as a potential constituent of the next generation of integrated systems, (Goranson, et al, 2002), could be incorporated as a special property of an environment. The same can be said about the ability of Enterprise Activities in CIMOSA for direct communication between them by exchanging events or messages. In this case, either the Business Environment is considered able to "mediate" for this communication, meaning that the Enterprise Activity is not seen as an elementary morphism any more, or it totally ignores this exchange. It is all a matter of modelling architecture but not of language semantics: the fundamental semantics framework is already there! Similar "environmental" constructions may give models the (now missing, (Goranson, et al, 1997) ability to represent their own state and causality (introspection). Finally, as also observed in (Goranson, et al, 2002), the insight into an object's internal mechanism demanded by level 4/5 integration can be served by suitable object decompositions (e.g. elementary objects and morphisms in the case of CIMOSA Functional Operations).

6 MODEL CORRECTNESS AS AN INVARIANT PROPERTY

Model correctness as introduced here is made possible by the generality of the object morphism characterisation. It also has an interesting invariance property: correctness should be maintained at two axes of the CIMOSA cube, derivation (from requirements to design-implementation) and genericity (from generic to partial-particular). For derivation: all models at different levels should be correct.

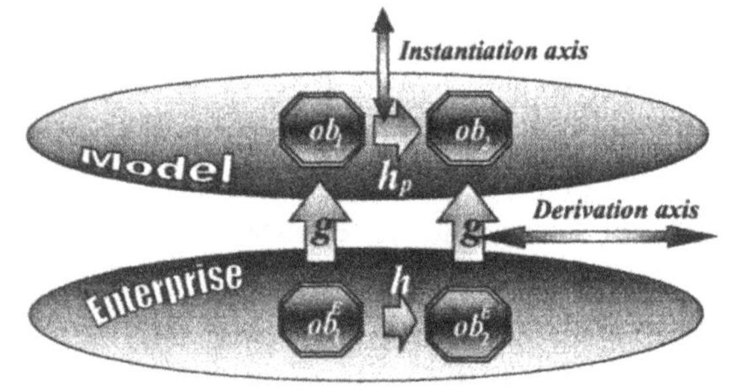

For genericity: the partial and generic levels should be able to provide classes of morphisms between objects which, if instantiated (particular level), should maintain correctness (attached diagram).

7 EPILOGUE

We argued the case that the cocomplete category of objects and their morphisms can be used as a semantic framework for enterprise modelling languages. The very general setting obtained in this way can be mapped to physical (real world) objects to allow for the physical layer semantics of those languages. At the same time, it can accommodate the formal features of architectures and associated languages such as CIMOSA or their suggested extensions (to accommodate agents for example). New properties of models, such as model correctness, can also be defined.

The author believes that far more can be achieved by the application of the full power of category theory to modelling. To this end, the present paper serves only as an introduction and a call for further work in the future.

8 REFERENCES

Bernus, P. Nemes, L. Morris, R. (1996), *"The meaning of an Enterprise Model"*, in Bernus, Nemes, (Eds.), Modelling and Methodologies for Enterprise Integration, Chapman & Hall.

CEN (1995), ENV 12204, *"Constructs for Enterprise Modelling"* TC 310/WG1.

CIMOSA Association, (1996), *"CIMOSA, Technical Baseline"*, private publication.

Ehrich, H.D. Goguen, J.A. Sernadas, A. (1990), *"A Categorial Theory of Objects as Observed Processes"*, in Foundations of Object-Oriented Languages, REX School/Workshop Proceedings.

Goranson, H.T. (1997), *"ICEIMT in Perspective – 92 to 97"*, in Enterprise Engineering and Integration, Proceedings of ICEIMT'97, Springer-Verlag.

Goranson, H.T. (Ed.) Huhns. M, Nell, J.G. Panetto, H. Tormo Carbó, G. Wunram, M. (2002), *"A Merged Future for Knowledge Management and Enterprise Modeling"*, this conf.

Hoare, C.A.R., (1985), *"Communicating Sequential Processes"*, Prentice-Hall.

Hollocks, B.W. (Ed), Goranson H.T., Shorter D.N., Vernadat F.B. (1997), *"Assessing Enterprise Integration for Competitive Advantage – Workshop 2, Working Group 1"*, in Enterprise Engineering and Integration, Proceedings of ICEIMT'97, Springer-Verlag.

Kotsiopoulos, I.L. (1993), *"Theoretical aspects of CIMOSA modelling"*, CIM Europe Conference, Amsterdam, in Kooij C., MacConaill, P.A., Bastos J. (Eds.),"Realising CIM's industrial potential", IOS PRESS.

Kotsiopoulos, I.L. (1996), *"Objects and Environments in Dynamic CIMOSA Models"*, in Bernus, P. Nemes, L. (Eds.), *"Modelling and Methodologies for Enterprise Integration"*, IFIP TC5 Working Conference on Modelling and Methodologies for Enterprise Integration, IFIP/IFAC Task Force, Heron Island, Australia, Chapman & Hall.

Kotsiopoulos, I.L. (1999), *"Railway Operating Procedures: Regulating a Safety-Critical Enterprise"*, Computers in Industry, 40.

Kotsiopoulos, I.L. (Ed.), Engel, T. Jaekel, F-W. Kosanke, K. Mendez, J-C. Ortiz Bas, A. Petit, Raynaud, P. (2002), *"Steps in Enterprise Modelling – A Roadmap"*, this conference.

Petit M (Ed.), Goossenaerts, J. Gruninger, M. Nell, J.G. Vernadat, F.B. (1997), *"Formal Semantics of Enterprise Models – Workshop 4, Working Group 2"*, in Enterprise Engineering and Integration, Proceedings of ICEIMT'97, Springer-Verlag.

Modelling of Distributed Business Processes

H. Grabowski, and Torsten Engel
Research Center for information technologies at the University of Karlsruhe, pde@fzi.de

Abstract: Today enterprises face the challenge to participate in enterprise networks. Business processes in these networks extend well beyond enterprise boundaries. Definition and optimisation of business processes requires adequate modelling languages, which support the modelling of cross-organizational information and material flows. This paper discusses innovative concepts, which extend existing modelling language concepts for these new requirements.

1 INTRODUCTION: NEW ORGANIZATIONAL FORMS

Today enterprises face dramatic changes: Internationalisation of market and competition relations, increasing complexity of products and services, shorter life cycles and individualisation of market and client requirements, dynamic changes and innovations of processes and organizational structures etc. These changes require that in the future more business partners must contribute to the value chain. New forms of collaboration over enterprise boundaries have to be developed and set up (Gora, Scheid, 2001). By leaving enterprise boundaries risk sharing, reduction of complexity and bundling of task-specific competencies can be used in a wider range (Specht, Kahmann, 2000). Two general organizational forms have been developed in the last years: the virtual corporation and the extended enterprise.

These modern organizational forms require a more detailed and usually short-term inter-organizational information exchange. In this context business-process models can decisively support the decision making in two ways: First, to bring potential partners together and second to form the business processes of the common enterprise in an optimal way (Kosanke, Zelm,

2002). But cross-organizational business processes, the *distributed business processes*, can nowadays only be modelled in an insufficient way.

2 STATUS QUO

According to DAVIDOW and MALONE (quoted in Brütsch, 1999), "the *virtual corporation* is a temporary network of independent companies, suppliers, customers even erstwhile rivals – linked by information technology to share skills, costs and access to one another's markets". A virtual corporation is usually created from corporate networks for a specific business task. An inter-enterprise networking across the value chain is usually termed an *extended enterprise*. The business in an extended enterprise is linked back through the supplier chain and forward into the distribution and customer chain (Brown, Zhang, 1999).

Both have in common that they imply crossing-boundary business activities. These organizational forms are the basis for an optimal configuration of business processes and they present the chance to optimise the value-chain across company boundaries (Brütsch, 1999). The efficient implementation of distributed business processes across several companies requires a best possible support by information systems. Various innovative concepts are developed from different vendors and research organisations to support the collaboration of companies and their different information systems at different locations. But they often focus on technical solutions like data transfer and they neglect the optimisation of the distributed business processes, which are integrated within the information systems.

Distributed business processes have to be continuously analysed and optimised during the evolution of a virtual organisation in order to maximally benefit from the collaboration. A prerequisite for an optimisation and realisation of distributed business processes is a common understanding and clear definition of the processes within the virtual organisation. Tools and methods of enterprise modelling have proved success for the analysis and optimisation of business processes in an individual company (Grabowski, Adamietz, 1998). But traditional enterprise modelling focused on business processes and organizational structures within an isolated company. Collaboration with other companies was merely described as information and material flow from and to other, usually unspecified companies. Optimisation across company boundaries requires an extended view on business processes: Business processes in virtual companies or extended enterprises can influence the work of several companies. Several aspects of business process analysis and modelling must be extended to support the concepts of virtual corporations and extended enterprises:

- *Analysis methodology*: Methodologies must be extended to integrate all companies participating in a virtual corporation or an extended enterprise in a consistent way.
- *Modelling language*: Modelling languages must semantically and syntactically support the modelling of distributed business processes.
- *Modelling tools*: Tools should support concurrent modelling.

In the EC-funded project BURMA-X (Business Relationship Management for the extended enterprise) (BURMA_X, http://) a new approach for analysing extended enterprises was developed. This approach integrates the analysis of cross-organizational aspects of an extended enterprise and also the consequences for the individual companies.

Many existing modelling tools often already support concurrent modelling. Often the tools were extended to client-server-applications with a central database, but the used modelling language usually was not adapted to the requirements of modelling distributed business processes. Following the requirements for a modelling language to make documentation and improvement of distributed business processes possible for modellers and users are discussed.

3 REQUIREMENTS FOR MODELING DISTRIBUTED BUSINESS PROCESSES

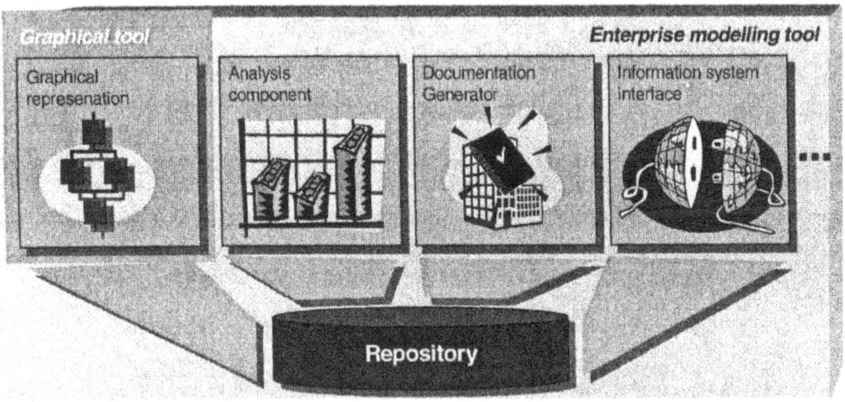

Figure 1: Advantages of an enterprise modeling tool

In contrast to graphical tools modelling tools have the decisive advantage of having the enterprise model in a repository. This allows to graphically displaying the model for further usage of the model for analysis, automated documentation, model-based adaptation of information systems etc. (Fig. 1). But current modelling tools very seldom offer the possibility to model cross-organizational aspects in a graphical way. If, they usually are not representing the model semantic in the repository.

An example for a typical, important aspect in modelling cross-organizational business processes is to identify the change of responsibility in a process. A widespread way to show this graphically is to place activity-constructs in columns ("swim lanes") representing a company (Fig. 2). An analysis or evaluation of the change of responsibility further requires, that the company-specific responsibility be also represented in the enterprise model repository.

Fig. 2 shows an extract from a distributed business process. An analysis of the number of changes of responsibility in this example can only be done if the companies "Deliverer", "Company X" and "Client" are represented in the modelling language as actual enterprise objects assigned to the relevant activities and not as simple graphical elements of a view.

The decisive difference between enterprise models for individual companies and enterprise models for virtual or extended enterprises is the coexistence of company-neutral and company-specific modelling constructs. An enterprise modelling language for virtual organisations therefore has to offer the additional possibility to assign relevant enterprise objects to a certain company represented in the model.

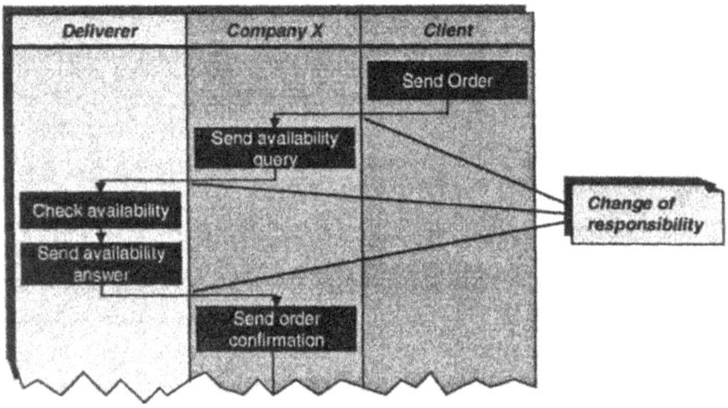

Figure 2: Example for graphical representation of change of responsibility within a distributed business process

4 BASIC CONCEPTS

In the course of the BURMA-X project mentioned above an Internet portal for the operation of extended enterprises is developed in order to facilitate development and production across company boundaries, provision of high-level after sales services, even if components have been produced by partners and by definition and easy execution of cross-organizational business processes. The communication platform integrates various services like cross-organizational product catalogues, stock control, knowledge management, contact search etc. that can be easily and quickly integrated (Fig. 3). In the same way new companies can easily be integrated into or replaced from the portal. A first prototype of the portal will be presented in summer 2002.

The portal development is based on a business process analysis of an existing extended enterprise within the BURMA-X consortium. In the course of the project the MERGE method and toolkit, which were developed at the research centre for information technologies (FZI, http://), were used. The MERGE method was adapted to the requirements of extended business processes analysis (Grabowski, Engel, 2002). Further, a first prototype of an extended version of the MERGE-Toolkit was used, which allows modelling of extended business processes. In this context the following basic concepts for an extended business processes modelling language were determined.

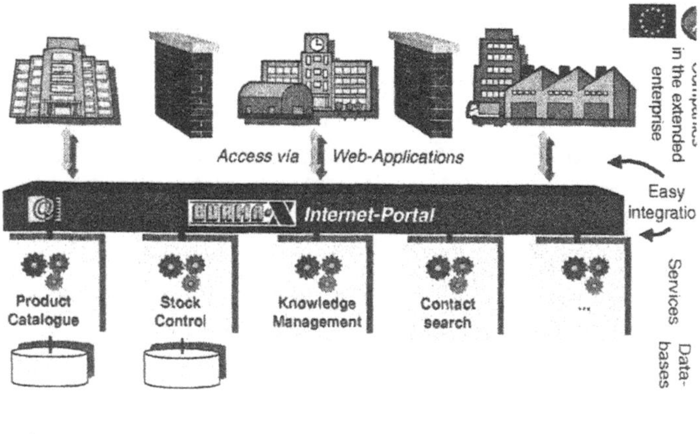

Figure 3: Architecture of the BURMA-X internet portal for extended enterprises

As mentioned above enterprise models for virtual companies or extended enterprises should offer the possibility to assign enterprise objects to certain companies. The basic prerequisite to model company-specific enterprise objects is the existence of an enterprise object class `company`. If an enterprise object in a model is connected to an enterprise object of the class `company`, it is identified as a specific object for this company. The responsibility of a company for a certain activity in a business process can now be modelled by assigning the activity to the company.

But an assignment of an activity to a company is often not detailed enough. The responsible organisation unit or situation in the company is also important both for internal co-ordination and to make the relevant contact clear to business partners (Kugeler, 2002). For that reason a modelling language should offer the possibility to model the organizational structure of a company and to assign an organizational unit of a certain company to an enterprise object in order to identify enterprise specific objects. Fig. 4 shows an extract of an organizational structure diagram created with the mentioned MERGE-Toolkit prototype.

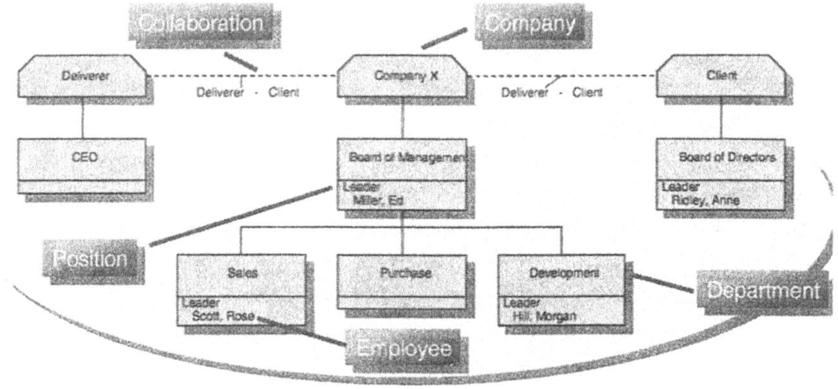

Figure 4: Example of an extended enterprise structure modeled with Merge

Typical examples for company-specific enterprise objects are:

- *Activities*: As mentioned above company-specific activities show the responsibility of a company to perform this action in a process. Further it is possible to identify at which points in a process companies have to exchange data and/or material respectively where they have to collaborate.
- *Resources*: As collaboration between companies is often involved with collaboration of information systems or resources in general, relevant resources should not only be identified but also assigned to the company, which is the owner of the resource.
- *Data objects*: The information exchange between companies requires the definition of data objects, which are to be exchanged. The data objects are the basis to model the information flow between the companies: Which data does a company send to a partner?

The last aspect, the cross-organizational data modelling, has a strong impact on connecting different information systems (Kugeler, 2002). In different companies data objects can have the same semantic meaning, but different attributes and data formats. Mapping of different data structures is an important aspect to enable collaborating companies to exchange data electronically. If two companies collaborate e.g. as client and supplier, the client will send orders to the supplier. The IT-systems of both companies will store data for orders. But they will use different formats or they will even store different data concerning an order. So the modelling of extended business processes requires transparency concerning

- The *differentiation* of equal data objects with different formats and
- The *transformation* between the formats within the processes.

5 SUMMARY AND OUTLOOK

Modern organizational forms like virtual or extended enterprises require an extended way of modelling business processes. Current modelling tools respectively languages do not fulfil these requirements. New modelling constructs are necessary to identify company-specific elements in a cross-company enterprise model.

The growing importance of concepts for cross-company information systems like EAI (Enterprise Application Integration) or SCM (Supply Chain Management) will increase the demand for modelling distributed business processes in the future. Virtual or extended enterprises will need an efficient way to co-ordinate the co-operation of the participating companies. Extended business-process models combine the experience of traditional business process modelling with special features for modelling the business processes of these new organizational forms.

6 REFERENCES

Browne, J. Zhang, J. (1999), *Extended and virtual enterprises – similarities and differences,* in: International Journal of Agile Management Systems 1/1, MCB University Press

Brütsch, D. (1999), *Virtuelle Unternehmen,* Zürich: vdf, Hochschulverlag AG an der ETH Zürich

BURMA-X, http://www.burma-x.de

FZI, http://www.fzi.de/pde

Gora, W.; Scheid, E.: (2001), *Organisation auf dem Weg zur Virtualität*, in: Gora, W.; Bauer, H.: „Virtuelle Organisationen im Zeitalter von E-Business und E-Government", Springer-Verlag

Grabowski, H. Adamietz, P. (1998), *Prozeßorientiertes Customizing von EDM/PDM-Systemen*, in: Informationsverarbeitung in der Konstruktion '98 – Prozeßketten für die virtuelle Produktentwicklung in verteilter Umgebung; Düsseldorf: VDI-Verlag

Grabowski, H., Engel T. (2002), *Business Process Analysis in Virtual Organisations*, in: Proceedings PDTEurope 2002; Sandhurst: Quality Marketing Services, 2002

Kosanke, K.; Zelm, M.: (2002), *Geschäftsprozeßmodelle für Wissensmanagement und Entscheidungsunterstützung*, in: Industrie Management 1/2002; Berlin: Gito-Verlag

Kugeler, M. (2002), *Supply Chain Management und Customer Relationship Management – Prozessmodellierung für Extended Enterprises*, in: Becker, J.; Kugeler, M.; Rosemann, M. (Hrsg.): Prozessmanagement: Ein Leitfaden zur prozessorientierten Organisationsgestaltung; Berlin: Springer-Verlag

Specht, D.; Kahmann, J. (2000), *Regelung kooperativer Tätigkeit im virtuellen Unternehmen,* in: Albach, H.; Specht, D.; Wildemann, H.: Virtuelle Unternehmen, Wiesbaden: Gabler

Needs and Characteristics of Methodologies for Enterprise Integration

Marc Hawa[1], Angel Ortiz Bas[2], and Francisco-Cruz Lario Esteban[2]
[1]*DMR, Spain,* [2]*Universidad Politecnica de Valencia, Spain,* aortiz@omp.upv.es

Abstract: Methodologies are one of the main elements in EI Projects. In this paper we present an aggregated and comparative analysis of the methodological aspects of several of the main EI proposals from the state-of-the-art of existing EI Methodologies (CIMOSA, PERA, IE-GIP among others). A definition of the set of characteristics that must be provided by an EI methodology is presented as well.

1 UNDERSTANDING ENTERPRISE INTEGRATION METHODOLOGIES

A generic methodology can be defined as "the system of methods and principles used in a particular discipline". This definition outlines two main issues. First of all, a methodology is necessarily linked to a particular discipline where it is useful; outside of this discipline, it may not be applicable. Within this work, we focus on IE as the particular discipline of interest. Secondly, the two basic elements of a methodology are methods and principles. Furthermore, Vernadat, (1996) defines a methodology as a "group of methods, models and tools". This definition is made from an EI perspective and identifies more specifically the three different elements that must be provided by an EI methodology: (1) methods, (2) models and (3) tools.

The authors define an EI methodology as a group of methods targeted to support a business architect in the development of EI projects, supported both by models (the so-called architectures) and tools. Methodologies, architectures and tools are indeed the three main engineering ingredients of EI projects, and are thus clearly separated in this definition. This has led the authors to define the MAT (Methodology – Architecture – Tool) concept (Fig. 1) as a general framework to guide EI business architects (Hawa, 2002). The MAT approach is still consistent with commonly accepted definitions in EI (e.g., "a set of instructions that offer the user a step by step guide to carry out all the necessary aspects for the execution of an IE project"(Williams, 1997), and enables business architects to select, and even combine, the specific methodology, architecture and tools that best suit their engineering endeavour.

The role of methodologies is further enhanced based on the fact that the real-world application of any EI methodology is indeed a complex, long and expensive process (Vernadat, 1996, Bernus, et al, 1996, Kosanke, 1997). It requires the co-ordination, understanding and mutual acceptance of multidisciplinary teams (Weston, 1997), and must deal with all the intrinsic complexity of the business entity it is applied to (human, knowledge, structural, operational, technological and financial capitals).

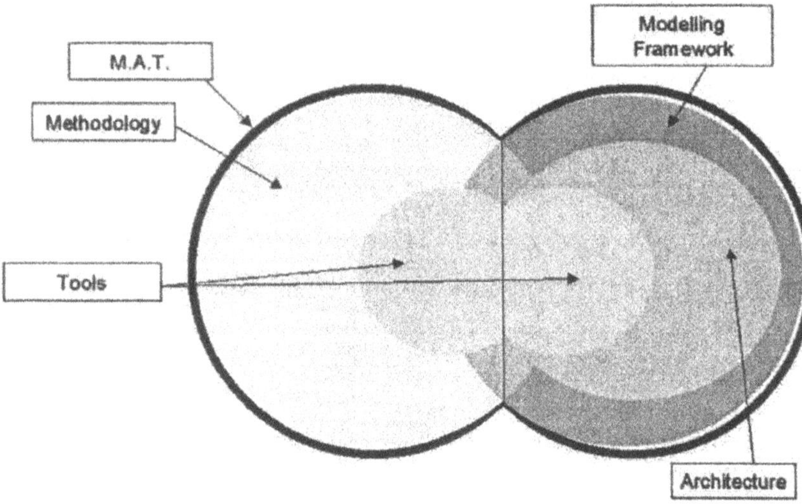

Figure 1: The MAT Concept

Within EI, the main advantage of methodologies lies in the assumption that they can be considered mostly generic, i.e. independent of particular applications and applicable in a large number of projects. Thereby, although the specific details will vary from one project to another, the general guidelines, methods and procedures can still be considered and defined as substantially generic (Williams, 1997, Williams, 1996). The results of the

IFAC/IFIP Task Force On Architectures for Enterprise Integration reinforce this understanding in IE, and stress its added value (Bernus, et al, 1996).

Although each EI methodology is intended to cover a wide range of projects, it must be clear that, on one hand, not all of them suit all real-world applications and, on the other hand, depending on the project considered, some of them stand as more appropriate than others. Based on this statement, there is a clear need to provide business architects with a list of criteria to help them in the assessment of the added value and adequacy of each potential methodology.

2 DEFINING THE SET OF CHARACTERISTICS TO ASSESS ENTERPRISE INTEGRATION METHODOLOGIES

We have developed an aggregated and comparative analysis of the methodological aspects of several of the main EI proposals from the state-of-the-art of existing IE Methodologies (see Annex 1):

- CIMOSA (Computer Integrated Manufacturing–Open System Architecture),
- GERAM (Generalised Enterprise Reference Architecture and Methodology),
- GRAI-GIM (Graphes à Réseaux et Activités Interreliés–GRAI Integrated Methodology), (Domeingts, 1984)
- ICEIMT'97 (International Conference on Enterprise Integration and Modelling Technology) Workshop 2, Working Group 1 Proposal (Kosanke, Nell, 1997).
- IE-GIP (Enterprise Integration– Business-Process Integrated Management - Spanish acronyms),
- PERA (Purdue Enterprise Reference Architecture),

This has led to the definition of the set of characteristics that must be provided by an EI methodology. We have classified those characteristics in Technical o Human oriented

2.1 Methodology Human Oriented Characteristics

1. To explain and justify its steps: Besides describing the steps to solve a problem, a methodology should explain and justify those steps.
2. To be comprehensive and easy to follow: It should be user-oriented and friendly for all project team members profiles (Bernus, et al, 1996)

3. To act as a guide: It should guide the work of the people involved in the IE project (Williams, 1996).
4. To provide different presentations and visions: Depending for example on the profile of the business architect or the problem to be addressed, the methodology needs to provide customised views of the project and/or enterprise (Williams, 1996).
5. To support change management: It must consider intrinsically the migration path concept (i.e. evolutionary path from AS-IS to TO-BE) and provide change management support.
6. To generate documentation: It should specify how to elaborate the project documentation (and even generate automatically this documentation), and among others a Master Plan, similar to the Master Plan of PERA or the deliverables of IE-GIP.
7. To provide the user with a manual: Similar to the Implementation Procedures Manual of PERA that helps the user to plan and defines the work to be done and how to do it.

2.2 Methodology Technical Oriented Characteristics

1. To be exact and precise: i.e. it must specify all the necessary steps of a project. Each phase should have clearly defined its objectives, inputs, delivery documents (deliverables) and roadmap of activities.
2. To cover the complete project lifecycle: The users won't adopt a methodology that doesn't provide support in all the different phases of the project. For some phases, the support may rely upon a specialised methodology (for example a methodology for software engineering). In such cases, the main methodology should provide clear interfaces with that specialised methodology.
3. To be flexible: Within the range of projects covered by the methodology, it must be flexible enough to allow users to adapt it to each particular project.
4. To be supported by a computer tool: The complexity of the application of a methodology requires the availability of a supporting software application. This is a key point and it represents a common drawback of many EI methodologies that lack of such computer-assistance.
5. To be open: It must be open to a rich variety of models and techniques, and not be bounded to a specific model or technique. This can be achieved for example through the evolutionary integration of new technologies, models and techniques (Bernus, et al, 1996).

6. To offer different alternative paths: It should provide different alternative strategies to solve a problem and to support in the selection of the most appropriate path.
7. To be located in a general dimension of continuous improvement: A project begins, is developed, and concludes. The day after the end of a project should be consider in the own methodology. Questions as continuous improvement, reuse of results, next steps, inheritance in next project, etc., should be dealt with by the methodology.
8. To cover project planning: It should give an appropriate methods and procedures to establish and assess the scope and necessary effort for the development of the project (Ortiz, 1999).
9. To support different project paradigms: Depending on the type of considered project, the selected architecture, the involved personnel, the results of previous projects, etc., it should be let the business architects choose the most suitable project paradigm (waterfall, concurrent engineering, etc.).
10. To negotiate in an efficient and effective way the project: For it, the project should be developed inside the budget, of the waited temporary window and of the existent restrictions.
11. To re-use previous efforts: It should allow the direct and easy reuse of the all of some of the results, deliverables and models developed in previous projects (Bernus, et al, 1996).

3 ASSESSING THE ADDED-VALUE OF METHODOLOGIES FOR AN ENTERPRISE INTEGRATION PROJECT

Based on the set of characteristics defined above, a template (table 1) has been developed to evaluate methodologies and assess their suitability to match the specific needs of a project. It is important to point out that this template is defined to assess a unique methodology. Normally, a comparative evaluation of several candidate methodologies is recommended, to select the one that best suits the reality of a real-world endeavour.

In this template *Project Weight* is a score for the project purposes (1, low importance to 5, high importance) and the *Project Threshold* is the minimum score to consider the Methodology suitable for the specific project's needs. *Methodology Support* means the score that we assign to the analysis capability of the Methodology. *Partial Evaluation* scores the specific characteristic for the specific methodology and *Final Evaluation* scores the Methodology (only if the Project threshold requirements have been overcame).

Table 1: Template for methodologies evaluation

Characteristic		Project weight	Project threshold	Methodology support	Partial evaluation
Human Oriented					
1	Explains and justifies its steps	w1	t1	s1	w1*s1
2	Is comprehensive and easy to follow	w2	t2	s2	w2*s2
3	Acts as a guide	w3	t3	s3	w3*s3
4	Provides different presentations and visions	w4	t4	s4	w4*s4
5	Supports change management	w5	t5	s5	w5*s5
6	Generates documentation	w6	t6	s6	w6*s6
7	Provides the user with a manual	w7	t7	s7	w7*s7
Technical Oriented					
8	Is exact and precise	w8	t8	s8	w8*s8
9	Covers the complete project lifecycle	w9	t9	s9	w9*s9
10	Is flexible	w10	t10	s10	w10*s10
11	Is supported by a computer tool	w11	t11	s11	w11*s11
12	Is open	w12	t12	s12	w12*s12
13	Offers different alternative paths	w13	t13	s13	w13*s13
14	Is located in a general dimension of continuous improvement	w14	t14	s14	w14*s14
15	Covers project planning	w15	t15	s15	w15*s15
16	Supports different project paradigms	w16	t16	s16	w16*s16
17	Negotiates in an efficient and effective way the project	w17	t17	s17	w17*s17
18	Re-uses previous efforts	w18	t18	s18	w18*s18
Final evaluation		Sum (wi*si)			

4 CONCLUSIONS

As one of the important part of the MAT Concept (Methodology – Architecture – Tool), we have focused on Methodologies to support EI Projects. We have analysed the main EI Proposal (CIMOSA, PERA, GRAI-GIM, GERAM, IE-GIP and ICEIMT'97 WS2WG1) and a set of characteristics to be provide for an IE Methodology has been identified. Additionally, we have developed a comparative analysis among the methodologies taking into account several key aspects. Finally, we have defined a method to evaluate IE Methodologies, this method consider that the real added value of a Methodology depends of the specific purpose and project. Therefore the proposed method allows business architect to identify the most appropriate IE Methodology considering the weight given to each characteristic, the support provide for each methodology to this characteristic and taking into account that is necessary to overcome the defined threshold.

5 REFERENCES

Bernus, P. Nemes, L. Williams, T.J. (1996), *Architectures for Enterprise Integration*, Chapman & Hall.

Doumeingts, G. (1984), *Methode GRAI: Methode de Conception des Systemes de Productique*, These d'Etat en Automatique, Universite de Bordeaux 1, Bordeaux, France.

Hawa, M. Ortiz A. Lario, F. Ros L. (2002), *Improving human capabilities in the development of enterprise engineering and integration projects through training base on multimedia technology*, Intern. Journal of Computer Integrated Manufacturing. To be published in June.

Kosanke, K. (1997), *Enterprise Integration – International Consensus: An Europe – USA Initiative*, in: K. Kosanke, J.G. Nell (Eds.), Enterprise Engineering and Integration: Building International Consensus, Springer-Verlag, pp. 64-74.

Kosanke K. Nell J.G. (Eds.), (1997), *Enterprise Engineering and Integration: Building International Consensus*, Springer-Verlag, pp. 64-74.

Ortiz, A. Lario, F. Ros, L. (1999), *IE-GIP: A Proposal for a Methodology to Develop Enterprise Integration Programs*, Computers in Industry. Vol. 40. pp. 155-171.

Vernadat, F.B. (1996), *Enterprise Modeling and Integration: Principles and Application*, Chapman & Hall.

Weston, R.H. (1997), *Enterprise Modelling and Integration – Towards Agile Manufacturing Systems*. in: K. Kosanke, J.G. Nell (Eds.), Enterprise Engineering and Integration: Building International Consensus, Springer-Verlag, pp. 348-358.Williams, T.J. (1996), *The needs of the Field of Integration*, Chapter 3. Architectures for Enterprise Integration, Chapman & Hall.

Williams, T.J. (1997), *PERA Methodology*, International Workshop in Business Integration, Valencia. Spain.

6 ANNEX I: COMPARATIVE ANALYSIS OF CONSIDERED METHODOLOGIES:

The following Methodologies have been analysed in terms of the characteristics identified in the template (table 1) with specific reference to the life cycle concept of pre-EN ISO 19349 and GERAM.

1. CIMOSA = CIM Open Systems Architecture
2. GERAM = Generalised Enterprise Reference Architecture and Methodologies
3. GRAI
4. IE-GIP
5. PERA = Purdue Enterprise Reference Architecture

References to the template characteristics are in brackets (Tx)

Aspects related to the company where the methodology is applied	
Supports the business entity concept? (T3/T9)	PERA, GERAM, IE-GIP
Supports the enterprise system concept? (T3/T9)	CIMOSA
Supports the domain concept? (T3/T9)	CIMOSA, IE-GIP
Provides identification of the company type (for example manufacturing, continuous, etc)? (T3/T9)	GERAM, CIMOSA, IE-GIP, PERA
Supports a company life cycle concept (T3/T9)	all.
Supports the life history concept? (T3/T9)	GERAM
Supports the relationship between company life cycle and project life cycle? (T5) (T3/T9)	GERAM, PERA, IE-GIP
Supports a product life cycle concept? (T3/T9)	CIMOSA, GERAM
Supports the relationship between company life cycle and product life cycle? (T3/T9)	CIMOSA, GERAM
Supports the business-process life cycle concept? (T3/T9)	CIMOSA
Supports an engineering environment concept? (T3/T9)	CIMOSA
Supports an operation environment concept? (T3/T9)	CIMOSA
Supports a continuous improvement concept? (T14)	CIMOSA, IE-GIP, GERAM
Aspects related to the development of the methodology	
Provides a methodology? (T1/T2/T8)	CIMOSA, PERA, IE-GIP, GRAI
Supports a structured Approach? (T1/T2/T8)	PERA, GRAI-GIM
Provides Methodology Functional abstraction (T4)	CIMOSA, IE-GIP
Supports a project concept (T9)	all
Supports a program concept (T3)	PERA, IE-GIP
Supports a project life cycle (T9/T15)	all
Supports the relationship between company life cycle and project life cycle? (T9/T15)	GERAM
Supports the life history concept (T3/T15)	GERAM
Supports a program life cycle concept (T9/T15)	PERA (with reservations)
Supports a project life cycle concept (T9/T15)	all

Supports a project management concept (T9/T15)	CIMOSA, GERAM
Supports a program management concept (T9/T15)	PERA, IE-GIP
Supports a simultaneous change processes concept (T5/T15)	CIMOSA, GERAM
Supports the macro and detail level (T3/T9)	IE-GIP
Supports the functional and temporal presentation (T4)	GERAM
Supports the engineering environment (T3/T9)	CIMOSA
Supports the operation environment (T3/T9)	CIMOSA
Supports the master plan (T3/T9)	PERA, IE-GIP, GERAM
Provides an implementation procedures manual (T1/T8/T19)	PERA
Supports continuous improvement (T14)	CIMOSA, IE-GIP, GERAM
Aspects related to human aspects of the methodology application	
Defines project teams (T3/T9/T15)	IE-GIP, GRAI-GIM, PERA
Provides user and technology orientation (T2)	GRAI-GIM
Provides implementation procedures manual (T7)	PERA
Supports meetings (T3/T9)	CIMOSA, GRAI-GIM, IE-GIP
Supports interviews (T3/T9)	CIMOSA, GRAI-GIM, IE-GIP
Supports knowledge management (T4)	all
Aspects related to the Change Management	
Supports change management (T5)	all
Describes current system (or as-is) (T4/T5)	all
Describes future system (or to-be state) (T4/T5)	all
Describes migration path (T4/T5)	all
Describes change process (T4/T5)	all
Describes simultaneous change processes (T4/T5)	GERAM, CIMOSA, PERA
Provides performance indicators (T4/T5)	GERAM, IE-GIP
Aspects related to the computer application of a methodology	
Provides a supporting tool? (T11)	IE-GIP, CIMOSA, GRAI-GIM
Supports the relationship between the supporting tool and the methodology? (T11)	IE-GIP,
Supports the relationship between language / modelling technique and methodology? (T11)	GERAM
Displays functional and temporal presentation? (T17)	GERAM
Supports in the master plan development? (T3/T9)	PERA, IE-GIP, GERAM
Provides an implementation procedures manual (T7)	PERA

Argumentation for Explicit Representation of Control within Enterprise Modelling and Integration

Bruno Vallespir, David Chen, and Guy Doumeingts
LAP/GRAI, University Bordeaux I - CNRS, France, vallespir@lap.u-bordeaux.fr

Abstract: The paper addresses the necessity and rationale to take enterprise control into account explicitly in order to ensure that the implemented architecture enables to get the level of performance required by the strategy of the company. To control a system that cannot be entirely formalised and modelled, it is necessary to provide a set of decision-makers with co-ordination links between them. Focusing on control means focusing on decision-making and on the role of man in modern industrial systems as well. Man-based decision-making oriented enterprise control is proposed as a complementary approach with respect to formalised views used in enterprise modelling and integration. The purpose of this paper is only to argue this proposition and not to give operational means to do so.

1 INTRODUCTION

Enterprise Engineering and Integration is a complex and interdisciplinary project in which not only technical aspects, but also human factors, control and decisional aspects must be considered (IFAC-IFIP, 1999). Within the current industrial challenge, control plays a major role for matching performances and then keeping a company competitive. However, the global control aspect is not well taken into account. From our point of view, most of existing approaches are information modelling oriented and the control is reduced to an information processing issue. This statement is not only ours, many others think that to match current industrial challenges, control needs to be in an equivalent position to information and communication (ICIMS, 1999).

Focusing on control means focusing on decision-making and on the role of man in industrial systems as well. It is necessary to avoid mechanistic approaches and to situate human decision into the system.

In this paper, firstly some theoretical arguments to justify the necessity of having a control view are formulated. Then basic concepts and principles to implement enterprise decision-making system are presented. The impacts of introducing human decision making to control industrial system are discussed.

2 THEORETICAL ARGUMENTS AND BASIC CONCEPTS

The traditional approach in system engineering coming from automation considers that a system has some significant states that can be formalised (i.e. explicitly identified and defined). Formalising these states enables to take them into account, i.e. to translate them into a model, which is used to control the system. To consider that the significant states can be formalised is worthy of remark. This assumption is the basis of automation. It leads the system to run "automatically" when it is in a formalised state or to be blocked or insensitive in other states. (In the largest meaning. An activity completely defined by a procedure, even if this activity is performed by a man, is considered here as to be "automatic".) However, this automatic control approach may be justified and used when (1) the product is cheap or the wasted raw material can be re-collected in case when the system is out of order; (2) the performances expected from the system are low so the decrease of the level of service is acceptable (waste of time for example); or (3) the risk to go out of the set of formalised states is low because the system is quasi-exhaustively modelled.

Many products we use everyday (domestic devices, PCs, etc.) belong to this class of systems. However, this is not the case for an industrial system: the system must not be damaged and cannot be reconstructed, the performances expected from the system do not allow any degradation of the result. Furthermore the system is weakly modelled, and this for two main reasons:

- It is generally possible to describe an operational process from a qualitative point of view but it is more difficult to quantify it. Very often, the lead-time is an average value with a large dispersion around it. This dispersion becomes larger when the number of human operators grows;

- The number of situations and events influencing the system cannot be counted and exhaustively identified beforehand. A large and complex system such as an industrial system is rarely in a nominal mode. Very often, the model of the system is a description of an ideal process corresponding rarely to the reality.

How to control such a system? More precisely, how to control the system when it is in a significant but not formalised state? A principle from cybernetics gives a piece of solution: a system is really controlled only if the control system proposes at least as many states (control system variety) as the controlled system (controlled system variety). Because this variety does not exist *a priori*, it is necessary to have inside the control system some variety generators, i.e. organs that are capable to react to a situation *at the time* the situation appears. The only organ known to be versatile and quick enough to play this role is the human being (Fig. 1).

The human being can play the role of variety generator if his or her freedom to make decision is high. This leads to a difficulty: a high degree of freedom can lead the decision-maker to have a behaviour that is not compatible with the overall objectives of the system. Then, the decisional freedom needs to be limited in some way (framed). The problem grows when the number of decision-makers grows.

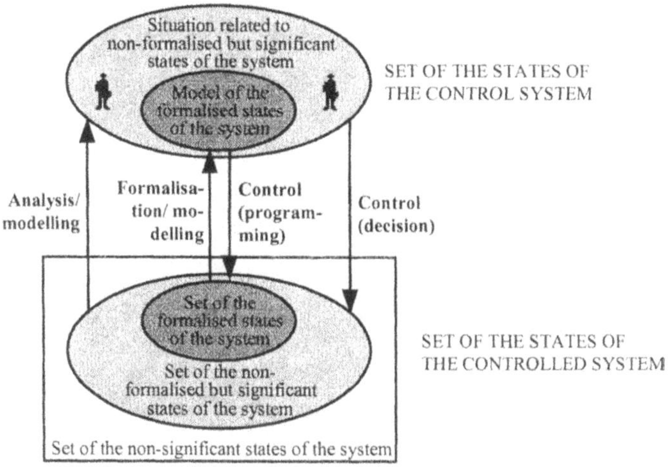

Figure 1: Increase of the control system variety by the presence of human beings

There is a need to consistently frame a set of local decisional freedoms dealing with specific objectives. The solution is *co-ordination* in the meaning that decisional freedoms are defined and delimited by a co-ordinator module. Generally, the co-ordinator is in the same situation as the co-ordinated modules: it is co-ordinated too. Then the complex, multi-objectives systems, such as industrial systems, are controlled by a hierarchical, multi-level structure, which owns the important property of decentralisation of decision-making (Doumeingts *et al.*, 1984). This aspect must not be confused with the organisation chart, which is concerned with tasks and responsibilities assignment.

At the level of decision-making itself, we can consider that the main items to define a frame for the decision-making are (1) the valued *objective* or *set of objectives* the decision centre has to match and (2) the *decision variables* enabling the decision centre to know under what constraints it can act.

At the detailed level, decision-making may be considered to the search for a position in a space defined by decision variables and closed (limited) by constraints in order to process the received information and to match the objectives (Fig. 2).

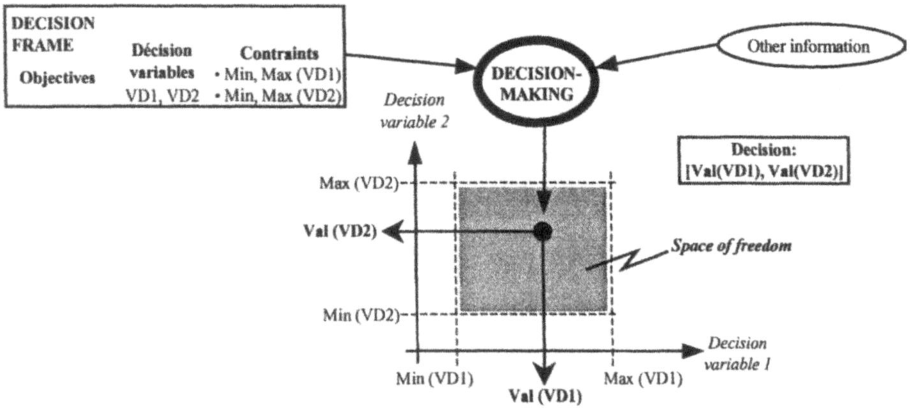

Figure 2: Principle of decision-making in conformity with a decision frame (here: only two decision variables are considered)

These concepts allow one to define a consistent decision-making environment rather than to show how to make an individual decision.

3 IMPACT ON CONTROL AND PERFORMANCE

An industrial system is an artefact, which must run in accordance with the objectives defined by the strategy of the company. This aspect has to be taken into account during the design phase and during the operating phase. Our interest for this section is the operating phase.

An industrial system has most of time a high degree of flexibility. This flexibility is absolutely necessary to match the "hard requirements" coming from the market. On the other hand, the potential risk is to use this flexibility in a bad way, leading to a mismatch with the objectives of the company. That is why control is today a key issue for reaching performances.

The main purpose of control is to ensure that operational activities run in a consistent way and that objectives coming from upper level (corporate strategy) are met. For that purpose, control must on one-hand co-ordinate tasks in relation to these objectives and to enable the deployment of these objectives into the structure and, on the other hand, follow up performances of operations in order to compare them to objectives. However, many industrial cases show that sometimes these follow-up activities are often not very usable or do not even exist at every level of the hierarchy and. Then, the main issue is to merge a top-down approach corresponding to the deployment of objectives and a bottom-up approach dealing with the follow-up of operations and aggregation of information (Fig. 3).

We find here the concept that a structure can really be controlled only if some feedback loops are implemented (in particular when the system is weakly modelled). The source of this feed-back is all the raw data coming from the operational processes and from the environment of the system. These raw data can be relevant for the lowest levels of control because these levels are close to the real system. However, they are not relevant for upper levels, for two reasons.

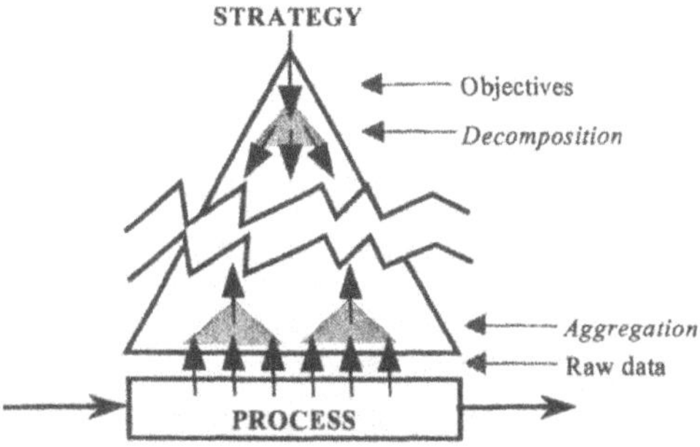

Figure 3: The issue of matching decomposition of objectives and aggregation of information

The first reason is the cognitive limitation of decision-makers. This limitation can be expressed in term of quantity of information. Beyond this limit, the decision-maker is submerged by the information: he or she is not able to interpret the supposedly useful information in order to make decision. The quantity of information is proportional to (1) the size of the domain in which the decision-maker is supposed to make decision and (2) the detail of the information handled by the decision-maker. Therefore information needs to be aggregated as it goes up to levels that have a broader and more conceptual view of the operational process in the structure, in order to provide the good granularity of information to decision-makers.

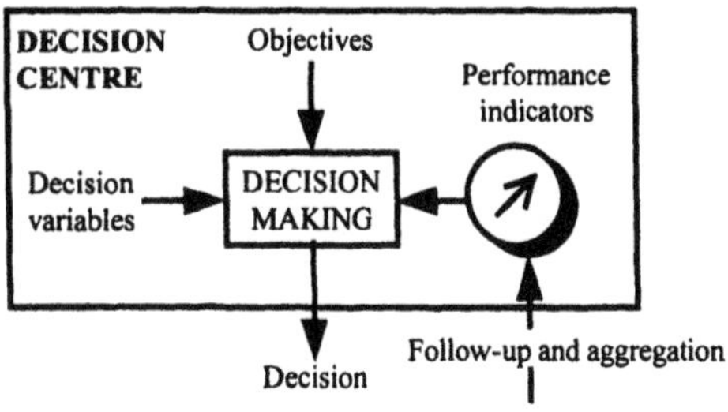

Figure 4: Decision centre and triplet {Objectives, Decision Variables, Performance Indicators}

The second reason is related to the fact that decision is mainly made on a basis of objectives and decision variables, so that information must be expressed in a way closely related to these two items (Doumeingts *et al.* 1992, 1995, 1998). This is the notion of *performance indicator* (Fig. 4).

Performance indicators must be consistent with objectives because it is necessary to compare the target performances (objectives) and performances reached (indicator) (*observability*). Performance indicators must be consistent with decision variables because these must have an effect on the performance concerned (*controllability*). Therefore the main issue is to ensure the internal consistency inside a decision centre in terms of the triplet of *objectives, decision variables* and *performance indicators*. This consistency is ensured if the performance indicators allow the verification of the achievement of objectives, and if they are influenced by actions on decision variables.

4 IMPACT ON DECISION-MAKING AND "MEN IN THE SYSTEM"

Very often, objectives on quality, costs and lead times are enunciated simultaneously. The main problem is that all these objectives are looked for at the same time. Control is always multi-objective. That is why the problem of decision-making is of a tremendous importance because, most of time, decision-making is not reducible to the optimisation of one criterion. Two systems, with the same resources, products, etc. may have a very different level of performances in accordance with the quality of decisions made inside their control system. That is why it is very important for an industrial system to create the best environment for decision-making.

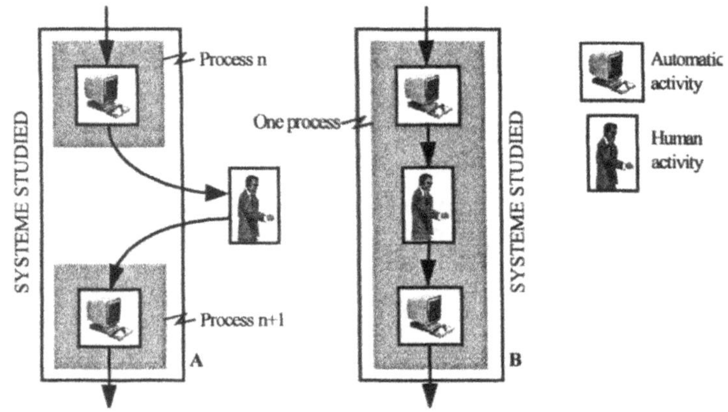

Figure 5: Presence of a men in processes: customer of the system and between two processes (A) or actor of the system and inside the process (B)

Today, the essential role of man inside an industrial system is not to operate any more but to decide. Therefore to focus on decision-making in system engineering is also to focus on the position of men in the system.

The first effect of this consideration is to avoid any model where men would be considered outside the system (men = customers or service providers). One of the main concepts making Enterprise Modelling and Integration different to techno-centred approach such as automatic control, computer sciences, etc. is that men are completely part of the system and be involved in processes of the system (Fig. 5).

The techno-centred approach remains relevant when important automatic processes exist. However, this is not often the case in industrial systems for the reasons discussed previously.

The second effect is that is not reasonable to expect to model and formalise the internal running of decisional activities. In a modelling activity, it is attractive to try to do so because these activities belong to the studied process. This trend could be problematic because:

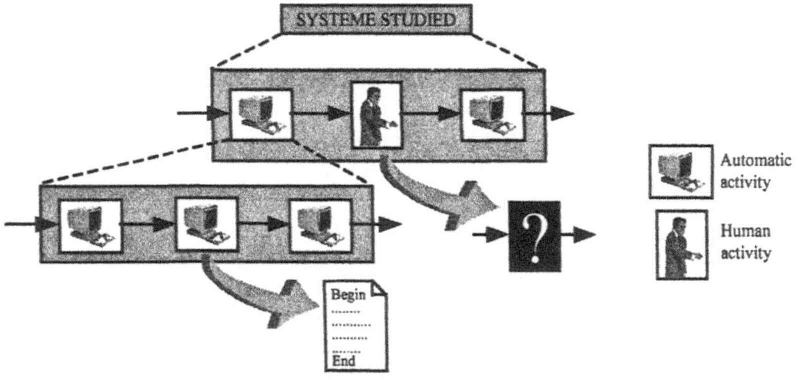

Figure 6: Modelling an "automatic" activity and a decisional activity

- Firstly, it will be generally disappointing because, despite the efforts to model these activities, men will present numerous behaviours outside the model. Therefore it is wiser to consider real decisional activities as black boxes whose the main inputs and outputs are known in order to situate them inside the process (Fig. 6);
- Secondly, it could be dangerous to formalise the activity in order to constraint men to a nominal behaviour. The capacity of the decision-maker to generate a variety of decisions could be partially or completely inhibited by this constraint. This aspect is perfectly illustrated in some structures where the standardisation of activities has led to the loss of responsibility and imagination in abnormal situations.

5 CONCLUSIONS

This paper has presented the necessity to take into account the control aspect in enterprise modelling and integration. This consideration has been argued and justified from a theoretical point of view. A set of basic concepts to implement decision centres with well defined triplets of information (Objectives, Decision Variable and Performance Indicators) to allow a system-wide consistent decision making have been proposed.

To take our statement into account, two levels of detail are concerned: (1) a local view enabling to define the environment of decision (information required, performance indicators, decision variables, etc.) and (2) a global view in order to define the architecture of decisions to implement co-ordination between decision centres and to ensure the deployment of corporate objectives throughout the industrial system.

6 REFERENCES AND BIBLIOGRAPHY

Doumeingts, G. (1984), *Méthode GRAI: méthode de conception des systèmes en productique. Thèse d'état,* University Bordeaux I.

Doumeingts, G. Vallespir, B. (1995), Les aspects humains dans la conception des systèmes de production - Proc. of 30th Congress of Société d'Ergonomie de Langue Française Biarritz, France.

Doumeingts, G., Vallespir, B. Chen, D. (1998), *Decision modelling GRAI grid,* Chapter in: Handbook on architecture for Information Systems, Peter Bernus, Kai Mertins, Gunter Schmidt, (Eds.). Springer-Verlag.

Doumeingts, G., Vallespir, B. Zanettin, M. Chen, D. (1992), *GIM: GRAI Integrated Methodology for designing CIM systems,* GRAI/LAP, University Bordeaux I, version 1.0.

ICIMS (2000), ICIMS-NOE Scientific meeting, Brussels, Belgium, November 24, 1999, in ICIMS NEWS, March.

IFAC-IFIP Task Force (1999), *GERAM: Generalized Enterprise Reference Architecture and Methodology*, Version 1.6.2, Annex A in IS 15704, Requirements for Enterprise Reference Architecture and Methodologies, ISO TC 184/SC5/WG1.

AUTHORS INDEX

A

Aguilar-Savén, R.S. 195
Ahn, G-J. 205
Akkermans, H. 71
Ang Cheng L. 127

B

Bernus, P. 127, 135
Byer, N. 183

C

Callot, M. 51
Cardoso, J. 303
Chen, D. 61, 273, 417
Chu, B. 205, 253
Cieminski, G. v. 167

D

delaHostria, E. 245, 283
Doumeingts, G. 417

E

Engel, T. 337, 399
Engwall, R. 245, 295

F

Fukuda, Y. 113

G

Garetti, M. 167
Ghenniwa, H.H. 313
Goossenaerts, J. 51
Goranson, H.T 7, 15, 37, 113, 253, 347
Grabowski, H. 399
Gruninger, M. 253
Gutierrez Vañó D. 217

H

Harrison, R 225
Hawa, M. 407
Heisig, P. 51
Huhns, M.N. 37, 83

I

Ivezic, N. 253

J

Jaekel, F-W. 235, 337
Jochem, R. 127, 347, 371

K

Kang, M. 245
Kosanke, K. 3, 51, 127, 245, 337
Kotsiopoulos, I.L. 337, 389
Krogstie, J. 51, 91
Kulvatunyou, S. 253

L

Labrou, Y. 253
Lario, F.C. 407
Lee, E.W. 113
Levi M.H. 147
Li, Y. 313
Lillehagen, F. 61, 91
Liu, H. 325

M

Macchi, M. 167
Masuoka, R. 253
Mendez J.C. 177, 245, 337
Miller, J. 303
Ming, H. 127

N

Nell, J.G. 15, 37, 113, 245, 347

O

Obrst, L. 325
Ortiz Bas, A. 217, 337, 407

P

Panetto, H. 37, 347, 381
Partridge, C. 101, 347
Payne, M. 265
Peng, Y 253
Petit, M. 337, 359
Poler Escoto, R. 61, 217
Preez, N.D du 61

R

Rabe, M. 235
Raynaud, P. 337
Reber, J.W. 295

S

Sempere Ripoll, F. 347
Shen, W. 245 313
Sheth, A. 253, 303
Shorter, D. 253, 347
Sims, J.E. 205
Stefanova, M. 101
Stephens, L.M. 83
Stojanovic, N. 51

T

Tolone, W.J. 205
Tormo Carbó, G. 37, 113, 217

V

Vallespir, B. 417
Vernadat, F.B. 25, 273

W

Webb, P. 159, 347
West, A.A. 225
Weston, R.H 127, 183, 225
Wiendahl, H-P. 167
Wilhelm, R.G. 205
Wilson, L. 325
Wray, R. 325
Wunram, M. 37

Z

Zelm, M. 61, 113, 347

GPSR Compliance
The European Union's (EU) General Product Safety Regulation (GPSR) is a set of rules that requires consumer products to be safe and our obligations to ensure this.

If you have any concerns about our products, you can contact us on

ProductSafety@springernature.com

In case Publisher is established outside the EU, the EU authorized representative is:

Springer Nature Customer Service Center GmbH
Europaplatz 3
69115 Heidelberg, Germany

www.ingramcontent.com/pod-product-compliance
Ingram Content Group UK Ltd.
Pitfield, Milton Keynes, MK11 3LW, UK
UKHW021103170926
15229UKWH00007B/28
* 9 7 8 1 4 7 5 7 5 1 5 0 5 *